BIOMATERIALS FOR MEMS

BIOMATERIALS FOR MEMS

Editors
M CHIAO & JC CHIAO

PAN STANFORD PUBLISHING

Published by

Pan Stanford Publishing Pte. Ltd.

Penthouse Level, Suntec Tower 3
8 Temasek Boulevard
Singapore 038988

E-mail: editorial@panstanford.com
Web: www.panstanford.com

British Library Cataloguing-in-Publication Data
A catalogue record for this book is available from the British Library.

BIOMATERIALS FOR MEMS

ISBN: 978-981-4241-46-5 (Hardcover)
ISBN: 978-981-4241-47-2 (eBook)

Printed in Singapore.

Preface

Microelectromechanical Systems (MEMS) technology has been developed for more than three decades. MEMS applications have proliferated that cover many industry sectors, including automotive, aerospace and consumer clectronics. MEMS for biological applications, or BioMEMS, and not received much attention until perhaps mid-1990s. It has been extremely challenging for engineers and scientists to cross discipline boundaries, as such, many existing relevant textbooks on MEMS focus on introducing MEMS technologies to non-MEMS researchers.

In recent years, research efforts have been made in finding biomedical applications for MEMS devices. As a result, MEMS engineers are facing new challenges in biomaterial selections that are suitable for miniaturization. Many research papers demonstrated micromachining of particular biomaterials and application of those biomaterials in microdevices. However, there is insufficient organized effort to systematically introduce biomaterials to MEMS researchers.

This book intends to serve as a guide for practicing engineers, researchers and students interested in MEMS devices using biomaterials and biomedical applications using these devices, This book highlights important features and issues of biomaterials that have been used in the MEMS and biomedical areas. The materials in this book come from earlier research papers and detailed references are included. The contents of this book provide an overview in various aspects to build up understanding in this multidisciplinary subject. This book only serves as an introduction and does not intentionally leave out other relevant work.

We are grateful to all the contributors and staff involved in the book's editing process.

Editors
M. Chiao and J.-C. Chiao

Biomaterials for MEMS, Edited by M. Chiao and J.-C. Chiao

www.panstanford.com
978-981-4241-46-5

Contents

Chapter One

Introduction

M. Chiao* and J.-C. Chiao†

*Department of Mechanical Engineering, University of British Columbia,
6250 Applied Science Lane, Vancouver, V6T 1Z4 Canada.
E-mail: muchiao@mech.ubc.ca
†Department of Electrical Engineering, Nedderman 538, Box 19016,
University of Texas at Arlington, Arlington, TX 76019-0016, USA.
E-mail: jcchiao@uta.edu

Biomaterials have been used in biomedical industry to produce implantable devices for the last several decades. Most of such devices have utilized conventional manufacturing processes. MEMS (Microelectromechanical systems) uses micromachining techniques to build miniature devices and can enable novel applications. Over the last decade, biomaterial-based MEMS devices have found applications in medicine. However, due to the cross-disciplinary nature of the research that often involves mechanical engineering, electrical engineering, chemistry, biology and material science, review literatures on the subject that could lead scientists in a specific field into the new biomaterial-based application world is often lacking. The goal of this book is to provide a review on new micromachining techniques of biomaterials and their applications.

1.1 INTRODUCTION

The definition of *Biomaterial* can be found in a Merriam-Webster's dictionary, dated 1966: *A natural or synthetic material (as a metal or polymer) that is suitable for introduction into living tissue especially as part of a medical device (as an artificial joint).* However, with the advances in technology, materials that are used *ex vivo*, but interacting with living organisms can be arguably a biomaterial. For example, blood contacting materials such as plasticized polyvinyl chloride (PVC) commonly used in blood bags, may be considered as a type of biomaterial. More recently, rapid diagnostic tools using miniaturized fluidic channels[1] combining with micro mixers, valves and pumps have the potential to fundamentally change the ways biochemical analysis have been performed. The materials involved in such devices or systems include silicon, glass, metals and polymers. These materials, either

Biomaterials for MEMS, Edited by M. Chiao and J.-C. Chiao

www.panstanford.com
978-981-4241-46-5

used for the substrates, passive parts or active elements, need to interact with biological components such as cells, proteins, DNAs, genes or reagents. The interfacing mechanisms and reactions between the biological components and the mechanical parts are not often discussed or taught in the curricula of conventional MEMS courses in the departments of electrical or mechanical engineering. Thus, the purpose of this book is to introduce biomaterials to engineers who are interested in MEMS and microfluidics, yet have no prior background in biomaterials.

Common biomaterials used in biomedical industries are summarized in Table 1.1. It is clear that these materials cover a wide range from metals, ceramics to polymers. Common factors need to be considered in biomaterial applications include: toxicity, surface fouling and mechanical strength, besides their functional characteristics and capabilities. For example, biomaterials used in dentistry need to have satisfactory mechanical strength and also adequate biocompatibility. The mechanical strength also has to be compatible with surrounding tissues to avoid damages due to motions. *In vivo* or *ex vivo* microfluidic devices, either for drug delivery implants or diagnostic tools, will have different biocompatibility concerns. It is important to maintain certain flexibility and deformability of the *in vivo* devices so the devices can stay inside the body without causing any tissue scarring. However, the devices have to be rigid enough to maintain a proper internal pressure or allow drug refill using a syringe. For *ex vivo* microfluidic platforms, although the entire system does not have toxicity concerns, the interfacing parts such as the substratum, channels and active sensing or actuation elements need to maintain a proper environment for cell proliferation and movement. Another example illustrates the importance of surface fouling. Long-term cardiovascular catheters for intravenous drug delivery is critical for patients under chemotherapy. However, sepsis as a result of bacterial contamination on the catheter walls, is a major failure of the treatment. Surface fouling, initiated by protein adsorption

Table 1.1 Common biomaterials and application fields.

Application	Material
cardiovascular	stainless steel; titanium; polyurethane; polytetrafluoroethylene (PTFE); polyethylene terephthalate (Dacron^{TM})[2]
dentistry	gold; calcium phosphate; porcelain; amalgam; glass ionomer[3]
controlled drug delivery	poly(D,L-lactide-co-glycolide) (PLGA);[4]
ophthalmology	silicone; hydrogel;[5]
orthopedics	titanium[6]
urology	polyurethane; silicone[7]
neurology	silicon[8]

on the catheter surfaces, is said to create a favorable environment for bacteria to grow. While the catheters should provide required strength to maintain their mechanical structures integrity, the prevention of surface fouling is also critical for device design.

1.2 MICROMACHINING OF BIOMATERIALS

MEMS is a technology that uses micromachining techniques to form microscale devices, such as accelerometers,[9] optical lens,[10] RF (radio frequency) devices[11,12] and microfluidic devices. Micromachining uses thin-film deposition and etching techniques developed by the IC (integrated circuit) industry to fabricate small mechanical parts monolithically on a substrate. Limitations, such as processing temperature and chemicals used in established IC fabrication processes, restrict materials to silicon, silicon dioxide, silicon nitride, aluminum and refractory metals. Conventional micromachining methods start with a bulk material, for example, a silicon wafer, following by thin-film deposition, photolithography and etching. The process can be repeated, with some materials forming sacrificial layers and structures, to form more complex 2-D or 3-D structures. The level of complexity of a micromachining process is often determined by the number of photolithography masks used in the processes.

Historically, micromachining process can be categorized into two groups: (1) *bulk micromachining* and (2) *surface micromachining*. Bulk micromachining is often involved in etching of a silicon wafer itself and use the single crystalline planes in silicon to form mechanical structures. Examples include early ink-jet printing nozzles, inertial sensors, as well as microfluidic valves and pumps. On the other hand, surface micromachining uses silicon substrates only as a mechanical support, the add-on thin-films, such as polycrystalline silicon (poly-si), silicon nitride and metals are the actual mechanical components. Examples include scanning optical mirrors, inertial sensors and radio frequency switches. Typically, bulk micromachining consumes more silicon real estate than surface micromachining, but may require less processing steps. Bulk-micromachined structures are generally simpler and provide generic advantages using single crystalline structures such as smoother planes, higher Young's Modulus and reliability. Surface micromachining, on the other hand, offers potentials for complex structures to satisfy nonconventional functionality and requirements, possibility to be integrated with complementary metal oxide semiconductor (CMOS) circuitry and ability to adapt to exotic materials.

Micromachining of biomaterials initially used bulk-micromachined silicon molds.[13] Silicon wafers were etched into micro-scale structures as molds as they offer precision in the molds by which the conventional metal drilling or quick prototyping methods may not offer. Casting, hot-embossing and micromolding have been used[14,15] to mold biomaterial (mainly polymers) to desired shapes. Hot-embossing and micro-molding are rather mature technologies and many foundry services exist to provide custom-made micromechanical polymeric parts. These techniques however have their own individual requirements for dimensions, thermal budgets and pressure tolerance. It is not likely one single method can suit all biomaterials.

A new technique using Polydimethylsiloxane (PDMS), a biocompatible silicone-based polymer to cast over pre-patterned PR (photoresist) has been demonstrated,[16] the results were surfaces with patterned biological material, such as anti-bodies, enzymes and cells.[17] Photosensitive resist can be patterned monolithically with micro- and nano-scale features. This is an inexpensive and rapid process suitable for research purposes to fabricate a microfluidic device in a short time. Companies have build business around the idea as well. On-chip pneumatic pumps and valves have been fabricated using this method (Fluidigm, South San Francisco, CA, USA) and demonstrated applications in rapid biological analysis.[18]

Other non-conventional techniques have also been successful in micromachining of biomaterials. For example, laser micromachining has been used to define microchannels for biosensor applications.[19] An industrial CO_2 laser was used to fabricate microchannels by ablation with channel widths of 300–350 μm and aspect ratios of 0.52 on co-polyester thermoplastic substrates. Three-dimensional laser rapid prototyping of polymer resin was also demonstrated.[20,21] Stereolithography of photopolymerized resin was used to make molds for micromolding, however, the method could be utilized to make the MEMS part directly. Raw resin was polymerized by scanning a focused laser beam to form the 3-D mold. The laser beam was scanned by computer-controlled mirror scanners. Another technique was demonstrated with the *thick-film* technology,[22,23] that has been used to form metal interconnects on printed-circuit boards (PCB) for decades. The aerosol deposition method produces thick films by particle densification from accelerated particles impacting onto a substrate at high speeds. The method can fabricate piezoelectric films that can be used as actuation elements or energy harvesting components in the micro devices. This technique produced devices with larger features ($>$ 100 μm) but was relatively inexpensive compared to other techniques that requires conventional IC photolithography. There have been efforts in industry as well as academia trying to adapt the printing process to produce polymeric features. Printing techniques offer a cost-effective and quick prototyping solution. Further utility of the printing technique incorporating biological reagents as the ink enables direct deposition femto-/nano-liter droplets onto conventional microarray substrates for high-throughput applications including rapid proteomics, viral, and bacterial screening.[24] The technique consists of a target template with small metal or metal-oxide patterns on one side of a transparent substrate and a layer of protein solution facing down to the microarray substrate. An excimer pulsed laser shines the light on the target and the metal or metal-oxide patterns act as absorbers locally heating the elements. The temperature variation around the absorber produces a fluidic jetting mechanism and transfers a droplet of the protein solution onto the microarray substrate. By precise timing of laser pulses and substrate/target step motion, the protein solution then can be patterned on the substrate. These commonly-used manufacturing methods have been implemented for conventional industries and matured with values in cost effectiveness and mass production. They are now being considered, with some adjustments and fine-tuning in procedures, processes, apparatus, and formula, for advanced bioMEMS and biomaterial fabrication and manufacturing.

1.3 BIOMEDICAL MICRODEVICES

MEMS community has been focusing on many biological applications, and the field of BioMEMS has grown rapidly and significantly. Some of the demonstrated devices and fabrication materials are summarized in Table 1.2. This summary only serves as an example of devices and materials that are related to the BioMEMS fields and not purposely excluding other materials that have been demonstrated in MEMS devices. Most of the MEMS applications focus on non-implantable applications, such as diagnostic tools or tissue engineering devices. These devices or applications have fewer concerns toward material uses and device architecutures. Implantable MEMS drug delivery devices have also been demonstrated.[25] Commercial efforts have been pursuit (MicroCHIPS, Inc., Bedford, MA, USA). Micro-scale silicon reservoirs filled with drugs are sealed by thin metallic films. The metallic film is connected to a battery. When a voltage is applied, electrochemical reaction causes corrosion on the film and drug can be released by diffusion. Another implantable MEMS device is a micro strain gauge for applications in orthopedics. Thin-film strain gauges are fabricated and connected to an RF module.[26] The powering and sensor data transfer are done using electromagnetic induction wirelessly. Microelectrodes have also been used to record or stimulate neurological activities in brain.[27] Bulk micromachined silicon needle arrays integrated with CMOS circuitry were used to record neuronal signal from the motor cortex and the signals were amplified and transmitted to a computer. The brain prosthetic system serves as a brain-computer interface to control artificial limbs with specific neuronal activities. Similarly, smaller arrays of microneedles can also be used with integrated circuits for electrical stimulations on neurons for tremor control or pain management.[28,29] Recently, researchers from ETH have demonstrated elegantly a MEMS device for ophthalmological applications.[30] A contact lens was attached with a thin-film metal inductive coil and metallic strain gauges. An ASIC chip was mechanically thinned down to 50 μm thick and attached onto the contact lens to regulate power and data transmission. The pressure in eyes then could be monitored continuously *in vivo* for a long period of time.

These implantable devices often face tremendous technical challenges. First of all, the size requirements for implantation as the body cavity may be small present a design and fabrication challenge for the functionalities required. MEMS technology offers great advantages to address this issue. Secondly, in order for the implant to stay inside the body in a long period of time, biocompatibility needs to

Table 1.2 Biomaterials used in MEMS and microfluidic devices.

	Diagnostic Tools	Actuators	Tissue Engineering
semiconductor material			
silicon	PCR[32]	microneedles;[33] surgical tools[35]	neuron growth[34]
polymers			
PDMS	PCR;[36] flow cytometry[37]	drug delivery;[38] pumps[41–43]	cell and neuron growth[39,40]
hydrogel			cell-interacting matrix[44]
polyimide	microneedles[45]		
other materials			
iron-based magnetic particles	magnetic tagging[46]	pumps[43]	
piezoelectric material (PZT; ZnO)	acoustic wave sensor[47]	pumps[48]	
titanium		microneedles[49,50]	

be addressed. The concerns include:

(1) bio-mechanical compatibility that ensures the mechanical form factor of the devices have no damage to the tissues;
(2) electrical compatibility that ensures the electricity or electromagnetic energy does not harm the tissues;
(3) thermal compatibility that ensures the local temperature does not rise to burn the tissues; and
(4) biocompatibility that ensures minimum biofouling.

Thirdly, many devices with sensors to measure physiological parameters or actuators to dispense drugs or deliver stimuli will have a need for control. Often electronic control mechanisms for either wireless or locally wired solutions are needed. Finally, integrated functionalities or capabilities will be required in order to achieve a small platform not only to fit the implantation site but also to address the implementation methods that are clinically beneficial to the patients. Many of these implemtnation methods are minimally invasive procedures or by natural orifice translumenal endoscopic surgery (NOTES),[31] for which the device designs also need to be mechanically compatible to the commonly used implementation tools.

1.4 ORGANIZATION OF THE BOOK

Facing the great challenges in engineering to satisfy the rigorous specifications for implantable devices, the combination of MEMS techniques and biomaterials provides a new direction. The advantages of miniaturization, integration, electrical functionalities and cost-effectiveness of manufacturing in MEMS, and the flexibility, deformability, capabilities and biocompatibility in biomaterials offer an attractive solution to the dire challenges mentioned before. However, due to such a relatively young research topic and fast-growing academic and industrial research societies, it is difficult to provide a conventional review of the established knowledge. In this book, we aim to illustrate some of the advanced technologies that involve the cross-disciplinary between biomaterials and MEMS technologies. The first half of the book, Chap. 2 to Chap. 6, introduces materials, devices and applications of biomaterials. The second half of the book, Chap. 7 to Chap. 10, describes biocompatibility issues and characterizing of biomaterials.

References

[1] Jessica Melin and Stephen R Quake. Microfluidic large-scale integration: the evolution of design rules for biological automation. *Annu Rev Biophys Biomol Struct*, 36:213–231, 2007.

[2] Xinwen Wang, Peter Lin, Qizhi Yao, and Changyi Chen. Development of small-diameter vascular grafts. *World J Surg*, 31(4):682–689, Apr 2007.

3 Richard S Schwartz and Ron Fransman. Adhesive dentistry and endodontics: Materials, clinical strategies and procedures for restoration of access cavities: A review. *J Endod*, 31(3):151–165, Mar 2005.

4 Harjit Tamber, Pal Johansen, Hans P Merkle, and Bruno Gander. Formulation aspects of biodegradable polymeric microspheres for antigen delivery. *Adv Drug Deliv Rev*, 57(3):357–376, Jan 2005.

5 Krzysztof Slowinski, Marta Misiuk-Hojlo, and Marek Szalinski. Influence of material on biocompatibility of intraocular lenses. *Polim Med*, 37(1):35–45, 2007.

6 George 2nd Eaninwene, Chang Yao, and Thomas J Webster. Enhanced osteoblast adhesion to drug-coated anodized nanotubular titanium surfaces. *Int J Nanomedicine*, 3(2):257–264, 2008.

7 Darren T Beiko, Bodo E Knudsen, James D Watterson, Peter A Cadieux, Gregor Reid, and John D Denstedt. Urinary tract biomaterials. *J Urol*, 171(6 Pt 1):2438–2444, Jun 2004.

8 Andreas Schneider, Luca Frasson, Tassanai Parittotokkaporn, Ferdinando M. Rodriguez y Baena, Brian L. Davies, and S. Ejaz Huq. Biomimetic microtexturing for neurosurgical probe surfaces to influence tribological characteristics during tissue penetration. *Microelectronic Engineering*, 86(4-6):1515–1517, Apr–Jun 2009.

9 Kevin King, S. W. Yoon, N. C. Perkins, and K. Najafi. Wireless MEMS inertial sensor system for golf swing dynamics. *Sensors and Actuators A-Physical*, 141(2):619–630, Feb 15 2008.

10 Chin-Pang-Billy Siu, Haishan Zeng, and Mu Chiao. Magnetically actuated MEMS microlens scanner for *in vivo* medical imaging. *Optics Express*, 15(18):11154–11166, Sep 3 2007. Optics Express.

11 Christopher D. Nordquist, Michael S. Baker, Garth M. Kraus, David A. Czaplewski, and Gary A. Patrizi. Poly-silicon based latching rf MEMS switch. *IEEE Microwave and Wireless Components Letters*, 19(6):380–382, Jun 2009.

12 Joan Lluis Lopez, Jaume Verd, Arantxa Uranga, Joan Giner, Gonzalo Murillo, Francesc Torres, Gabriel Abadal, and Nuria Barniol. A CMOS-MEMS rf-tunable bandpass filter based on two high-q 22-mhz polysilicon clamped-clamped beam resonators. *IEEE Electron Device Letters*, 30(7):718–720, Jul 2009.

13 Fardad Chamran, Yuting Yeh, Hong-Seok Min, Bruce Dunn, and Chang-Jin Kim. Fabrication of high-aspect-ratio electrode arrays for three-dimensional microbatteries. *Journal of Microelectromechanical Systems*, 16(4):844–852, Aug 2007.

14 T. Lomas, A. Wisitsoraat, F. Chevasuvit, and A. Tuantranont. A precision hot embossing mold fabricated by high-resolution powder blasting with polydimethylsiloxane and su-8 masking technology. *Journal of Micromechanics and Microengineering*, 19(3), Mar 2009.

15 P. P. Shiu, G. K. Knopf, M. Ostojic, and S. Nikumb. Rapid fabrication of tooling for microfluidic devices via laser micromachining and hot embossing. *Journal of Micromechanics and Microengineering*, 18(2), Feb 2008.

16 Y. N. Xia and G. M. Whitesides. Soft lithography. *Annual Review of Materials Science*, 28:153–184, 1998.

17 Hyundoo Hwang, Gyumin Kang, Ju Hun Yeon, Yoonkey Nam, and Je-Kyun Park. Direct rapid prototyping of pdms from a photomask film for micropatterning of biomolecules and cells. *Lab On a Chip*, 9(1):167–170, 2009.

18 Elizabeth A Ottesen, Jong Wook Hong, Stephen R Quake, and Jared R Leadbetter. Microfluidic digital pcr enables multigene analysis of individual environmental bacteria. *Science*, 314(5804):1464–1467, Dec 2006.

19 Shau-Chun Wang, Chia-Yu Lee, and Hsiao-Ping Chen. Thermoplastic microchannel fabrication using carbon dioxide laser ablation. *J Chromatogr A*, 1111(2):252–257, Apr 2006.

20 A. Bertsch, S. Jiguet, P. Bernhard, and P. Renaud. *Microstereolithography: A review.* Materials Research Society, 506 Keystone Drive Warrendale, PA 15086-7573, USA, 2003.

21 Kohki Mukai, Shinya Kitayama, Yasunobu Kawajiri, and Shoji Maruo. Micromolding for three-dimensional metal microstructures using stereolithography of photopolymerized resin. *Microelectronic Engineering*, 86(4-6):1169–1172, Apr-Jun 2009.

22 Xuan-Ya Wang, Chi-Yuan Lee, Cheng-Ren Peng, Pei-Yen Chen, and Pei-Zen Chang. A micrometer scale and low temperature pzt thick film MEMS process utilizing an aerosol deposition method. *Sensors and Actuators A-Physical*, 143(2):469–474, MAY 16 2008.

23 F. M. Dauchy and R. A. Dorey. Patterned high frequency thick film MEMS transducer. *Integrated Ferroelectrics*, 90:42–52, 2007.

24 J A Barron, H D Young, D D Dlott, M M Darfler, D B Krizman, and B R Ringeisen. Printing of protein microarrays via a capillary-free fluid jetting mechanism. *Proteomics*, 5(16):4138–44, Nov 2005.

25 Gabriela Voskerician, Matthew S. Shive, Rebecca S. Shawgo, Horst von Recum, James M. Anderson, Michael J. Cima, and Robert Langer. Biocompatibility and biofouling of MEMS drug delivery devices. *Biomaterials*, 24(11):1959–1967, 2003.

26 R. Melik, N. K. Perkgoz, E. Unal, C. Puttlitz, and H. V. Demir. Bio-implantable passive on-chip RF-MEMS strain sensing resonators for orthopaedic applications. *Journal of Micromechanics and Microengineering*, 18(11):115017, 2008.

27 Q. Bai and K. D. Wise. Single-unit neural recording with active microelectrode arrays. *IEEE Transactions on Biomedical Engineering*, 48(8):911–920, 2001.

28 T. Ativanichayaphong, J. W. He, C. E. Hagains, Y. B. Peng, and J.-C. Chiao. A combined wireless neural stimulating and recording device for pain management study. *Journal of Neuroscience Methods*, 170,(1):25–34, 2008.

29 S. Roy, L. A. Ferrara, A. J. Fleischman, and E. C. Benzel. *BioMEMS and Biomedical Nanotechnology*, chapter MEMS and Neurosurgery, pages 95–123. Springer, 2007.

30 M. Leonardi, E. M. Pitchon, A. Bertsch, P. Renaud, and A. Mermoud. Wireless contact lens sensor for intraocular pressure monitoring: Assessment on enucleated pig eyes. *Acta Ophthalmologica*, 87(4):433–437, 2009.

31 S. Bergman and W. S. Melvin. Natural orifice translumenal endoscopic surgery. *Archives of Physical Medicine and Rehabilitation, October 2008.*, 88(5):1131–1148, 2008.

32 M. Carles, T. Lee, S. Moganti, R. Lenigk, K. W. Tsim, N. Y. Ip, I. M. Hsing, and N. J. Sucher. Chips and qi: microcomponent-based analysis in traditional chinese medicine. *Fresenius J. Anal Chem*, 371(2):190–4, Sep 2001.

33 W. H. Smart and K. Subramanian. The use of silicon microfabrication technology in painless blood glucose monitoring. *Diabetes Technol Ther*, 2(4):549–59, 2000.

34 G. Zeck and P. Fromherz. Noninvasive neuroelectronic interfacing with synaptically connected snail neurons immobilized on a semiconductor chip. *Proc Natl Acad Sci USA*, 98(18):10457–62, Aug 2001.

35 K. J. Rebello. Applications of MEMS in surgery. *Proceedings of the IEEE*, 92(1), 2004.

36 Naveen Ramalingam, Hao-Bing Liu, Chang-Chun Dai, Yu Jiang, Hui Wang, Qinghui Wang, Kam M Hui, and Hai-Qing Gong. Real-time pcr array chip with capillary-driven sample loading and reactor sealing for point-of-care applications. *Biomed Microdevices*, May 2009.

37 Donald Wlodkowic, Shannon Faley, Michele Zagnoni, John P. Wikswo, and Jonathan M. Cooper. Microfluidic single-cell array cytometry for the analysis of tumor apoptosis. *Anal Chem*, 81(13):5517–23, Jul 2009.

[38] Urs O. Häfeli, Amir Mokhtari, Dorian Liepmann, and Boris Stoeber. *In vivo* evaluation of a microneedle-based miniature syringe for intradermal drug delivery. *Biomed Microdevices*, Apr 2009.

[39] A. J. Blake, T. M. Pearce, N. S. Rao, S. M. Johnson, and J. C. Williams. Multilayer pdms microfluidic chamber for controlling brain slice microenvironment. *Lab Chip*, 7(7):842–9, Jul 2007.

[40] N. Bontoux, L. Dauphinot, T. Vitalis, V. Studer, Y. Chen, J. Rossier, and M.-C. Potier. Integrating whole transcriptome assays on a lab-on-a-chip for single cell gene profiling. *Lab Chip*, 8(3):443–50, Mar 2008.

[41] Jungyul Park, Il Chaek Kim, Jeongeun Baek, Misun Cha, Jinseok Kim, Sukho Park, Junghoon Lee, and Byungkyu Kim. Micro pumping with cardiomyocyte-polymer hybrid. *Lab Chip*, 7(10):1367–70, Oct 2007.

[42] Kee Suk Ryu, Kashan Shaikh, Edgar Goluch, Zhifang Fan, and Chang Liu. Micro magnetic stir-bar mixer integrated with parylene microfluidic channels. *Lab Chip*, 4(6):608–13, Dec 2004.

[43] Scott McDonald, Tingrui Pan, and Babak Ziaie. A magnetically driven PDMS micropump with microball valves. *Conf Proc IEEE Eng Med Biol Soc*, 4:2650–3, 2004.

[44] Biancamaria Baroli. Hydrogels for tissue engineering and delivery of tissue-inducing substances. *J Pharm Sci*, 96(9):2197–223, Sep 2007.

[45] Shinji Ichimori, Kenro Nishida, Seiya Shimoda, Taiji Sekigami, Yasuto Matsuo, Kenshi Ichinose, Motoaki Shichiri, Michiharu Sakakida, and Eiichi Araki. Development of a highly responsive needle-type glucose sensor using polyimide for a wearable artificial endocrine pancreas. *J Artif Organs*, 9(2):105–13, 2006.

[46] Kyu Sung Kim and Je-Kyun Park. Magnetic force-based multiplexed immunoassay using superparamagnetic nanoparticles in microfluidic channel. *Lab Chip*, 5(6):657–64, Jun 2005.

[47] Christopher D. Corso, Anthony Dickherber, and William D Hunt. An investigation of antibody immobilization methods employing organosilanes on planar zno surfaces for biosensor applications. *Biosens Bioelectron*, 24(4):811–7, Dec 2008.

[48] C. I. Shen and H. K. Liu. Fabrication and drive test of piezoelectric pdms valveless micro pump. *Journal of the Chinese Institute of Engineers*, 31(4):615–623, 2008.

[49] Kazuyoshi Tsuchiya, Naoyuki Nakanishi, Yasutomo Uetsuji, and Eiji Nakamachi. Development of blood extraction system for health monitoring system. *Biomed Microdevices*, 7(4):347–53, Dec 2005.

[50] E. R. Parker, M. P. Rao, K. L. Turner, C. D. Meinhart, and N. C. MacDonald. Bulk micromachined titanium microneedles. *Journal of Microelectromechanical Systems*, 16(2):289–295, April 2007.

Chapter Two

Spider Silk as a MEMS Material

J. Bai*, W. C. Chu†, J.-C. Chiao‡ and M. Chiao*,§

*Department.of Mechanical Engineering, The University of British Columbia, Canada
†School of Environmental Health, The University of British Columbia, Canada
‡Department of Electrical Engineering, University of Texas at Arlington, USA
§E-mail: muchiao@mech.ubc.ca

A thin-film made of dragline spider silk is produced and fabricated at room temperature using a spin-cast technique; the film's micromachining, optical and mechanical characterization is obtained. The mechanical strength of spider silk film is 5 to 10 times greater than the reported strength of hydrogel. Magnetic spider silk thin-film is also made by photolysis of iron pentacarbonyl mixed with spider silk in solution. A free-standing microcantilever beam is fabricated and actuated under a magnetic field.

2.1 INTRODUCTION

Due to the limitations of polysilicon, in particular, its lack of flexibility in biomedical applications and its expensive processing cost, numerous studies have been undertaken to investigate alternate materials for microelectromechanical processing and applications. Therefore, polymeric materials like polymethylsiloxane (PDMS) and hydrogel have been used to fabricate micromechanical components such as microfluidic valves and micropump diaphragms. These materials do offer more economical fabrication and elasticity than polysilicon does; however, their mechanical strength is not high enough for many biomedical applications.[1]

On the other hand, spider silk, a natural biopolymer, has been shown to possess good biomaterial qualities and can potentially be engineered to change its mechanical and chemical properties to fit certain application requirements. Spider silk is one of the strongest natural materials with a high tensile strength and toughness. The amount of energy required to break spider silk is three times higher than Kevlar[18] and more than 25 times higher than steel.[8] These preliminary results have led to much research on fabricating spider silk for artificial ligaments, tendons,

Biomaterials for MEMS, Edited by M. Chiao and J.-C. Chiao

www.panstanford.com
978-981-4241-46-5

surgical sutures or biodegradable membranes, bullet proof vests and structural applications. Moreover, silk based scaffolds for bone formation offer options to address the limitation of existing materials, such as collagen (which has poor mechanical properties), poly(lactiv-glucolic) acid (which induces inflammation), and hydroxypatite (which is not completely biodegradable).[32] Biomembranes made of spider silk film can also be used for pumping or as valves in microfluidic channels. Spider silk's robust structure and modifiable surface chemistry make it an ideal candidate as a chemical sensor or actuator. Recombinant spider silk proteins in plants and goat milk have been successfully demonstrated[16,27] to increase the commercialization potential of spider silk dramatically.

Dragline silk is a type of spider silk made up of two different but similar proteins (fibroins). In the secondary structure of these proteins, alanine-rich regions organize into beta-sheets which form the core of crystalline structures held together by hydrogen bonds. The glycine rich regions are less ordered. It is this combination of crystalline and amorphous regions that is responsible for the strength and extensibility of spider silk.[1,15,33] Regenerated spider silk is first harvested from spiders, dissolved into solvents and then re-spun through an orifice. Mechanical properties, such as the toughness of the regenerated silk, rely largely on the assembly process of the proteins during re-spinning and drying. However, the tensile strength of native silk is found to be three times higher than that of regenerated spider silk.[30]

Many microelectromechanical systems (MEMS) require a magnetic aspect to be incorporated in the function; therefore, studies to incorporate magnetic property components into MEMS structures have been undertaken.[7] For example, thin-film NiFe attached to a polysilicon micro cantilever beam was shown to actuate under a magnetic field[13] More recently, polydimethylsiloxane (PDMS) was mixed with NdFeB particles to form a magnetic membrane used in a micro pump.[41] Several magnetic oxide nanoparticles, including Fe_2O_3 and magnetite, have been synthesized with particle sizes between 4–16 nm by using microemulsion, electrochemical deposition, and other methods.[11,35] Composite films consisting of iron-iron oxide have also been achieved through chemical vapour deposition of iron pentacarbonyl.[19] More importantly, polymer coated magnetic nanoparticles have been fabricated using thermal deposition in the presence of ammonia and polymeric dispersants.[2,3] However, most of these magnetic incorporation methods require high temperature, high pressure or very expensive instrumentation to aid in the fabrication.

This chapter will describe a new magnetic spider silk composite fabricated using a regenerated spider silk matrix and iron pentacarbonyl ($Fe(CO)_5$) at room temperature and atmospheric pressure. Nanoparticles of iron oxides were made via photolysis of iron pentacarbonyl using UV light instead of thermal or chemical decomposition. Fourier Transformed Infrared Spectroscopy (FTIR) was used to characterize the interaction of iron pentacarbonyl with spider silk proteins. A micro cantilever beam made of iron oxide nanoparticles in thin-film spider silk was fabricated. The micro beam was actuated using magnetic field.

2.2 THIN-FILM SPIDER SILK PREPARATION

Spider dragline silk harvested from the golden orb weaving spider *N. clavipes* was obtained based on a traditional forced silking technique.[30,10,29] The silk was dissolved in a 1,1,1,3,3,3 hexafluoro-2-propanol (HFIP) solution with a ratio of either 1% w/w or 0.5% w/w.[30] The solution was allowed to dissolve for approximately 24 hours at ambient temperature before spin-coating onto a silicon wafer. A 10 ml droplet was spun at approximately 500 rpm for 30 seconds and then air dried for further thin-film mechanical measurements.

Iron spider silk thin film: Iron pentacarbonyl was purchased from Sigma Aldrich. Liquid iron pentacarbonyl has a low boiling point and is sensitive to UV light; therefore it has to be refrigerated and shielded from UV light. The mixture was diluted in 1:1 v/v (spider silk solution vs. $Fe(CO)_5$) ratio. The mixture was then deposited onto the silicon substrate using a pipette. The sample was then exposed under an Entela model UVG-54 handheld short wave UV lamp. This was performed for 2 to 3 hours. A thin film was formed from the droplet.

2.3 CHARACTERIZATION

2.3.1 SEM and EDS

Scanning electron microscopy was used to examine the quality of thin-films. Figure 2.1 shows a cross-sectional view of a thin spider silk film of two dilutions: 1% and 0.5%. It can be seen from the SEM image that porosity exists throughout the film, and there is no significant porosity difference between the two diluted solutions (1% and 0.5% w/w). These voids are presumably left behind by evaporation of the HFIP solvent (1,1,1,3,3,3 hexafluoro-2-propanol) during spinning and at the drying stage.

SEM images were also taken to examine magnetic spider silk film deposited on a silicon substrate (Fig. 2(a)). Figure 2.2 shows a SEM image of 1% v/v silk mixed with $Fe(CO)_5$. The sample appeared to exist in two phases - as a film and as a porous/fibrous-like area. It is not entirely clear why the formation of the

Figure 2.1. Thin film cross section of (a) 1% w/w (b) 0.5% w/w spider silk.

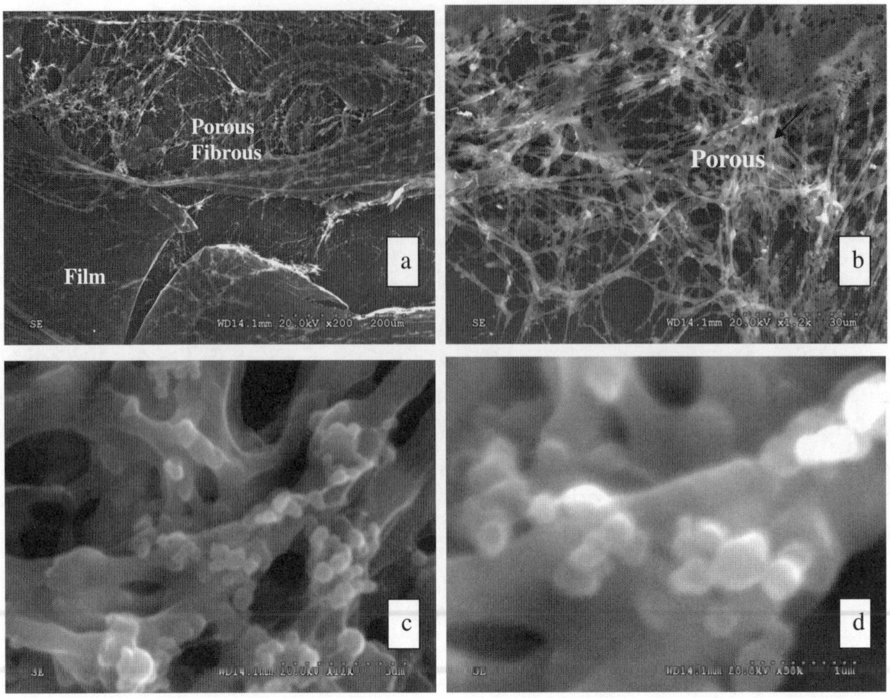

Figure 2.2. SEM image of 1% w/w silk mixed with $Fe(CO)_5$ with 1:1 v/v.

fibrous-like structure occurs. However, since both HFIP and $Fe(CO)_5$ are volatile solutions with a mixing ratio of 1:1 v/v, more than 99.5% of the final film solution will evaporate prior to spin casting. As the solvent quickly evaporates, it mimics the solvent extraction method used in creating porous materials.[17] Spin casting spider silk solution without $Fe(CO)_5$ (0.5% w/w ratio of spider silk with 99.5% of HFIP), produced a thin film without fibrous structure. Similar effects can be seen for more highly diluted film solutions. Therefore, these fibrous structures only form when $Fe(CO)_5$ is mixed in the film solution and after CO ligands have left the thin solid films from photolysis.

Figures 2.2(c) and 2.2(d) show higher magnification SEM images, where spheres can be seen on the fiber-like material. These particles are approximately 0.1 – 1 μm in diameter. They appear to be brighter than the rest of the film. The brighter area in the figures probably represents a higher concentration of Fe. It is suspected that because heavier elements generate more back-scattered electrons, they tend to appear brighter in SEM pictures. This conclusion is bolstered by the fact that biomaterial on its own is difficult to examine under SEM, as it is not conductive. Therefore any material that is dimmer in colour is expected to have lower Fe concentration. Conversely, it is also necessary to keep in mind that due to the spherical shape of the particles, more backscattered electrons will also be produced for imaging. Therefore, this brightness could also result from particle shape rather than conductivity.

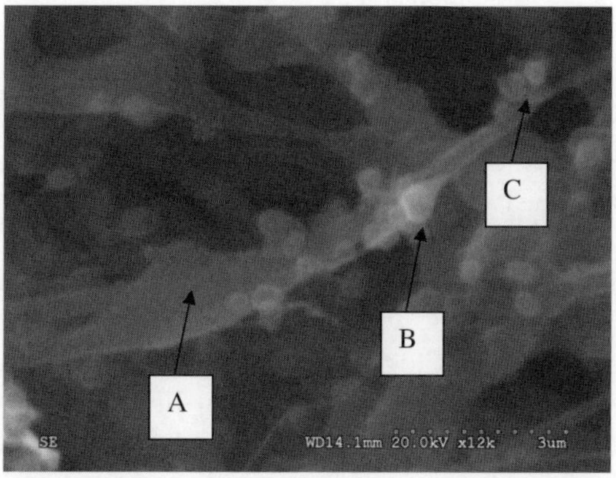

Figure 2.3. Image of the Area Used by X-ray EDS.

Table 2.1 Fe wt% found using X-ray EDS at Three Locations Indicated in Fig. 2.3.

	A	B	C
FE(wt%)	2.07	1.52	3.55

In order to determine the nature of the spheres, EDS was used to examine the elemental composition and concentrations in films made of iron pentacarbonyl doped spider silk and iron pentacarbonyl alone. For comparative purposes in this study, the UV exposure method outlined for iron spider silk film was performed on both films. The sample areas are shown in Fig. 2.3 and the EDS results are tabulated in Table 2.1. EDS was focused on the centres of both types of particles under SEM. It is important to note that EDS examines a small volume that is at least 1 mm in diameter. In the case of spheres from iron spider silk samples, it is difficult to separate the data from the spheres versus the data from the fibrous material underneath. This will be addressed in later sections. The initial concentration of iron contained in the iron spider silk mixture was calculated to be 13.9 wt%, however, EDS shows 1.52 wt% iron in the solid film. A much higher concentration of iron (36.2 wt%) was found using the pure $Fe(Co)_5$. Significant amounts of nitrogen and carbon are also found in the film since nitrogen is one of the basic building blocks of amino acids, thus indicating the presence of spider silk. The small concentration of fluorine is a result of HFIP. Similar concentrations of fluorine for regenerated spider silk samples have previously been reported.[29]

Two types of iron oxide occur naturally, Fe_2O_3 and Fe_3O_4, both having ferromagnetic properties.[39] Nano iron oxide particles of 10-30 nm diameter can be produced thermally from $Fe(CO)_5$.[14] The orange/copper colour of spheres observed from photolysis of iron pentacarbonyl have an iron concentration of 36 wt%, which is similar to that if Fe_2O_3, also orange/copper in color.

2.3.2 TEM

Magnetic spider silk film was investigated using TEM, where the size and contour of the spheres could be much more defined. The spherical particles varied in diameter, approximately between 150 to 650 nm. A select few were as large as 1.4 μm. The particles were not dispersed in any orderly fashion. They did however appear to connect and form chains or clusters (see Figs. 2.4(a) and 2.4(b)). Note that, particularly in Fig. 2.4(b), the spheres appear to blend in with each other forming fibrous structures, with the more transparent materials between the spheres appearing to join them together. It is possible, then, that the spheres not only exist on the surface, but also within the fiber-like material. However, in areas that are more concentrated with spider silk material (see Fig. 2.4(c)) spheres exist in fewer numbers and smaller sizes. These smaller particles could be a result of Fe particles not being able to bind with oxygen as easily, making diffusion within a higher concentrated spider silk region more difficult.

2.3.3 FTIR

Fourier Transformed Infrared Spectroscopy (FTIR) was used to characterize the molecular structure of the spin-on spider silk film. Previously, FTIR measurements

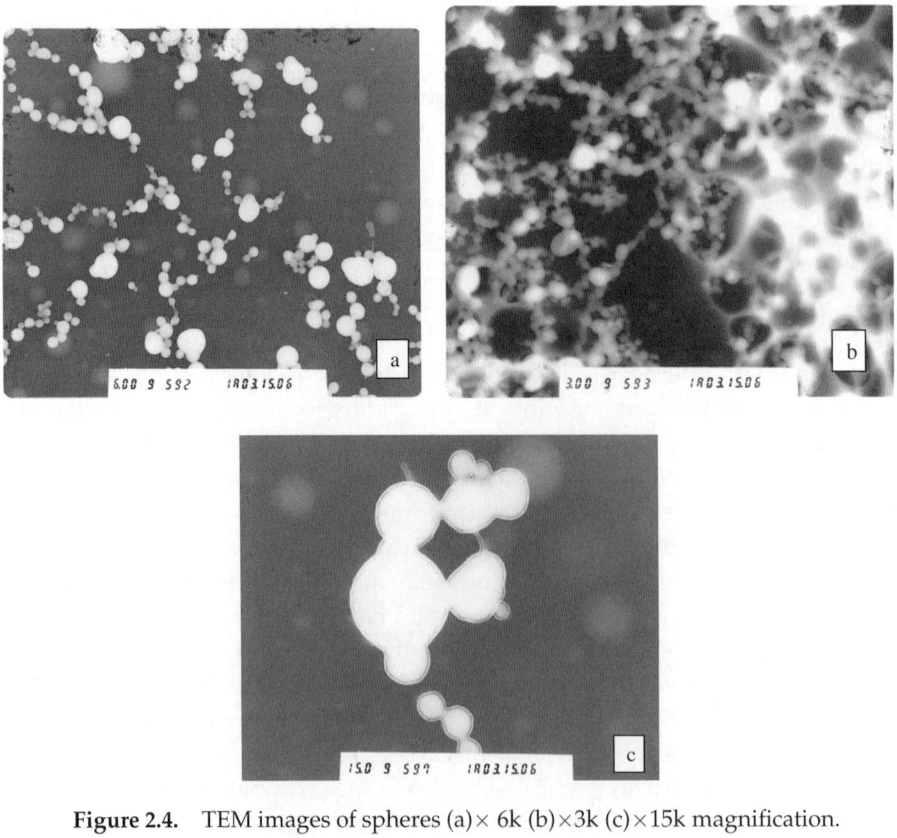

Figure 2.4. TEM images of spheres (a)× 6k (b)×3k (c)×15k magnification.

of native spider silk fibers have shown peaks corresponding to Amide I, II and III vibration modes. These peaks correspond to the arrangement of polypeptides present in the silk's secondary structure. The material properties of spider silk have been shown to be strongly linked to its secondary structures.[10,33] FTIR Peaks at wave numbers 1652 and 1669 cm^{-1} in the Amide I region have been assigned to α-helical and β-sheet backbone conformations respectively.[31] A peak at 1540 cm^{-1} in the Amide II band corresponds to disordered conformations,[6] while peaks at 1230 and 1240 cm^{-1} in Amide III are associated with β-sheets and disordered conformations.[45] The FTIR spectra are shown in Fig. 2.5. The top spectrum represents the iron spider silk sample and the bottom spectrum represents the pure regenerated spider silk sample.

As shown in Fig. 2.5, the FTIR spectrum of the regenerated spider silk thin film obtained from a spin-on process shows peaks at 1660, 1542 and 1242 cm^{-1}. The pattern of the three peaks, ranging in intensity from moderate to strong, is similar to that reported to indicate the presence of α-helical and random conformation components. The β-sheet conformation, which is suggested to be responsible for the high stiffness and strength of the material[29] was not found to be present in this work. This indicates a change in the material's secondary structure in comparison to the native fiber state.

Peaks below 1450 were not resolved and lacked distinctiveness, thus they are not addressed here. The Amide I and Amide II peaks are shown at wave numbers 1657 and 1537. The vibration mode in the spectral region between 1600–1700 cm^{-1} predominantly originated from the C = O stretching of Amide I group.[21] Based on the position within this spectral region, 1657 cm^{-1} represents a 20% presence of α-helix in the sample's secondary structure.[46,47] 1537 falls within the Amide II

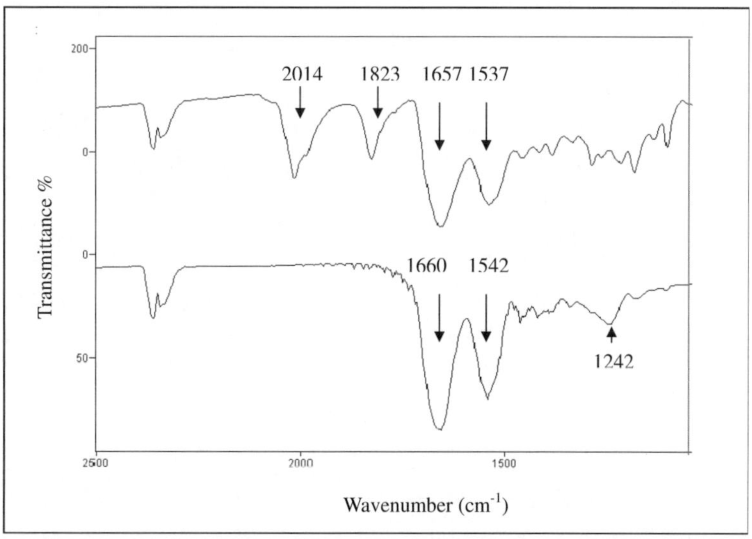

Figure 2.5. FTIR of iron spider silk sample. Top spectrum is iron spider silk; bottom spectrum is the pure regenerated spider silk.

spectral region of 1500–1600 cm^{-1}. This region represents N-H bending and C-N stretching of the backbone of Amide II.[24] Spectral 1537 also represents an α-helix secondary structure.[22] These peaks exist in the pure regenerated spider silk FTIR spectrum; therefore their presence here also represents the secondary structure of spider silk in the iron spider silk film. Peak 1242 in the pure spider silk sample has become resolved with the iron spider silk sample. This could be explained by the fact that 1242 represents a random conformation, and was not a high intensity peak in the pure spider silk sample. The resolved peak in the iron spider silk sample therefore does not represent a finding of any new significance.

The spectral peak found at 2014 cm^{-1} represents the presence of Fe-CO vibration assigned for $Fe(CO)_5$, confirming the presence of this iron compound in the solid film. The photolysis reaction to convert $Fe(Co)_5$ to iron oxide was monitored with FTIR and illustrated in Fig. 2.7. Mechanistically, one of the Fe-CO bonds will first be broken by one photon upon UV irradiation.[5] As the photon cleaves one bond, the CO is broken free from what is now an unstable molecule of $Fe(CO)_4$.[25] Two possible reactions of $Fe(CO)_4$ could then follow. One is that the CO will evaporate, leaving the unstable $Fe(CO)_4$ molecule to naturally decompose where the Fe and the CO can completely separate. However, due to the matrix of proteins, the CO might not be able to escape but instead be trapped again by the iron. Therefore, it is also possible due to the close proximity of CO to $Fe(CO)_4$, that $Fe(CO)_5$ will be reformed as it is a more stable state for the molecules. As can be seen in Fig. 2.6, within 40 minutes of irradiation all the Fe-CO bonds had been broken.

Figure 2.5 also shows a peak at 1823 cm^{-1} which lies in the region normally associated with bridging CO groups.[5] It suggests the presence of multiple CO

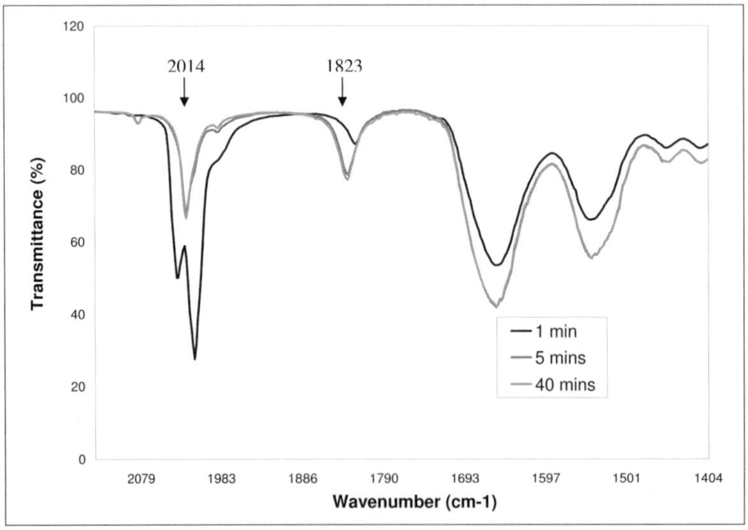

Figure 2.6. FTIR spectrums of iron spider silk sample after different exposure time. For color reference, see page 253.

bands in an unresolved envelope. The value of 1823 is very close to that of 1828, which has previously been reported for another compound, triiron dodecacarbonyl ($Fe_3(CO)_{12}$).[5] This peak could possibly show the existence of two types of Fe bindings taking place. After initial photolysis, it is possible that several unstable $Fe(CO)_4$ molecules, having just lost their own CO, share CO ligands with their respective neighbours to form $Fe_3(CO)_{12}$ due to their close proximity to each other. This could be an additional reason why a high concentration of carbon is found within the particles produced by pure. $Fe(CO)_5$ Moreover, C = O side groups are also found in the major spider silk amino acids such alanine and glycine.[42] Therefore, the existence of C=O side groups could provide an explanation as to why, in some areas, formation of iron spider silk spheres is observed. As the spider silk dissolves in HFIP, the hydrogen bonds between the amino acids break, thus allowing the spider silk to dissolve into a solution. It is likely that $Fe(CO)_4$ can bind to the C = O group within the amino acids, and form $Fe_3(CO)_{12}$ with other $Fe(CO)_4$ molecules, as well as with amino acids.

To further support the hypothesis related to peak 2014 and 1823, iron spider silk samples were tested against time using FTIR methods. Here, spectra were taken at different times after deposition of iron spider silk solution. Again, a potassium bromide plate was used as the substrate. The results are shown in Fig. 2.6. The peak of 2014 decreased in transmittance very quickly during the first 5 minutes of UV exposure. This shows the breaking of Fe-CO bonds within $Fe(CO)_5$. Evaporation of $Fe(CO)_5$ could also have contributed to this decrease. Once passed the initial 5 minutes, the reaction appeared to reach a plateau. This result confirms with the hypothesis (illustrated in Fig. 2.7) that a certain level of $Fe(CO)_5$ will always remain in the sample.

In contrast, the peak of 1823 increased in transmittance during the first five minutes of UV exposure. This increase supports the hypothesis that formation of $Fe_3(CO)_{12}$ is likely to have occurred. After five minutes, the reaction appeared to also have reached a plateau. These three FTIR spectra show that the initial minutes of UV exposure is a crucial state. Once that is passed, the reaction significantly

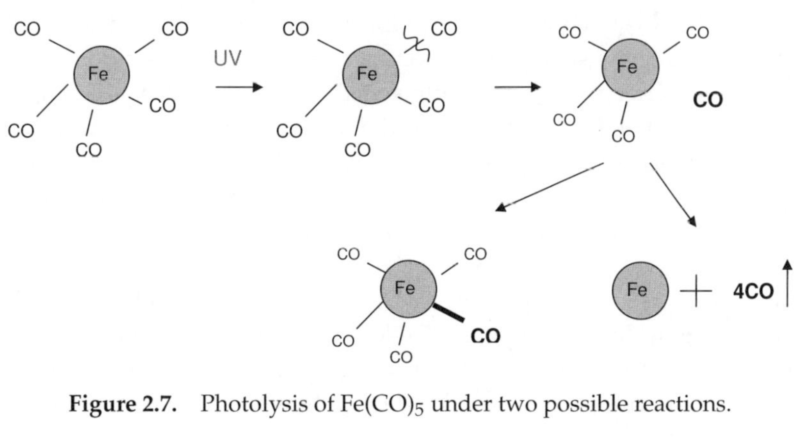

Figure 2.7. Photolysis of $Fe(CO)_5$ under two possible reactions.

slows down. This is shown by overlapping of the five minute spectrum with the 40 minute spectrum.

Finally, a small shift in wave number at the 1823 peak is shown between the one minute and five minute spectra. Since the wave number shifts when the sample is in different states (such as liquid, solid or gas), it is likely that the shift here is a result of the sample not being completely dried when the spectrum was taken after one minute of UV exposure.

For the magnetic spider silk material, the formation of the observed spheres is a direct result of mixing iron pentacarbonyl with the regenerated spider silk solution. The spheres contain a concentration of Fe that is slightly higher than the rest of the material. Fibrous and porous material containing a low concentration of iron also formed as a result of this mixed solution. In order to further explore the reason behind this phenomenon, an FTIR was performed to study any binding that could have occurred between the iron particles and the spider silk proteins.

2.3.4 Squid

A SQUID was used to characterize the magnetic hysteresis of the material. Figure 2.8 illustrates the M vs. H hysteresis plot for a solid iron spider silk sample at 300 K. The saturation magnetization is found at 1.4×10^{-5} emu/g, which is 105 times smaller than the previously reported level for polymer coated iron particles.[2] This is not surprising, as the iron wt% in the reported particles is around 21%; whereas the amount of Fe in this sample is only around 1.5–3 %. The remanence ratio of 16.6% and coercivity of 30 Oe was also found. These two values are in the same range as the ones reported for polymer coated iron particles.

2.3.5 Micromachining and Mechanical Testing of a Spider Silk Microbridge

The mechanical testing results of the film performed by a nano indentation machine [mts] are summarized in Table 2.2. The elastic modulus and hardness decrease as the % w/w ratio of silk to solvent decreases. This is likely due to the difference in fibroin concentration. However, an advantage of the 0.5% w/w solution is its film consistency, reduction in surface roughness, and its increased sensitivity to the effects of spin-coating speed. As shown in Fig. 2.1(a), the film porosity appears to be less prominent in the upper region compared to the lower region of the cross section. Mechanical testing data were mostly collected in the upper region of the film thickness (the first 500 nm), and this could be a contributing factor to the high Young's modulus value, which is not common in porous materials.

Data on the viscoelastic properties (loss tangent) of the film were obtained for the first time in this experiment. The loss tangent showed no dependency on % w/w dilution in the range used and was found to be similar to the loss tangent of human bone specimens (0.05).[38] The data collected using the nano indentation

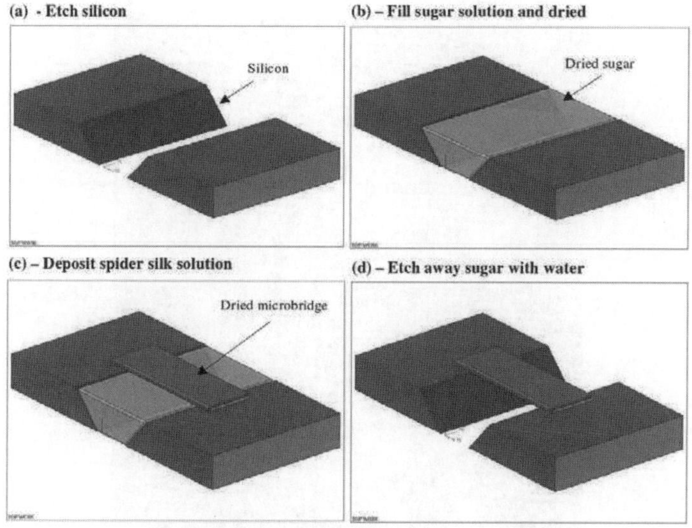

Figure 2.8. Sacrificial etching process steps to fabricate spider silk microbridge. For color reference, see page 253.

Table 2.2 Mechanical testing of thin film spider silk

Sample	Elastic Modulus (GPa) [std]	Hardness (GPa) [std]	Viscoelasticity (loss tangent) [std]
1% w/w	7.251[0.494]	0.276 [0.005]	0.0436 [0.0065]
0.5% w/w	4.41 [0.342]	0.182 [0.012]	0.182 [0.012]

machine takes into consideration the substrate effect on thin film tests. This has been previously documented in many studies.[4,26] Using a widely accepted rule of thumb, the intrinsic property of the film can be reasonably detected with minimal effect from the substrate when the indentation depth is kept to 10% of the film thickness.[37] Only consistent mechanical property data within this region is used in this work. In addition, the thicknesses of the films used are in the range of 5–10 μm, which should improve the accuracy of the nano indentation data.

In order to demonstrate a micromachined spider silk microstructure, a modified surface micromachining process with a sacrificial etching technique[36] was carried out to fabricate a free-standing spider silk microbridge. Figure 2.8(a), shows a silicon substrate that is anisotropically etched to form a cavity. Sugar solution (50 wt%) was poured into the cavity and air dried to form a sacrificial layer (Fig. 2.8(b)). The spider silk solution was then deposited using a 2 ml needle syringe and air dried (Fig. 2.8(c)). Finally, the silicon substrate was immersed in water to "etch" away the sugar (Fig. 8(d)), leaving the free-standing spider silk microbridge shown in Fig. 2.8(d). Figure 2.9(a) shows a SEM microphoto of the spider silk microbridge ($800 \times 800 \times 40\ \mu m^3$).

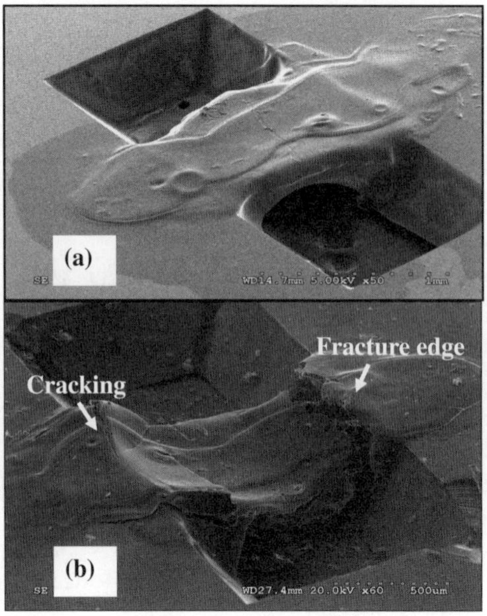

Figure 2.9. SEM photo of the spider silk microbridge (800 × 800 × 40) before (a) and after (b) a fracture test.

A nano indentation machine was used for a loading-unloading test on the spider silk microbridge and the results are shown in Fig. 2.10. The microbridge was first loaded up to 5 mN and unloaded. A second loading-unloading cycle with a larger maximum force of 100 mN was then applied and released. The 100 mN loading-unloading cycle overlaps with the 5 mN loading-unloading cycle when the initial displacement is small. However, as the unloading did not follow the path of the loading curve, it can be concluded that plastic deformation occurred with the loading of 5 mN.

A fracture test was carried out on the spider silk microbridge by the nano indentation machine. The loading force was increased gradually and the displacement was recorded until the beam fractured. The spider silk microbridge is assumed as a fixed-fixed beam structural member with a rectangular cross section. Since the load is applied on to the beam through a large silicon chip to prevent a local indentation, the beam experiences a distributed load. The maximum moment experienced by the two fixed ends of the beam can be modeled as [48]:

$$M = F \times L/12$$

where M is the maximum moment at the fixed-ends, F is the loading force applied by the indentation tip and L is the length of the undeformed beam. The normal stress can then be calculated using [48]:

$$\sigma = M \times c/I$$

where σ is the maximum stress, c is half of the thickness of the beam, and I is the moment of inertia of the cross section area of the beam. The strain is calculated

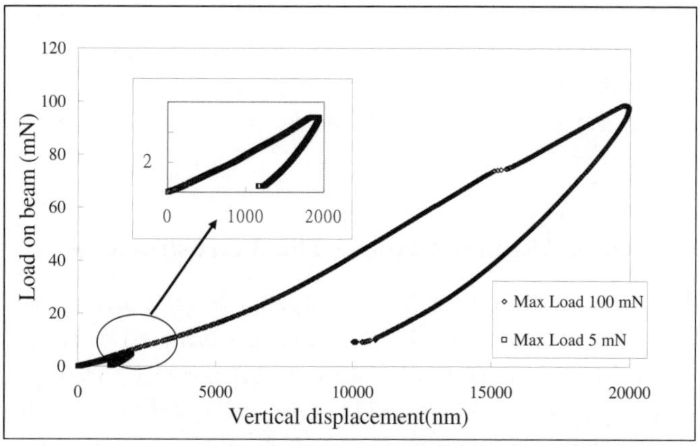

Figure 2.10. Loading and unloading tests. Two maximum loads were reached and released.

following the Pythagorean Theorem:

$$l^2 + B^2 = C^2$$

where, B is the vertical distance traveled by the nano indentation tip, and l is the original half length of the beam, C is the new half length of the beam. Therefore, the change in beam length, Δ is equal to C/l. The strain is then calculated by Δ/l. Figure 2.9(b) shows a SEM microphoto of the spider silk microbridge ($800 \times 800 \times 40\ \mu m^3$) after a fracture test. The fractures occurred at the fixed-ends where maximum moment and maximum normal stress occurred.

Figure 2.11 shows a stress-strain curve obtained from the fracture testing. Three microbridges were tested. The Ultimate Tensile Strength (UTS) was found to be in the range of 65.6–85.1 MPa with a strain range of 1.84–2.4%. One main

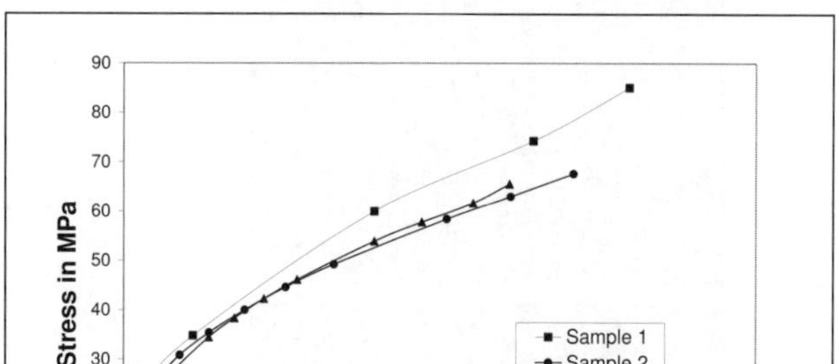

Figure 2.11. Stress and strain curve of a spider silk microbridge. The UTS' were measured as 65.6, 67.6, and 85.1 MPa, for the three samples with microbridge widths of 0.9, 0.8 and 0.4 mm respectively. The other dimensions are identical between the three samples.

contributing factor in the differences observed between the three sets of fracture data is due to the variations in the dimensions of the microbridges. The volumetric change of the microbridge depends on the time duration in which it is in contact with water during the release step. A precise time control in the release step is difficult to achieve.

2.3.6 Actuation of a Magnetic Spider Silk Microstructure

Spider drag-line silk is dissolved in a 1,1,1,3,3,3 hexafluoro-2-propanol (HFIP) solution with a ratio of 1% w/w.[49,50] Ni particles with an average diameter of 2.6 μm were added to the spider-silk solution at a ratio of 15% w/w. A spin-on process (30 seconds at 500 rpm) was used to form a Ni/spider-silk film on a silicon substrate. Figure 2.12(a) shows the top view of the thin-film Ni/spider silk. While porosity is evident throughout the film, this porosity can be reduced by decreasing the weight ratios between the Ni particles and the spider-silk solution. The spider silk matrix acts as a structural support, and the morphology of the Ni/spider-silk film is dominated by the packing preferences of the Ni particles, which are likely a result of Van der Waals forces between the particles. As shown in Fig. 2.12(b) spider silk has a smoother surface and a lighter color than the Ni particles.

In order to demonstrate a Ni/spider-silk microstructure, a modified surface micromachining process using a sacrificial etching technique was carried out to fabricate a free-standing spider-silk microbridge. A plastic substrate was preformed with a cavity using a conventional drilling process. A sugar solution (50% wt) was poured into the cavity and air dried to form a sacrificial layer. The Ni/spider-silk solution was then deposited using a 2 μl needle syringe, and air

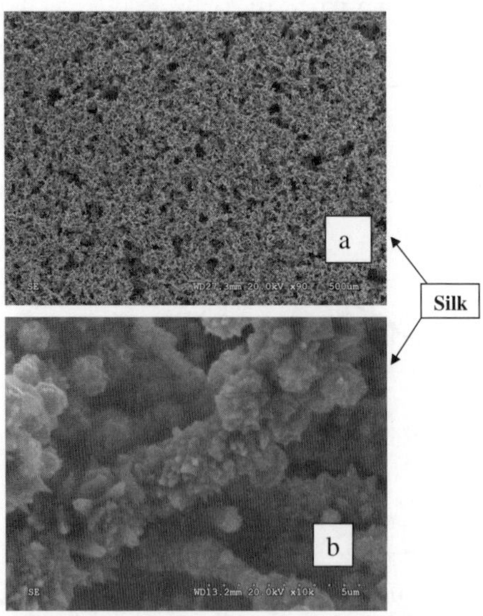

Figure 2.12. SEM imaging of Ni-spider silk film (a)×90 top view (d)×10k top view.

dried. Finally, the plastic substrate was immersed in water to "etch" away the sugar, leaving the free-standing Ni/spider-silk microbridge shown in Fig. 2.13 (3.5 mm $\times$ 1.3 mm $\times$ 0.2 mm).

A conventional *M* (magnetization) versus *H* (magnetizing field) hysteresis loop was measured and plotted for the Ni/spider silk, as shown in Fig. 2.14. A Quantum Design RSO SQUID magnetometer was used to characterize the Ni/spider silk film at 300°K. The saturation magnetization (M_S), coercivity (H_C), and remanence ratio (M_r/M_S) were found to be 0.028 *emu/g*, 74 O_e and 13.4%, respectively. These measurements were performed on a 14% w/w (Ni:spider-silk solution) sample. Bulk Ni, on the other hand, has M_S, H_c and M_r/M_S values of 55 emu/g, 100 O_e and 5%, respectively [51]. It can be seen that the M_s of the Ni/spider silk is significantly lower than that of the bulk Ni, indicating that the Ni/spider silk is less magnetic. However, the values of H_c and M_r/M_S for the Ni/spider silk are comparable to those of the bulk Ni, indicating that the Ni/spider silk material retains ferromagnetic properties.

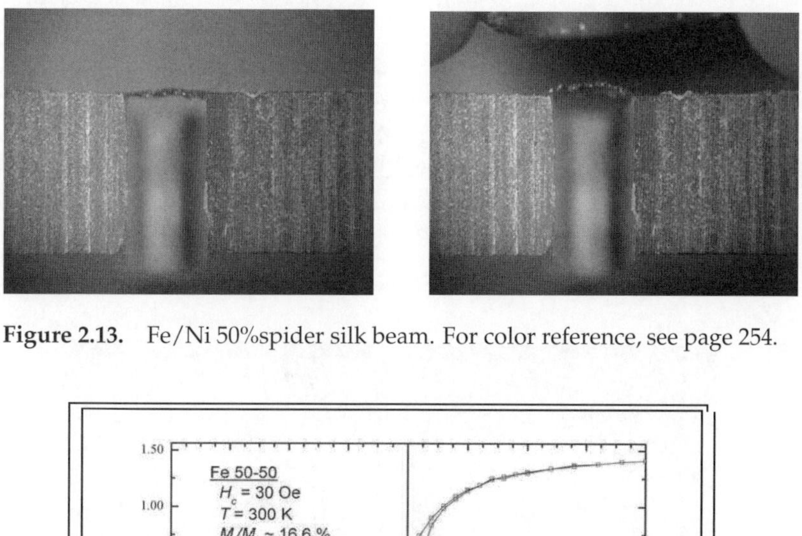

Figure 2.13. Fe/Ni 50%spider silk beam. For color reference, see page 254.

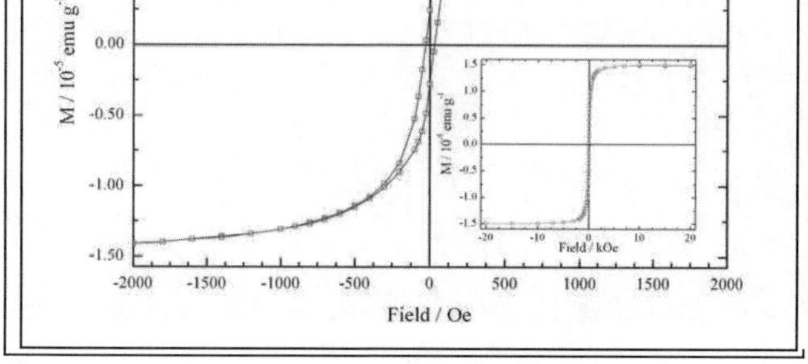

Figure 2.14. Hysteresis curve for 1:1 v/v iron spider silk sample. For color reference, see page 254.

A static bending test was performed using a permanent magnet. A Ni/spider-silk microbeam was positioned at one end of a fixture, and two permanent magnets were attached to a glass rod. As the glass rod was slid in and out of the fixture from the opposite end, the distance between the magnet and the microbeam changed, and thus varing the magnetization strength. A gaussmeter was used to measure the magnetization field strength as a function of the distance between the magnets and the Ni/spider-silk microbeam sample. A scalpel was used to remove one fixed end of a microbeam and create a cantilever beam (Fig. 2.15) The movement of the beam was captured by a camera on a microscope, as shown in Fig. 2.15. The vertical displacement of the cantilever beam tip was measured by image software. Figure 2.16 shows the static vertical displacement of the cantilever beam tip as a function of the external field strength. Three samples prepared at 75 min, 7.5 hrs and 24 hrs of drying time after release in water were tested. The error bars of the three data sets overlap each other, except at 1.3 kO_e magnetic induction. The Ni/spider silk beam had a displacement of 1 mm under a static magnetic induction of 1.8 kO_e. Overall, no significant differences can be seen across the bending results for their allotted drying time, provided that there is a minimum drying time of 75 minutes.

Previously reported regenerated spider silk in a fiber form had an UTS of 320 MPa and a breaking strain of approximately 10–14%[28]. The material properties (UTS and breaking strain) obtained in this work are significantly lower. This difference could be explained by the lack of a post-drawing process involved in the spin-on formation of the spider-silk thin film, as compared to fiber formation with repeated post-drawing steps.[28,30] FTIR (Fourier transform infrared spectroscopy)

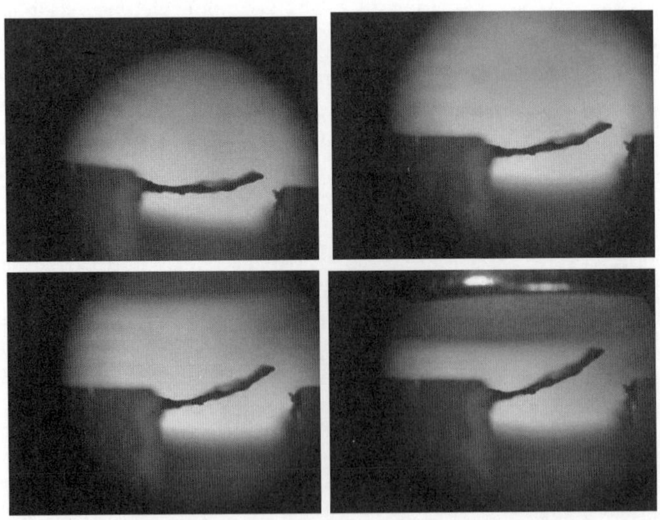

Figure 2.15. Images used to measure bending movement of the beam under increasing magnetic field. Sample size 3.25 × 0.98 × 0.17 mm (*Length* × *Width* × *Thickness*). For color reference, see page 255.

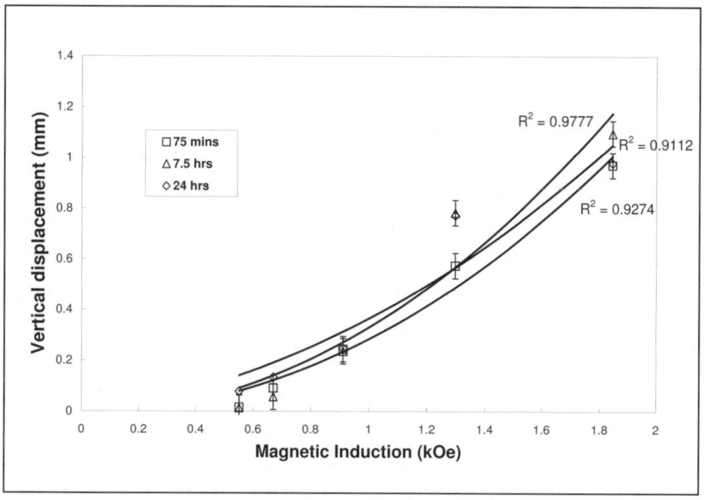

Figure 2.16. Bending distance as magnetic induction (field) increases.

spectra of regenerated spider silk in a thin-film form have been measured[12]leading to the suggestion that the β-sheet conformation, which is responsible for the high stiffness and strength of spider silk,[29] is missing from thin-film spider silk. This result indicates a dramatic change in the material's secondary structure compared to its native fiber state. Furthermore, the significant decrease in mechanical property shown in Ni/spider-silk film silk is perhaps due to the porosity of the material, which would allow stress concentration to occur around the Ni particles. The material property could be improved if the dilution of Ni/spider silk was reduced. However, in this work the near maximum dilution was used in order increase the material's magnetic properties.

Furthermore, contact with water during the release step also affects the strength of the microbridge. Previous studies have shown that water causes changes in the arrangement of the silk fibroins and causes super-contraction of the silk fibers, which can decrease the UTS and increase the brittleness of the material.[30,31,40] When regenerated fibers were stretched in water, the UTS was reduced to 25 MPa.[30] It is unlikely that our silk structures would super-contract as native fibers do; however, the immersion in water likely does enable some rearrangement of the protein-polymer backbone. The regenerated Ni/spider silk studied in this work has a comparable UTS to PDMS (with UTS in the range of 2.4–7 MPa [mit]), a popular material in MEMS.

2.4 CONCLUSIONS AND OUTLOOK

This chapter has presented regenerated Ni/spider silk as a material for MEMS. Thin-film formation using a spin-on process and a modified surface micromachined Ni/spider-silk microbridge were demonstrated. Static and dynamic bending tests on the Ni/spider-silk microbeams were performed. Mechanical testing of a spider-silk microbridge was carried out using a nano indentation machine and showed a UTS of 2.1 MPa. Magnetic membranes made of the Ni/silk film can be used for pumping or as valves in the microfluidic channels. Enhancement of the mechanical properties of the Ni/spider-silk material could involve the use of smaller particles, a sacrificial etching process that does not use water and possible processes that could rearrange the spider silk protein secondary structure during the thin-film formation step.

Our results allow us to come to several conclusions about the magnetic spider silk. The concentration of Fe in the iron spider silk material obtained by using the current method is low, resulting in low magnetization of the material. It has previously been shown that the spheres have a higher concentration of Fe then the rest of the film. In addition, in areas where there appears to be a high concentration of spider silk material, the formation of spheres is more inhibited.

Overall, this leads to the hypothesis that a lower concentration of spider silk solution should form more densely packed particles with a higher concentration of Fe. This is premised on the idea that when there is a higher concentration of spider silk in the film, the amino acids prefer to bind with themselves to form spider silk film. However, with less concentrated spider silk solution, the amino acids are more likely to bind with iron to form spheres, because smaller amounts of spider silk amino acids are available to form films. Based on this hypothesis, 0.25% w/w and 2% w/w regenerated spider silk solutions are mixed with iron pentacarbonyl to form iron spider silk films.

2.5 ACKNOWLEDGMENTS

This work was supported by the Natural Science and Engineering Council (NSERC) DISCOVERY grant and the Canada Foundation for Innovations (CFI) of Canada.

References

1 G. Altman , Silk-based biomaterials, *Biomaterials*, **24**, 401 (2003).

2 N. A. D. Burke, H. D. H. Stover, F. P. Dawson, J. D. Lavers, P. K. Jain, and Oka H., Preparation and characterization of polymer-coated magnetic nanoparticles. *IEEE Transactions on Magnetics*, **37**(4), 2660–2662 (2001).

3 N. A. D. Burke, H. D. H. Stover and F. P. Dawson, Magnetic nanocomposites: preparation and characterization of polymer-coated iron nanoparticles, *Chemistry of Materials*, **14**, 4752–4761 (2002).

4 S. Chen, L. Liu and T. Wang, Size dependent nanoindentation of a soft film on a hard substrate, *Acta Materialia*, **52**, 1089–1095 (2004).

5 M. J. Clarke, A. I. Cooper, S. M. Howdle and M. Poliakoff, Photochemical reactions of organometallic complexes impregnated into polymers: Speciation, isomerization, and

hydrogenation of residual alkene moieties in polyethylene, *J. Am. Chem. Soc.*, **122**, 2523–2531 (2000).

[6] Fukushima Yasumasa., Secondary structural analysis in the solid state for analogous sequential polypeptides of glycine-rich sequence of spider dragline silk. Polymer Bulletin [Online]. **45**(3), 237–244 (2000).

[7] M. R. J. Gibbs, E. W. Hill and P. J. Wright, Magnetic materials for MEMS applications, *J. Phys. D*, **37**, 237–244(2004).

[8] J. M. Gosline, P. A. Guerette, C. S. Ortlepp and K. N. Savage, The mechanical design of spdier silks: From fibroin sequence to mechanical function, *The Journal of Experimental Biology 202*, 3295–3303 (1999).

[9] R. C. Hibbler, *Mechanics of Materials*, 4th ed. New Jersey: Prentice Hall (1999).

[10] D. H. Hijirida, Kinh Gian Do, C. Michal, Shan Wong, D. Zax and L. W. Jelinski, 13C NMR of Nephila clavipes major ampullate silk gland, *Biophys. J.*, **71**, 3442–3447 (1996).

[11] T. Hyeon, S. S. Lee, J. Park, Y. Chung and H. B. Na, Synthesis of highly crystalline and monodisperse maghemite nanocrystallites without a size-selection process, *J. Am. Chem. Soc., 123* 12798–12801, Dec 26 (2001).

[12] J. T. Bai, W. Ma, R. Z. Chu, L. Wang, C. Silva, J. C. M. Chao and M. Chao, Regnerated spider silk as a new biomaterial for MEMS, *Biomedical Microdevices*, **8**(4), 317–323 (2006).

[13] J. W. Judy, R. S. Muller and H. H. Zappe, Magnetic microactuation of polysilicon flexture structures. *Journal of MEMS*, **4**, 162–169 (1995).

[14] M. N. A. Karlsson, K. Deppert, B. A. Wacaser, L. S. Karlsson and J.-. Malm, Size-controlled nanoparticles by thermal cracking of iron pentacarbonyl, *Applied Physics A: Materials Science and Processing*, **80**, 1579–1583 (2005).

[15] S. Kubik, High-performance fibers from spider silk, *Angewandte Chemie*, **41**, 2721 (2002).

[16] A. Lazaris, S. Arcidiacono, Y. Huang, J. Zhou, F. Duguay, N. Chretien, E. A. Welsh, J. W. Soares and C. N. Karatzas, Spider silk fibers spun from soluble recombinant silk produced in mammalian cells, *Science* **295**, 472–476 (2002).

[17] L. Li, F. Yan and G. Xue, Preparation of a porous conducting polymer film by electrochemical synthesis-solvent extraction method, *J. Appl. Polym. Sci.*, **91**, 303–307 (2004).

[18] S. F. Y. Li, A. J. McGhie and S. L. Tang, Comparative study of the internal structures of kevlar and spider silk by atomic force microscopy, in *40th National Symposium of the American Vacuum Society*, **15–19**, 1891–1894 (1994).

[19] T. Maruyama and Y. Shinyashiki, Iron-iron oxide composite thin films prepared by chemical vapor deposition from iron pentacarbonyl, *Thin Solid Films*, **333**, 203–6, 2011/2023 (1998).

[20] MIT Open Courseware, *Design of Medical Devices and Implants.* Chapter 8 (2004).

[21] C. Mouro, C. Jung, A. Bondon and G. Simonneaux, Comparative Fourier transform infrared studies of the secondary structure and the CO heme ligand environment in cytochrome P-450cam and cytochrome P-420cam, *Biochemistry*, **36**, 8125–8134 (1997).

[22] O'Hagan, Ryan (MTS Application Engineer). Nano indenter - high load testing method. E-mail communication.

[23] A. Naidja, C. Liu and P. M. Huang, Formation of protein-birnessite complex: XRD, FTIR, and AFM analysis, *J. Colloid Interface Sci.*, **251**, 46–56 (2002).

[24] K. Onoda, H. Mino, Y. Inoue and T. Noguchi, An FTIR study on the structure of the oxygen-evolving Mn-cluster of Photosystem II in different spin forms of the S(2) state, *Photosynth. Res.* **63**, 47–57 (2000).

[25] J. M. Parnis, M. G. K. Thompson and L. M. Ashenhurst, Matrix isolation of electron bombarded gases containing Fe(Co) 5: An FTIR absorption study of neutral and anion decomposition products, *Journal of Physical Chemistry A*, **107**, 7390–7395 (2003).

26 R. Saha and W. D. Nix, Effects of the substrate on the determination of thin film mechanical properties by nanoindentation, *Acta Materialia*, **50**, 23–38 (2002).

27 J. Scheller, K.-H. Guhrs, F. Grosse and U. Conrad, Production of spider silk proteins in tobacco and potato, *Nat. Biotechnol.*, **19**, 573–577 (2001).

28 A. Seidel, O. Liivak, S. Calve, J. Adaska, G. Ji, Z. Yang, D. Grubb, D. B. Zax and L. W. Jelinski, Regenerated spider silk: Processing, properties, and structure, Macromolecules, **33**, 775–780 (2000).

29 A. Seidel, O. Liivak and L. W. Jelinski, Artificial spinning of spider silk, *Macromolecules*, **31**, 6733–6736 (1998).

30 Z. Shao, Structure and behavior of regenerated spider silk, *Macromolecules*, **36**, 1157 (2003).

31 Z. Shao and F. Vollrath, Effect of solvents on the contraction and mechanical properties of spider silk, *Polymer*, **40**, 1799–1806 (1999).

32 Susan Sofia, Mary Beth McCarthy, Gloria Gronowicz and David L. Kaplan, Functionalized silk-based biomaterials for bone formation, *Journal of Biomedical Materials Research*, **54**, 139–148 (2001).

33 J. Stephens, Effects of electrospinning and solution casting protocols on the secondary structure of a genetically engineered dragline spider silk analogue investigated via fourier transform Raman spectroscopy, *Biomacromolecules*, **6**, 1405 (2005).

34 X. Sun and X. Dong, Magnetic properties and microstructure of carbon encapsulated Ni nanoparticles and pure Ni nanoparticles coated with NiO layer, *Mater. Res. Bull.*, **37**, 991–1004 (2002).

35 S. Sun, H. Zeng, D. B. Robinson, S. Raoux, P. M. Rice, S. X. Wang and G. Li, Monodisperse MFe2O4 (M = Fe, Co, Mn) nanoparticles, *J. Am. Chem. Soc.*, **126**, 273–279 (2004).

36 W. C. Tang, T.-H. Nguyen, M. W. Judy and R. T. Howe, Electrostatic-comb drive of lateral polysilicon resonators, in *5th International Conference on Solid-State Sensors and Actuators and Eurosensors III*, **25–30**, 328–31 (1990).

37 T. Y. Tsui, C. A. Ross and G. M. Pharr, Nanoindentation hardness of soft films on hard substrates: Effects of the substrate, in *Proceedings of the 1997 MRS Spring Symposium*, **3**, 57–62 (1997).

38 De Silva, Leandro Macedo Soares, V. Ebacher, D. Liu, H. McKay, T. R. Oxland and R. Wang, Elasticity and viscoelasticity of human tibial cortical bone measured by nanoindentation, in *2005 Materials Research Society Spring Meeting*, 91–96 (2005).

39 Wikipedia Contributors. Wikipedia, the free encyclopedia. [Online]. 2006(June/21), Available: http://en.wikipedia.org/wiki/Iron_(III)_oxide

40 Z. Yang, O. Liivak, A. Seidel, G. LaVerde, D. B. Zax and L. W. Jelinski, Supercontraction and backbone dynamics in spider silk: 13C and 2H NMR studies, *J. Am. Chem. Soc.*, **122**, 9019–9025 (2000).

41 C. Yamahata and M. A. M. Gijs, Plastic micropumps using ferrofluid and magnetic membrane actuation, in 17th*IEEE International Conference on Micro Electro Mechanical Systems*. Technical Digest, pp. 458–461 (2004)

42 Yong Zhang, Luru Dai and Zhong-Can Ou-Yang, Amino acid composition-dependent elasticity of spider silk, in *Progress in Statistical and Nonlinear Physics: Proceedings of the First Cross Taiwan-Strait Conference on Statistical Physics*, 2516–2522 (2004).

43 C. Mouro, C. Jung, A. Bondon and G. Simonneaux, Comparative Fourier transform infrared studies of the secondary structure and the CO heme ligand environment in cytochrome P-450cam and cytochrome P-420cam, *Biochemistry*, **36**, 8125–8134, Jul 1 (1997).

44 A. Naidja, C. Liu and P. M. Huang, Formation of protein-birnessite complex: XRD, FTIR, and AFM analysis, *J. Colloid Interface Sci.*, **251**, 46–56, Jul 1 (2002).

45 J. Stephens, Effects of electrospinning and solution casting protocols on the secondary structure of a genetically engineered dragline spider silk analogue investigated via fourier transform Raman spectroscopy, *Biomacromolecules*, **6**, 1405 (2005).

46 C. Mouro, C. Jung, A. Bondon and G. Simonneaux, Comparative Fourier transform infrared studies of the secondary structure and the CO heme ligand environment in cytochrome P-450cam and cytochrome P-420cam, *Biochemistry*, **36**, 8125–8134, Jul 1 (1997).

47 A. Naidja, C. Liu and P. M. Huang, Formation of Protein-Birnessite Complex: XRD, FTIR, and AFM Analysis, *J. Colloid Interface Sci.*, **251**, 46–56, Jul 1 2002.

48 R. C. Hibbler, *Mechanics of Materials*, 4th ed. New Jersey: Prentice Hall (1999).

49 Z. Shao, Structure and behavior of regenerated spider silk, *Macromolecules*, **36**, 1157 (2003).

50 J. Bai, T. Ma, W. Chu, R. Z. Wang, L. Silva, C. Michal, J. C. Chao, M. Chao, Regnerated spider silk as a new biomaterial for MEMS, *Biomedical Microdevices*, **8**(4), 317–323 (2006).

51 X. Sun and X. Dong, Magnetic properties and microstructure of carbon encapsulated Ni nanoparticles and pure Ni nanoparticles coated with NiO layer, *Mater. Res. Bull.*, **37**, 991–1004 (2002).

Chapter Three

Biodegradable Elastomeric Polymers and MEMS in Tissue Engineering

Richard Tran, Jagannath Dey, Dipendra Gyawali,
Yi Zhang and Jian Yang*

Department of Bioengineering, University of Texas at Arlington, 501 West First Street, Arlington, Texas 76019, USA
**E-mail: jianyang@uta.edu*

3.1 INTRODUCTION

Within the past decade, researchers in the field of tissue engineering have recognized the need for new materials with soft and elastic properties. As a result, many groups have focused on the synthesis, characterization, and application of materials with a wide range of biodegradable and elastomeric properties.[1] The combination of these polymers with Micro–Electro–Mechanical Systems (MEMS) technologies has sparked a new area of research with increasing practical applications.[2] The following chapter discusses important design criteria for creating polymers with elastomeric properties, recently researched biodegradable elastomers, and the use of MEMS in combination with biodegradable elastomers in tissue engineering applications.

3.1.1 Tissue Engineering

Currently, the only effective and permanent treatment to restore lost tissue function is transplantation. Although the success rate for organ replacement therapy has improved, the number of patients awaiting transplantation continues to increase, and the supply of transplantable organs does not meet the current demand.[3] In addition, complications can occur from chronic immune rejection and the required life–long immunosuppressive drug regimen. Due to the growing demand for transplantable organs, a heavy burden is placed on the healthcare industry and the national economy. For example, patients suffering from liver failure cost the United States over $9 billion annually since 1992.[4]

Biomaterials for MEMS, Edited by M. Chiao and J.-C. Chiao

www.panstanford.com
978-981-4241-46-5

Better alternatives need to be developed that are less invasive and more cost effective to provide the needed tissue.[5] As defined by Langer and Vacanti, tissue engineering, or regenerative medicine, is "an interdisciplinary field that applies the principles and methods of engineering and life sciences toward the understanding and development of biological substitutes to restore, maintain, and improve human tissue functions." By combining the fundamental principles and methods from chemistry, engineering, and biological sciences, the major goal of tissue engineering is to restore damaged or diseased tissue.[1]

The field of tissue engineering has progressed for almost 30 years. Due to the great potential of this field, much attention has been attracted to help overcome major healthcare needs.[6] Research groups in the field have attempted to recreate a variety of mammalian tissue. For example, ectodermal-, endodermal-, and mesodermal-derived tissue such as the nerve, cornea, skin, liver, pancreas, cartilage, bone, muscle, urethra, bladder, and blood vessels have been investigated.[7–15]

The foundation of tissue engineering relies on four key elements: cells, scaffolds, signals, and bioreactors.[16,17] In the general scheme for tissue engineering, cells are seeded onto a three–dimensional (3D) scaffold, a tissue is cultivated *in vitro*, then proper signals are supplemented to the system, and finally the construct is implanted into the body as a prosthesis.[17] The general scheme for the key elements involved in the tissue engineering paradigm is illustrated in Fig. 3.1.

The cells used in tissue engineering applications can be isolated from either an autologous, allogenic, or xenogenic source. The cells may be tissue specific, stem cells, or progenitor cells. Scaffolds, which provide a substrate for cell growth, can be composed of either a natural or synthetic material, and fabricated into a fibrous, foam, hydrogel, or capsule architecture. Signals can be introduced to enhance cell proliferation, differentiation, and vascularization of the construct. Bioreactors mimic the conditions inside the body, and provide many benefits towards a successful design. For example, bioreactors allow for an increase in

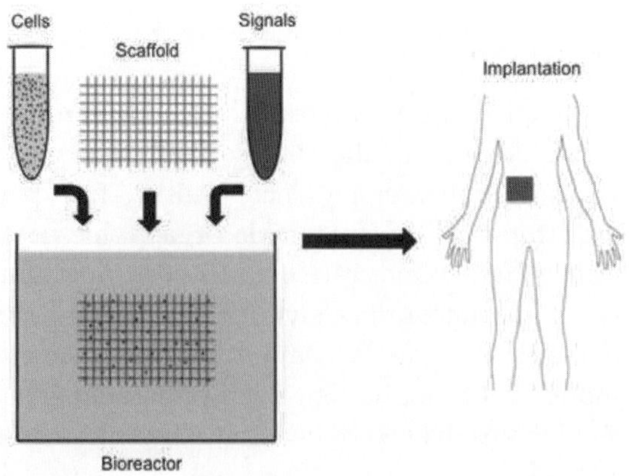

Figure 3.1. The key elements involved in the classic tissue engineering paradigm. For color reference, see page 255.

Table 3.1 The controllable parameters from the key elements of the tissue engineering paradigm.

Cells	Scaffold	Bioreactor
Source	Architecture	Nutrients/Oxygen Content
Type	Materials	Growth Factors
Density	Pore Size/Shape	Dynamic Flow Rate
Genetic Manipulation	Bioactive Molecules	Tension/Compression
Gene Expression	Mechanical Properties	Pulsatile Stress
	Degradation Rate	Shear Stress

the volume of cells that can be cultured *in vitro*, enhance mass transport, and add mechanical cues to stimulate cell differentiation and growth.[17,18] Thus, controlling the parameters from the key elements of the tissue engineering paradigm can ultimately influence the outcome of a cell–scaffold–bioreactor system (Table 3.1).

Despite much of the recent success in tissue engineering, key challenges remain to be addressed. Along with the difficulty in finding an appropriate cell source, the lack of suitable scaffolding biomaterials and the current graft engineering design strategies challenge the success of the field. For example, one major obstacle limiting the success of tissue engineering is compliance mismatch. The current scaffolds cannot be fully integrated with their surrounding tissues because of their incompliant molecular structures and mechanical properties. Thus, further consideration in regards to matching scaffold mechanical properties to the native tissues must be taken into consideration.

3.1.2 Mechanical Considerations for Tissue Engineering Scaffolds

All the tissue cells in the body are located in a unique 3-D extracellular matrix (ECM) environment. The ECM supplies important biochemical signals and functions, facilitates nutrient and waste exchange, guides cellular organization and differentiation, and provides mechanical integrity to the cells.[16,19] In order to sufficiently emulate the natural ECM, a successful scaffold design should include several key requirements. The ideal scaffold should be biocompatible, biodegradable, have an interconnected pore structure, possess a large surface area, allow for adequate cell loading, encourage cell attachment and proliferation, facilitate nutrient and waste exchange, and possess the appropriate mechanical properties for the intended target application.[6,20] Materials used in scaffold fabrication can be divided into four groups: metals, ceramics, polymers, and composites.[21] The content of this discussion will be limited to synthetic polymers. Unlike other materials, synthetic polymers have received great attention because of their controllable material properties such as strength, processability, degradation, microstructure, and permeability.[18]

Many of the soft tissues in the body have soft and elastomeric properties.[22] In order to successfully engineer these tissues, the use of a mechanically compliant biodegradable scaffold will be required. Engineered scaffolds must be strong enough to withstand the mechanical demands asserted upon them when implanted into the body, and must be able to retain their mechanical properties over time.[23] The utilization of an elastomeric scaffold is advantageous in that it can sustain and recover from multiple deformations without causing irritation to the surrounding tissue in a mechanically demanding environment.[24,25]

Another advantage of elastomeric scaffolds is their ability to be used with mechanical conditioning regimens to promote improved tissue formation. By gradually transferring stress from the degrading synthetic matrix to the newly forming tissue, scaffolds with applied cyclic mechanical strains have been shown to increase collagen and elastin production in vascular smooth muscle cells, and enhance the mechanical properties of the tissue engineered constructs in cardiac applications.[26,27] Research groups have also shown that mechanical signals aid in the development of tissue engineered cartilage.[28]

3.2 DESIGN CRITERIA FOR BIODEGRADABLE ELASTOMERIC POLYMERS

In order to fabricate constructs with the appropriate mechanical properties, many important design criteria must be met when creating the starting materials for the intended target application. The following section will discuss the design requirements and concerns that should be taken into consideration when creating an elastic material for soft tissue engineering applications.

3.2.1 *Polymerization Mechanisms*

The two main forms of polymerization for elastic polymers are polycondensation and polyaddition reactions. Polycondensation reactions have stepwise growth kinetics, and are characterized by the formation of by-products during synthesis. For example, a diol can be reacted with a diacid to produce a polyester with water as a by-product.

Polyaddition reactions display chain-growth kinetics, and require the use of an initiator. Chain initiation, propagation, and termination are steps that characterize a polyaddition reaction. Through this general mechanism, the average molecular weight of the polymer increases during the reaction. High molecular weight polymers and/or crosslinked polymers can be produced in a polyaddition reaction.

3.2.2 *Methods to Incorporate Elasticity*

The two methods to incorporate elasticity are physical crosslinking and chemical crosslinking. Certain segments of polymer chain will form a crystalline structure, which will serve as a means for physical crosslinking. In the case of polyurethanes, the clusters of hard segments act as "pseudo cross-links", and allow the material to behave as an elastomer.[29] When the temperature is raised, the hard segment clusters disassociate, and the material can be made to flow. When subsequently cooled, the clusters reform and the material will again exhibit elastomeric properties. Some ABA triblock co-polymers will also show elastomeric properties.[30] For example, the thermal liable crosslinks in an ABA triblock co-polymer can aggregate to form physical crosslinking between polymer chains.

Chemical crosslinking joins the polymer chains together into a network linked by covalent bonds. Unlike physical crosslinks, the chemical crosslinks are generally irreversible, and display greater mechanical strength and elasticity. It is well known that natural ECM components such as collagen and elastin are crosslinked polymers. The crosslinking provides these natural materials with their elastic nature. Due to this phenomenon, researchers have utilized the concept of crosslinking in the creation of elastomers to meet the versatile needs in tissue engineering and other biomedical applications.

For polycondensation mechanism, in order to create a polymer with a 3D elastomeric network structure, at least one of the monomers chosen should be multifunctional. In addition to providing the needed functional groups for chain extension, a multifunctional monomer provides valuable unused functional groups, which can be used in later post-processing to create a 3D crosslinked network. Thus, by creating crosslinked network structure, a material with elastic properties can be obtained.

3.2.3 Design Concerns

The three major concerns when designing a biodegradable elastomer for biomedical applications are the biocompatibility, mechanical properties, and degradation rate of the material. In the following section, a brief introduction of these three properties and how they affect each other are discussed.

3.2.3.1 Biocompatibility

Biocompatibility is a term used to describe the ability of a material to perform with an appropriate host response in a specific application. For the materials used in biomedical applications, the biocompatibility should always be put as the first concern. There are several factors that can affect the biocompatibility of a material. For example, the hydrophilicity or hydrophobicity of a material can greatly influence its biocompatibility. It has been demonstrated that the degree of hydrophilicity/hydrophobicity should be balanced to achieve optimal cell affinity.[31,32]

The acidity of a material can also influence its overall biocompatibility. Certain functional groups located on the polymer chain have the ability to greatly change the pH of the surrounding area. In addition to the chemistry of the bulk material causing pH changes, certain materials will degrade into acidic products to alter the pH of the immediate area. This deviation in pH from the body's normal values can create a cytotoxic effect, which can later lead to adverse reactions.

The use of certain chemicals during the synthesis of elastomers can also cause biocompatibility issues. In the case of elastomers created through a polyaddition mechanism, one factor to influence the material's biocompatibility is the toxicity of initiator used during synthesis. In other situations, redox initiators and photoinitiators used to crosslink the polymer have been shown to be toxic to cells when used in large concentrations. Thus, the amount of initiator used for crosslinking should be strictly controlled, and the residual remaining initiator not used during the crosslinking mechanism should be removed.

3.2.3.2 Important Mechanical Properties

During the mechanical test of a polymer, the stress–strain curve obtained is used to define many important parameters of a material's mechanical properties. The stress–strain curve is a graphical representation of the relationship between the amount of stress applied and the resulting strain of the sample.

The tensile strength of a material is the maximum amount of tensile stress that can be subjected to the material before failure. Normally for an elastomer, there is no yield point, and as a result, the peak stress of the stress–strain curve should appear at the break point. The compressive strength is usually obtained experimentally by means of a compression test, and is the value of uniaxial compressive stress reached when the material fails completely.

The elongation, also known as the stretch ratio, is a measure of the largest deformation of the material before failure during the tensile test. A higher elongation indicates the capability of a material to deform. The elastic modulus is also a very important parameter, and is used to determine the stiffness of a material. Depending on the type of mechanical test being performed, the three different types of modulus that can be obtained are the Young's modulus, shear modulus, and bulk modulus. The Young's modulus is the most commonly obtained for an elastomer in biomedical applications. It is defined as the ratio of stress over strain, and can be derived from the slope of the initial linear region of the stress–strain curve. The recovery from deformation is also a parameter used to characterize elastomers. Many of the tissues in the body are fully elastic within a certain deformation. The time to recover from deformation for elastomers should also be considered when characterizing elastomers.

3.2.3.3 Degradation Rate

In most cases, the degradability of an elastomer is due to the hydrolyzable bonds in the polymer network. *In vitro* degradation studies are always used to predict the degradation rate.[33] Enzymatic degradation and oxidative degradation are also two possible ways for the degradation of elastomer.[34,35] In some previous works, *in vitro* enzymatic degradation studies were carried out to evaluate the property of the polymer.[36]

The process of hydrolysis is mainly dependent upon the amount of water penetration into the network structure. Normally, a more hydrophobic material will have a lower degradation rate. Moreover, a lower glass transition temperature

(T_g) will also affect the degradation rate due to the increased water diffusion rate into the material. In order to prevent any changes in the elasticity of a material, it is important to maintain the T_g below the normal body temperature. In the case of an elastomer, both the T_g and mechanical properties are affected by the degree of crosslinking. A higher crosslinked elastomer will normally have a slower degradation rate, stronger mechanical strength and smaller elongation rate.

3.3 BIODEGRADABLE ELASTOMERIC POLYMERS

3.3.1 Polyesters

The use of elastomers in medical applications originates back to the beginning of the rubber industry. Since then, numerous materials have played a major role in medical technology.[37] Polyesters are the most widespread category of polymers used in biomedical applications. The ester bond is important because it allows for degradation through hydrolytic cleavage in the presence of water. Unlike enzymatic degradation, this form of degradation is advantageous because of the minimal site–to–site and patient–to–patient variations.

A polymer used in tissue engineering applications should show good degradability and biocompatibility when presented *in vivo*. Due to these requirements, glycolic and lactic acid based poly(α–hydroxy acids) such as poly-L-lactide acid (PLLA) and poly(lactic-*co*-glycolic acid) (PLGA) have gained attention in the past few decades as suitable polyesters for various medical applications. Their use can be seen in drug delivery systems, scaffolds for tissue regeneration, resorbable sutures, staples, and orthopedic fixation devices.[38]

However, these α–hydroxyl acid polymers are inappropriate for soft tissue applications because of their stiff nature. Due to this major drawback, researchers are advancing towards a new category of polyesters whose mechanical properties can be tuned for particular soft tissue engineering applications such as blood vessels, heart valves, ligaments, and tendons. Polyesters that possess elastic properties to meet the requirements for soft tissue engineering are shown in Table 3.2. The following section will focus on the polyester elastomers that have been used in the field of soft tissue engineering.

3.3.1.1 Polyhydroxyalkanoates (PHAs)

In the early 1920's, the bacteria *bacillus megaterium* was recognized for producing poly(3–hydroxybutyrate) (PHB), which is the most common polymer among the polyester class. Since then, more than 150 different monomer combinations have been used in the formation of different polymers within the PHA family.[39] Four different pathways have been revealed for the synthesis of PHA through the process of biosynthesis, which has been mentioned in detail elsewhere.[40] Due to advancements in the field of genetic engineering, researchers have also used plants as the production house for PHB-related polymers.[41]

Table 3.2 Mechanical properties of polyester elastomers in recent research.

	Mechanical Properties		
Polymer Name	Youngs Modulus(MPa)	Elongation at Break(%)	Tensile Strength (MPa)
PHB	2500	3	36
P4HB	70	1000	50
PGS	0.056–1.2	40–448	0.2–0.5
PGSA	0.048–1.37	47–170	0.54–0.5
POC	1.85–13.98	117–502	2.93–11.15
PEC	0.25–1.91	140–1505	0.51–1.51
PPSC	0.6–1.23	226–432	0.87–2.12
POM	Not reported	3.86–14.34	7.32–25.6
PAMC	0.05–1.8	55–450	0.29–0.88
CUPE	4.14–38.35	222.66–337.558	14.6–41.07

Several groups have also reported the chemical synthesis of poly(3–hydroxyalkanoates) (P(3HB)) through the process of a ring opening of β–butyrolactone (BL) in the presence of aluminum, zinc, and tin based catalysts.[42–44] However, these reactions did not yield high molecular weight polymers. To overcome this limitation, Hori *et al.* utilized the distannoxane complexes as an excellent catalyst for the ring-opening polymerization of (*R*)–b–butyrolactone ((*R*)–BL) and BL to produce P[(*R*)–3HB] and P(3HB) of high molecular weights and in high yields.[45]

By using different combinations of various monomers, researchers have successfully produced PHAs with a wide range of mechanical properties and degradation profiles. For example, poly(3–hydroxybutyrate) is a stiff polymer with a Young's Modulus of 2500 MPa and 3% elongation where as poly(4–hydroxybutyrate) is an elastic polymer with a Young's Modulus of 70 MPa and 1000% elongation. In terms of their biocompatibility, PHA elastomers are biosynthetic polymers and require serious consideration on their purity.[46]

In the early 1990's, Akhtar *et al.* reported a prolonged acute inflammatory response and severe chronic inflammatory response from PHA films implanted *in vivo*.[47] William *et al.* proposed the idea of using a depyrogenation technique through the use of an oxidizing agent that resulted in the reduction in the amount of endotoxins. In addition, William and co-workers also understood the problems associated with the use of solvents while extracting the polymer. The group found that a higher purity could be obtained if the polymer was extracted with hexane or acetone instead of the traditional chlorinated solvents. In order to support their study, the research group performed *in vivo* tests by placing several different types of implants such as microspheres, tubes, and pellets subcutaneously in mice. The histological results revealed the formation of a thin fibroblast capsule (four to six cell layers), and the absence of macrophages at the implant sites.[46]

Medical device companies have extensively investigated P4HB due to its potential as scaffold material for engineering various tissues. For example, Tepha Inc. is evaluating this member of the PHA family in order to meet all the standards set by the US Food and Drug Administration. In two independent studies lead by Stock and co-workers, it has been demonstrated that this elastomer is a potential candidate for engineering heart valves and for blood vessel augmentation.[48,49]

3.3.1.2 Poly(glycerol–sebacate) (PGS)

In the late 1990's, Nagata and co-workers reported their work on the synthesis and characterization of polymer based on sebacic acid and glycerol. By reacting glycerol and sebacic acid, they achieved their goal in creating an environmentally friendly plastic that can be degraded by soil bacteria.[50] PGS is synthesized through a polycondensation reaction, which produces degradable ester bonds throughout the polymer backbone to solve any degradation issues. However, material property challenges still remained due to the non-elastic nature of the polymer.

In 2002, Wang *et al.* realized that the monomers of this polymer are biocompatible, which opened the door for its use in biomedical applications.[24] After further study of the synthesis procedure and chemical structure, Wang and co-workers realized that Nagata *et al.* were using a 2:3 molar ratio of glycerol and sebacic acid in the reaction, which resulted in the total consumption of all the available functional groups. In contrast, Wang *et al.* divided the reaction into two steps to preserve some of the functional groups for later processing. Thus, a 1:1 molar ratio of glycerol and sebacic acid was used to obtain a linear pre-polymer. This initial step preserved the pendant hydroxyl groups in the PGS pre-polymer, which was later used to form an elastic 3D crosslinked network through ester bond formation.[24] Through this novel idea, a new trend of scaffolding materials was initiated in the field of soft tissue engineering.

PGS is a soft (Young's Modulus of 0.282 ± 0.025 MPa) and elastic (elongation of $267\pm59.4\%$) material that has potential for engineering soft tissue such as arteries, veins, and nerves.[24,51] A study by Sundback *et al.*, utilizing PGS for neural reconstruction, showed that PGS can be a good scaffolding material with a desirable biocompatibility.[51] The study also proved that a normal morphology and acceptable growth rate of Schwaan cells could be obtained when compared to PLGA, which is widely used in neural reconstruction.

In a study by Chen *et al.*, the mechanical properties of PGS were evaluated by varying the degree of crosslinking in order to match the mechanical properties of myocardial tissues. They further demonstrated that PGS is bioresorbable through hydrolysis and enzymatic degradation. The degradation rate of PGS can be fine tuned in order to meet the requirement for the construction of heart patches.[52]

3.3.1.3 Poly(glycerol sebacate)acrylate (PGSA)

In 2007, Nijst *et al.* created a photocurable elastomeric polymer based on the previously made PGS. The group incorporated vinyl functional groups into the polymer backbone by acrylating the available hydroxyl groups of the PGS pre-polymer with acryloyl chloride. Due to the presence of these vinyl groups, the polymer achieved a 3D crosslinked network structure through an ultraviolet crosslinking mechanism, which eliminated the long and harsh post polymerization conditions used during the PGS synthesis. This increased polymer's potential to encapsulate cells or temperature–sensitive biomolecules.[36] It was also reported that the mechanical properties of polymer could be tuned according to the degree of acrylation (Table 3.2). Furthermore, co-polymerizing the polymer with PEG diacrylate was also shown to modulate the mechanical properties of the polymer.[36]

The prime interest of developing PGSA was to proliferate and differentiate stem cells into the desired tissue by encapsulating them in the porous matrix of the polymer. Interestingly, human embryonic stem cells (hESCs) encapsulated in the polymer matrix and allowed to grow for seven days showed a colonial organization expressing the Ki67 protein and all three germ layers.[41]

3.3.1.4 Poly(diol–citrate) (POC)

In 2004, Yang *et al.* designed a citric acid based polyester elastomer to avoid the long and harsh processing conditions seen in the synthesis of PGS. The group used citric acid as multifunctional monomer to react with different aliphatic diols ranging from 3–16 carbon chains in 1:1 molar ratios.[53] By controlling the post polymerization conditions, Yang and co-workers demonstrated the ability to tune the mechanical properties and degradation profiles of the elastomer, which allowed for customizable properties for a specific application.

As seen in Table 3.2, a wide range of mechanical properties was reported in order to meet the specific needs for engineering tissues such as cartilage, small

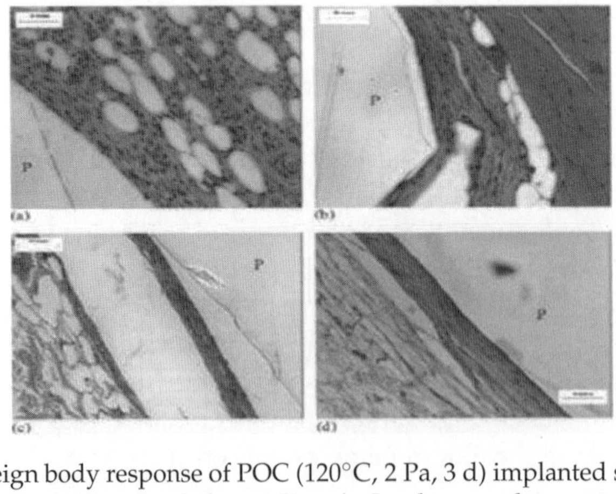

Figure 3.2. Foreign body response of POC (120°C, 2 Pa, 3 d) implanted subcutaneously in female Sprague–Dawley rats (scale bar = 50 μm). Implants and surrounding tissues were harvested after (a) 1 week; (b) 1 month; (c) 2 months; (d) 4 months implantation for H&E staining. "P" represents polymer section. Reprinted from Biomaterials, 27, Jian Yang *et al.*, Synthesis and evaluation of poly(diol citrate) biodegradable elastomers, 1889–1898, 2006, with permission from Elsevier. For color reference, see page 256.

diameter blood vessels, and tendons. The preliminarily biocompatibility tests confirm that the elastomer is a "cell friendly" material. Figure 3.2 shows H&E stained pictures of the foreign body response of POC subcutaneously implanted into female Sprague–Dawley rats[54] suggesting the biocompatibility of POC.

It was seen that POC supported the proliferation of human aortic smooth muscle cells and endothelial cells. Furthermore, the degraded products and the polymer are non-toxic in nature. Although the polymer is insoluble in water, the degraded products are soluble in water and can easily be eliminated from the body.[55] It has been shown that the 1,8–octanediol used in the reaction can be partially traded with N–methyldiethanolamine (MDEA) in order to increase the degradation rate, hydrophilicity, and mechanical properties of the elastomers.[54]

3.3.1.5 Poly(PEG–co–CA) (PEC)

Although Yang *et al.* already proposed the use of poly ethylene glycol as the diol to create poly(diol citrates), Ding *et al.* later published their work involving a similar synthesis procedure, but with a different monomer.[54] PEG200 was incorporated into the polymer to create a water-soluble elastomer with a rapid degradation profile. It has been shown that this polymer is highly elastic in nature (elongation of 1500%) with a tensile strength of >1.51 MPa. However, the primary objective of developing this material was to design drug-carrying devices. Its utility for tissue engineering applications is yet to be proved. Biocompatibility tests have not been reported for this elastomer, but the monomers (citric acid and PEG) have already been approved by the FDA for medical uses.[53]

3.3.1.6 Poly((1,2–propanediol–sebacate)–citrate) (PPSC)

After realizing the outstanding work of the previous groups, Lei *et al.* proposed synthesizing an oligomer terminated with an alcohol from the monomers sebacic acid and 1,2–propanediol. In addition to further linking these oligomers, citric acid was incorporated to fine tune the mechanical and degradation properties. The elastomer produced showed desirable mechanical properties with low water retention and a rapid degradation profile. However, biocompatibility tests and application oriented studies of this elastomer are not yet reported.[56]

3.3.1.7 Poly(1,8–octanediol malate) (POM)

A recent study by Wan *et al.*, proposed a thermoset elastomer based on 1,8–octanediol and malic acid utilizing a polycondensation reaction between the carboxylic acid and alcohol. This elastomer displayed a tensile strength of 7.32 ± 0.63 to 25.6 ± 1.42 Pa, a compressive Young's modulus of 0.12 ± 0.02 to 0.25 ± 0.01 KPa, and an elongation of <15%. However, no permanent deformation was reported after 500 press loading and release cycles with 30% maximum strain. The POM elastomer showed a linear degradation profile in PBS at 37°C, where the majority of mass loss (up to 90 %) was reported by week 13. The study was targeted for the regeneration Annulus Fibrosus (AF). POM not only shows good AF cell growth, but also cell penetration into the scaffold. However, in order to support significant cell growth, POM required more than 2 days crosslinking at 120°C under vacuum (2 Pa).[57]

3.3.1.8 *Poly(alkylene maleate citrates) (PAMCs)*[58]

In our lab, we have created a new family of elastomers named poly(alkylene maleate citrates) (PAMCs). The synthesis of PAMC pre-polymers can be achieved within 2–6 hours without the use of any harsh processing conditions. A random polycondensation reaction between 1,8–octanediol, citric acid, and maleic acid has been carried out to form a pre-polymer with a vinyl functional group incorporated into the polymer backbone. Furthermore, C3–C16 diols can be utilized as the alcohol, and different vinyl–containing monomers such as fumaric acid, maleic acid, and maleic anhydride can be used to incorporate the vinyl group.

For the first time, a family of elastomers has been created to utilize a dual crosslinking mechanism. The dual crosslinking mechanism allows the polymer to be C–C crosslinked through redox or ultraviolet mechanisms, and further crosslinked through a polycondensation reaction with the addition of heat. The dual crosslinking mechanism is highly advantageous in that the polymer can be quickly crosslinked through the vinyl group, and the mechanical properties and degradation profile of the polymer can be fine tuned through the post polymerization process to meet the requirements for a variety of soft tissue engineering applications.

The PAMC family of elastomers has a wide range of mechanical properties (Young's Modulus of 0.05–1.8 MPa) and elasticity (elongation of 45–450%) that can be modulated through different monomer concentrations, photoinitiator concentration, polymer concentration while crosslinking, and the dual crosslinking mechanism. This wide range of mechanical properties has a huge potential for the regeneration of various types of soft tissues.

A preliminary biocompatibility test confirms that PAMC support 3T3 fibroblast and smooth muscles cell proliferation and adhesion. Once inside the host, the degraded products of the elastomer are biocompatible as confirmed by *in vivo* biocompatibility tests. PAMC samples subcutaneously implanted for 1 week produced a slight acute inflammation response, which was expected and consistent with the introduction of foreign materials into body (Fig. 3.3). After 4 weeks of implantation, the number of infiltrating cells declined and a thin fibrous capsule was formed surrounding the implanted PAMC indicating that the degradation products of PAMC are not toxic.

The pre-polymers of PAMC are in a liquid or gel state, which allows the polymer to be applied onto any contour of the skin as seen in Fig. 3.4. In addition, PAMCs can be quickly polymerized within 3 minutes to seal any defects on the immediate surface. This PAMC property can be shows its potential in wound dressing and tissue sealant applications. Furthermore, pre-poly(octamethylene

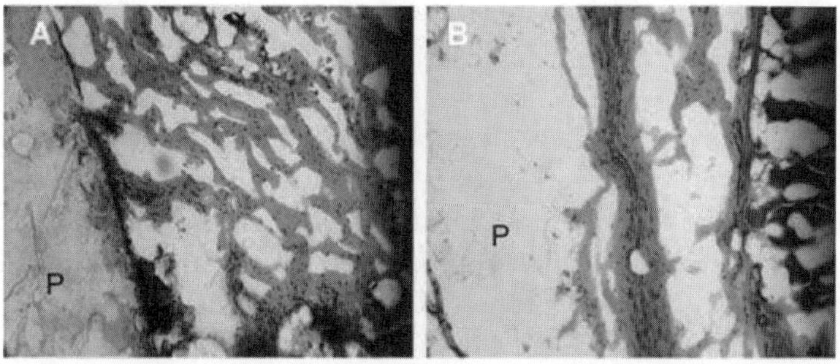

Figure 3.3. Foreign body response of PAMC (POMC 0.8, 20 mins. UV crosslinking) implanted subcutaneously in female Sprague–Dawley rats. Implants and surrounding tissues were harvested after (a) 1 week and (b) 4 weeks. "P" represents polymer section. For color reference, see page 256.

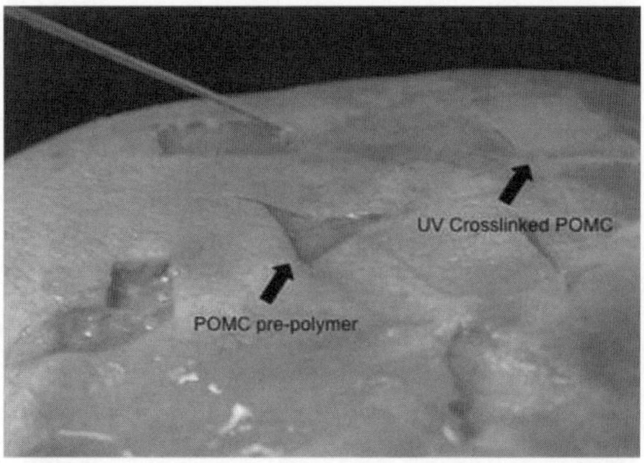

Figure 3.4. Pre-PAMC and ultraviolet crosslinked PAMC on porcine skin. For color reference, see page 257.

poly(ethylene glycol) maleate citrate) (PPEGMC) and pre-poly(octamethylene poly(ethylene glycol) maleate (PPEGM) are two members of the PAMC family that are soluble in water, and can be used as a injectable crosslinkable polymers for *in situ* applications. Our lab has also been extensively investigating the fabrication of nanogels, microgels, and hydrogels with various PAMC members for the potential use as a carrier for temperature sensitive drugs and cells.

3.3.1.9 *Crosslinked Urethane-Doped Polyesters*[59,60]

The design and development of a soft, strong, and completely elastic (100% recovery from deformation) material for tissue engineering still remains a challenge. Our lab has also recently developed a new generation of elastomeric polyesters, named crosslinked urethane–doped polyesters (CUPEs). The rationale behind the synthesis of CUPE was to combine the totally elastic properties of crosslinked polyesters with the mechanical strength of polyurethanes to create a family of strong and elastic polymers. The tensile strength of CUPE was as high as 41.07±6.85 MPa with corresponding break strains of 222.66±27.84%. The Young's Modulus ranged from 4.14±1.71 MPa to 38.35±4.5 MPa.

Pre-CUPE polymers were synthesized with different feeding ratios of pre-POC:1,6–Hexamethyl diisocyanate (HDI) (1:0.6, 1:0.9, and 1:1.2, molar ratios). The varying ratios of the POC to HDI regulated the amount of urethane bonds inserted in the polyester backbone, which was shown to influence the thermal, mechanical, and degradation properties of the polymer. For example, the T_g was shown to increase with respect to the concentration of the diisocyanate used during synthesis. CUPE polymers containing higher ratios of diisocyanate were also found to be stronger than CUPEs with lower diisocyanate content for similar post polymerization conditions.

In addition to diisocyanate feed ratios, the post-polymerization conditions also affected the properties of CUPE polymers. With increasing post-polymerization time and temperature, CUPE polymers displayed increased strength and stiffness along with a reduced ultimate strain value, which was attributed to the increased degree of crosslinking between the polymer chains.

For *in vitro* cell culture on CUPE films, scanning electron microscope (SEM) results indicated that seeded 3T3 fibroblasts and smooth muscle cells (SMCs) maintained their phenotype while proliferating on the surface of CUPE films. From the Methylthiazoletetrazolium (MTT) assay, it could be seen that a larger number of cells initially attached to the CUPE films as compared to control PLLA films, and the rate of growth and proliferation was comparable to that on PLLA. Subcutaneous implantation of CUPE films and scaffolds in Sprague–Dawley rats

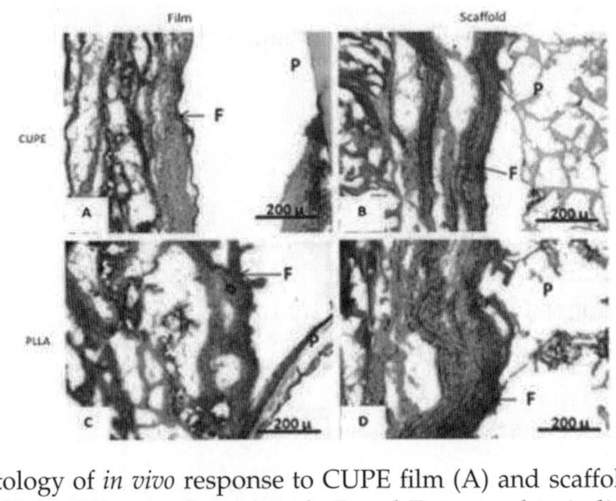

Figure 3.5. Histology of *in vivo* response to CUPE film (A) and scaffold discs (B). PLLA films (C) and scaffolds (D) served as control. P and F are used to indicate the regions of polymer and fibrous capsule respectively. All images were taken at 10x magnification. On the 1 week samples, although all samples were covered by a well defined fibrous capsule, CUPE implants were consistently surrounded by a thinner fibrous capsule as opposed to PLLA implants. In the case of the 4 week implants, overall, all the implants appear to trigger similar extent of tissue responses. For color reference, see page 257.

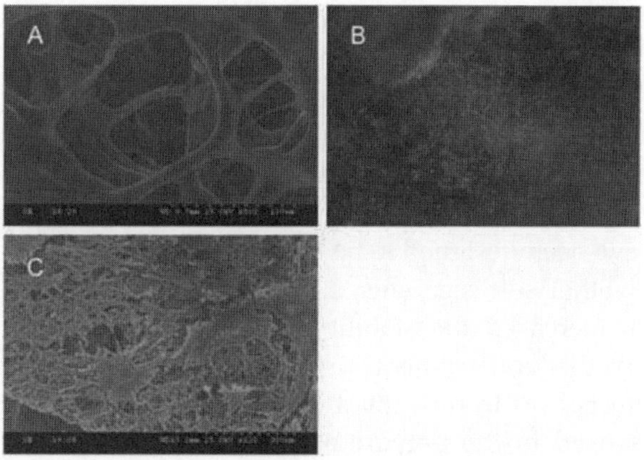

Figure 3.6. SEM images of scaffold surface indicate the presence of well defined pores (A) and even cell distribution of cells on the scaffold (B and C). For color reference, see page 258.

was performed to evaluate the foreign body response to CUPEs. Tissue samples were explanted at 1 week and 4 week time points, and examined histologically using H&E staining. At the 1 week time point, the fibrous capsules surrounding the CUPE implants were thinner than those surrounding the PLLA implants (Fig. 3.5). At 4 weeks, the fibrous capsule thickness was reduced for both the CUPE and PLLA implants, and was found to be comparable for both polymers. These results indicated that CUPE had a weaker acute inflammatory response and a similar chronic inflammatory response compared to PLLA.

We have also fabricated thin 3-D porous soft and elastic scaffold sheets (150 μm thick) by a simple freeze–drying method (Fig. 3.6(A)). Based on a scaffold–sheet tissue engineering strategy, we proposed the use of CUPE scaffold sheets for tissue regeneration. The thin scaffold sheets allowed even cell seeding, growth, and distribution (Figs. 3.6(B) and 3.6(C)), as the cells did not have to penetrate too deep within the scaffold. Soft scaffolds would also facilitate scaffold assembly into various shapes through folding, rolling, trimming, and bending. The mechanical strength of CUPE scaffolds would allow surgical handling and bioreactor training for the seeded scaffolds.

3.3.2 Polyurethanes

Polyurethanes are segmented block co-polymers, which consist of a soft and hard segment. The soft segment is composed of a macrodiol, and the hard segment is a combination of a diisocyanate and a chain extender. Typically, the macrodiol is usually a difunctional polyester or polyether segment, and a low molecular weight diol or diamine is used as a chain extender. The segmented architecture is responsible for the unique mechanical properties of polyurethanes. The partially crystallized hard segments act as virtual crosslinks to give polyurethanes their high tensile strength and elasticity.

Polyurethanes are a class of polymers which have been extensively used as biomedical materials since the 1960's.[61] In addition to good biocompatibility, their controllable and diverse mechanical properties make them ideal biomaterials.[62] The typical applications of polyurethanes in medicine over the past years have included pacemaker leads, catheters, artificial heart prostheses, and coatings for silicone breast implants.[61,63] These applications require that the material remain stable inside the body for long periods of time. Subsequently, all traditional polyurethanes have been designed to be biostable and not degrade easily *in vivo*.

By using polyether soft segments, a more hydrolytically stable material was produced, which increased the stability of the polymer in an *in vivo* setting. However, the polyether soft segments proved to be more susceptible to oxidation. The oxidative effects lead to unwanted degradation of the material. Due to the toxic pre-cursors used in the polyurethane synthesis, the degradation of these biostable polyurethanes would cause the release of carcinogenic compounds inside the body. For example, toluene diisocyanate is one of the most commonly used diisocyanates in the synthesis of biostable polyurethanes. Upon degradation of the urethane bonds, it results in the formation of toluene diamine, which has been shown to be carcinogenic. The effect of oxidation and subsequent degradation of the polyether–urethanes led to the development of oxidation and hydrolysis resistant polycarbonate based polyurethanes.

Due to these complications, the interest in the hydrolytically unstable polyester based urethanes has increased over the last decade. Currently, the primary degradable polyurethanes used as a biomaterial in tissue engineering include polyester–urethanes, polyether–urethanes, and polyester–ether urethanes. Alternatively, hydrolytically labile bonds may be introduced in the hard segment to control the degradation rate of the polyurethane to suit a particular application.[64–66] Faster degradation rates can also be obtained by making the polyurethane degradable, both hydrolytically and enzymatically.[67] The different types of biodegradable polyurethanes are discussed in the following sections.

3.3.2.1 *Polyester–urethanes*

Polyester–urethane is a term used to describe a polyurethane comprising of a polyester based soft segment. Different polyesters such as poly(L–Lactide) (PLA), poly(ε–caprolactone) (PCL), poly(vinyl alcohol) (PVA), and poly(glycolic acid) (PGA) have been used by various researchers for the synthesis of polyester–urethanes with different properties.

3.3.2.2 *PCL–based polyester urethanes*

Poly(ε–caprolactone) (PCL) diol has been used by various researchers to synthesize polyester–urethanes with a wide range of properties. Different polyester–urethanes can be obtained by varying the molecular weight of the PCL–diol, the ratio of hard segment and soft segment, and the properties of monomers used in the synthesis.[64] The low glass transition temperature of PCL (T_g −60°C) allows the polymer to be in an amorphous or semi-crystalline state at use temperature, and is partially responsible for the strength and elasticity of the PCL based polyurethanes.

The molecular weight of the PCL used for synthesis affects the mechanical properties of the synthesized urethane. With all other parameters remaining the same, many researchers have noted a trend of increasing initial modulus and tensile strength when higher molecular weight PCL is used as the diol.[64,65,68] This phenomenon has been attributed to increasing phase separation leading to greater crystallinity of the higher molecular weight PCL soft segments. A wide range of mechanical properties were also obtained by replacing the PCL soft segment with a PCL–PLA co-polymer segment.[69] By varying the PLA to PCL content of the co-polymer used as the soft segment, the properties of the polyurethane could be varied from a very stiff, inelastic polymer to a soft, elastic elastomer.

Other factors that affect the mechanical properties of the materials include the choice of diisocyanate and chain extender. Skarja *et al.* synthesized two different PCL–urethanes using both HDI and lysine diisocyanate (LDI).[64,65] The greater reactivity of the HDI resulted in higher molecular weight polyurethanes compared to those synthesized with LDI. At low soft segment molecular weights, the HDI based polyurethanes displayed a greater degree of phase separation when compared to the polyurethanes synthesized using LDI. Although mechanical tests were not performed on the HDI based urethanes, the greater degree of phase separation displayed better tensile properties when compared to the LDI based polyurethanes which is due to the better packing of the hard segments. This phenomenon was illustrated in more recent work.[70] However, at higher soft segment lengths, the effect of the diisocyanate on the mechanical properties of the polyurethane was not significant.

The chain extender used during the synthesis is another means of modifying the mechanical properties of the polyurethane. Chain extenders are usually low molecular weight difunctional polyamines, polyfunctional polyamines, or polydiols that are used to increase the molecular weight of the polyurethane. In addition to incorporating ester bonds in the soft segment, the incorporation of the appropriate chain extenders containing hydrolytically labile ester linkages is a technique that has been exploited by researchers to increase polyurethane degradation rates.[64,65,70,71] Furthermore, researchers have also synthesized amino acid based chain extenders which are susceptible to enzymatic degradation.[29,67,72]

These amino acid based chain extenders in combination with non-toxic diisocyanates such as lysine diisocyanate are expected to result in non-toxic and biocompatible hard segment degradation products upon implantation. The effect of the chain extender on mechanical properties mainly depends on the structure and the reactivity of the chain extender used. Chain extenders with pendant side chains may impede hard segment packing as opposed to aliphatic chain extenders, thereby resulting in inferior mechanical properties.[68]

Tatai *et al.* demonstrated the effect of the reactivity of chain extender on the final properties of the polyurethane. Less reactive chain extenders resulted in polyurethanes with lower molecular weight and poorer mechanical properties when compared to polyurethanes synthesized using more reactive chain extenders. Different chain extenders have also been used to specifically tune the mechanical properties of polyurethanes. Due to the use of aliphatic diisocyanates, the biodegradable polyurethanes lack the stiffness of those made with aromatic diisocyanates. In order to overcome this drawback, Hirt *et al.* introduced poly(hydroxybutyric acid)–co–(hydroxyvaleric acid) (PHB/PV) as a chain extender into a polyurethane with PCL–diethylene glycol–PCL triblocks and LDI as diisocyanate.[73] The PHB/PV chain extender crystallized quickly to form glassy domains thereby causing better aggregation of the hard domains. The hard segment aggregation increased the phase segregation, which resulted in a stiffer and stronger polyurethane.

3.3.2.3 *Poly(ester–ether) urethanes*

Although the incorporation of ester bonds in the soft segment made the polyurethane hydrolytically unstable, the rate of degradation was still found to be slow. This was primarily due to the fact that the soft segment aliphatic polyesters such as PGA, PLA, and PCL were inherently very hydrophobic. It was hypothesized that the incorporation of poly(ethylene glycol) (PEG) in the soft segment would increase the hydrophilicity of the polyurethanes, and accelerate the hydrolytic degradation.

Triblock co-polymers of PCL–PEG–PCL were first used by Cohn *et al.* for the synthesis of polyurethanes.[74] Further developments were made by Guan *et al.* who synthesized poly(ester–ether) urethanes using the triblock co-polymer as a soft segment, a hard segment comprising of 1,4–butane diisocyanate, and putrescine as the chain extender.[75] By varying the ratio of PCL and PEG in the soft segment, the studies showed that the mechanical properties of the polyurethane were improved. Reducing the lengths of the PEG segments, and increasing the length of the PCL segments increased the degree of crystallinity to improve the mechanical properties of the polyurethane.

Ciardelli *et al.* also observed a similar phenomenon. The higher molecular weight of the tri-block was able to increase the degree of soft segment crystallinity due to reduced interruption by the hard segments.[71] The effect of different ratios of the hydrophobic PCL and hydrophilic PEG on the overall hydrophilicity of the polyurethane has also been studied in detail.[76] A further increase in the degradation rates can be achieved by using hydrolytically and enzymatically labile chain extenders like phenylalanine diester in the hard segment of polyurethanes with PCL–PEG–PCL soft segments.[64]

Polyurethane materials have also been used for cardiac reconstruction for congenital heart defects. Guan *et al.* designed a polyester urethane (PEU) based on 1,4–butane diisocyanate, PCL, and putrescine which could be fabricated into highly porous scaffolds using a thermal induced phase separation technique.[77] These biodegradable PEU scaffolds were implanted in the heart of adult rats in which a surgical defect had been introduced. The PEU scaffolds were found to permit greater cellular infiltration with minimal inflammation.[78] These PEU

Table 3.3 A comparison of the thermal and mechanical properties of PTMC and its co-polymers.

		Thermal Properties		Mechanical Properties		
Polymer Name	MW	$T_g(°)$	$T_m(°)$	Modulus (MPa)	Peak Stress (Mpa)	Break Strain (%)
Low MW PTMC	42,100	−15	–	2.94	0.49	160
High MW PTMC	324,000	−19	–	6	12	830
DLLA 21 TCM 79[a]	358,000	−9	–	5	2	270
DLLA 80 TCM 20[a]	718,000	33	–	1900	46	7
DLLA 50 TCM 50[a]	644,000	11	–	16	10	570
CL 23 TCM 77[a]	246,000	−25	–	3.9	0.8	103
CL 75 TCM 25[a]	183,000	−55	5.7	4.6	0.1	236
CL 90 TCM 10[a]	184,000	−22	49.9	252	40	906

[a]Denotes the mol % of each monomer in the co-polymer.

scaffolds were also fabricated into tubular constructs to evaluate their effectiveness as vascular grafts. The tubular scaffolds were evenly seeded with mouse derived smooth muscle cells using a rotational vacuum seeding technique. Tubular constructs prepared by this method had burst pressure and suture retention values which closely matched that of native arteries.[79]

In addition to vascular engineering, biodegradable polyurethanes based on methylene diphenyl diisocyanate (MDI) as the diisocyanate component have also been used as scaffold materials for ligament tissue engineering. PHB/PV based polyurethanes have also been investigated as bioresorbable nerve guide materials.[80] The nerve guides fabricated form these polyurethanes were degradable and supported nerve regeneration with reduced inflammatory response.

3.3.3 Polycarbonates

Polycarbonates are a family of block co-polymers, which are characterized by the presence of a carbonate bond in the backbone of the polymer chain. To date, the two major classes of biodegradable polycarbonates that have been extensively studied for biomedical applications are the co-polymers of poly(1,3–trimethylene carbonate), and tyrosine derived polycarbonates. The latter family of materials has a glass transition temperature ranging from 52–90°C, making it a rigid material at room temperature. Since the scope of this discussion is limited to elastomeric materials, tyrosine derived polycarbonates have been omitted.

Poly(1,3–trimethylene carbonate) (PTMC) is an amorphous polymer, which was first synthesized by Zhu *et al.* through a bulk ring open polymerization of 1,3–trimethylene carbonate in the presence of catalysts.[81] PTMCs were found to be rubbery materials at room temperature, and displayed low glass transition temperatures ranging from −26°C to −15°C. Further properties of PTMCs are covered in Table 3.3.

In addition to moderate elastomeric properties, the utility of homopolymeric PTMC as a temporary implant material was hampered by its slow degradation. Over a 30 week period, the PTMC samples suffered a mass loss of only 9%. However, subcutaneously implanted PTMC samples were rapidly degraded *in vivo*, and the implanted samples beyond a 3 week period could not be detected macroscopically.[82] This was attributed to hydrolytic resistance and enzymatic cleavage of the carbonate bonds. Co-polymerization with other polymers, primarily degradable polyesters like poly(D,L lactide) (PDLLA) and poly(ε–caprolactone) (PCL), have been employed to improve the degradability and the mechanical properties of polycarbonates.[82–84]

The potential of these co-polymers as scaffolds for heart tissue engineering and synthetic nerve guides for nerve regeneration have also been evaluated.[85–87] It was also found that high molecular weight PTMC was very flexible and tough due to the excellent ultimate stress and strain characteristics. These mechanical properties were attributed to strain induced crystallization of the polymeric network upon application of high deforming stresses. This phenomenon has also been studied extensively elsewhere.[88]

3.3.3.1 Poly(D,L Lactide–co–1,3–trimethylene carbonate)

Both high molecular weight and low molecular weight co-polymers of amorphous poly(D,L–lactide) and PTMC have been studied by different researchers.[84,89,90] In all cases, the co-polymer synthesis was carried out through a bulk ring open polymerization of the different monomer ratios with stannous octoate as the catalyst. High molecular weight poly(DLLA–co–TMC) co-polymers were obtained by varying the reaction conditions to reduce the degree of transesterification. The low molecular weight co-polymers were synthesized using 1,3–trimethylene carbonate with a T_g value of −26°C.[81,89]

As expected, the co-polymerization was able to produce co-polymers, which had properties that were intermediate to those of the PDLLA and PTMC homopolymers. The T_g temperatures of the co-polymers ranged from −16 – 56°C depending on the percentage of lactide units in the chain. Co-polymers with a greater lactide percentage had higher T_g temperatures, and were generally stiffer. The mechanical properties of the different polymers ranged from weak elastomers to stiff and rigid materials. TMC–DLLA co-polymers with a higher TMC content exhibited high elongation at break (600–800%), but were weak (tensile strength ≤ 2 MPa) and underwent irreversible deformations upon extension. A higher DLLA content made the polymers strong (tensile strength 28–33 MPa), but brittle with low elongation at break (6–7%).[84]

Intermediate co-polymers containing similar molar percentages of both TMC and DLLA exhibited good elastomeric behavior. For example, a co-polymer containing equal molar percentages of DLLA and TMC (50%:50% molar ratio) displayed a tensile strength of 10 MPa and an elongation up to 570%. The dependence of mechanical properties on the polymer molecular weight was also evident. The higher molecular weight co-polymers displayed better mechanical properties than their low molecular weight counterparts.[84,90] All the co-polymers were found to be hydrophobic because of the hydrophobicity of the starting monomers. The *in vitro* degradation was influenced by the percentage of the monomeric constituents. For example, the polymers with a higher TMC content underwent surface degradation, while the higher lactide containing polymers underwent bulk degradation.[89]

Pego *et al.* conducted a detailed investigation of the degradation behavior of these co-polymers over a 111–week period. During this study, the group examined the effect of the monomer composition on the *in vitro* degradation profiles. The effect of degradation on the thermal and mechanical properties of the various co-polymers was also examined.[91] Even though PDLLA had a higher number of ester bonds, the co-polymers showed faster degradation rates compared to PDLLA and PTMC. This phenomenon was attributed to the lower T_g of the co-polymers when compared to PDLLA, which made the chains more mobile. The added mobility increased water uptake, and allowed the ester bonds to be more accessible in the co-polymers. The loss of mechanical properties, mass loss, and water uptake could all be co-related to the loss of molecular weight of the polymers as the hydrolytic scission of their chains progressed over the evaluation period.

The co-polymers containing a high percentage of TMC were not resorbed over the evaluation period, and the high DLLA constituting co-polymers were resorbed within 11 months. An extensive yearlong study was also conducted by Pego *et al.* to evaluate the degradation profile of these polycarbonates *in vivo*.[82] It was determined that the polymers degraded faster under the influence of the physiological environment, as opposed to the long *in vitro* degradation times. This was caused by the catalytic degradation of the carbonate moieties similar to that reported earlier.[81] To further support the theory of enzymatic breakdown of the carbonate bond, poly TMC was totally resorbed within 3 weeks *in vivo*, while a negligible mass loss was shown over a 2 year period *in vitro*.

TMC–DLLA co-polymer films were seeded with rat cardiomyocytes to determine their suitability for cardiac applications. Seeded cardiomyocytes adhered and proliferated very well on the TMC–DLLA films.[86,87] Porous scaffolds made from TMC and DLLA co-polymers were prepared by a compression molding of the salt–polymer precipitate, and followed by salt leaching. Although the TMC–DLLA co-polymer films showed good cardiomyocytes growth, no cell culture results on TMC–DLLA co-polymer scaffolds have been reported.[86,92]

3.3.3.2 Poly (ε–caprolactone–co–1,3–trimethylene carbonate)

As mentioned previously, co-polymerizations were found to be a suitable method to modulate the degradation rates of elastomeric polymers based on PTMC. DLLA was employed as a co-monomer to increase the degradation rate of the produced elastomers. However, for certain tissues repair applications such as synthetic nerve guides, it is desirable to use a material which is elastic and has a slower degradation rate.

Pego *et al.* hypothesized that the co-polymerization of ε–caprolactone and 1,3–trimethylene carbonate could yield a co-polymer, which could degrade slower than the TMC–DLLA co-polymers and retain their elasticity over a longer time period.[83] The rationale behind this idea was that poly (ε–caprolactone) is a semicrystalline polyester which degrades very slowly. High molecular weight (Mn>100,000) poly(caprolactone–co–trimethylene carbonate) co-polymers were synthesized through a ring open polymerization of the co-monomers in the presence of stannous octoate as a catalyst.

The glass transition temperatures of the co-polymers varied from −15°C to −60°C, depending on the molar percentage of each co-monomer in the melt. Co-polymers with higher caprolactone content had lower T_g values. As the caprolactone content increased, the co-polymers ranged from amorphous to semicrystalline in nature. As observed with the TMC–DLLA co-polymers, increasing the molar percentage of TMC in the melt produced weak polymers with low tolerance to deformation. Both *in vitro* and *in vivo* degradation studies were conducted to understand the mechanism of degradation of the co-polymers obtained.[82,91]

From the *in vitro* results, it was found that the TMC–CL based co-polymers degraded much slower than the TMC–DLLA based co-polymers.[91] The semicrystalline samples with a high CL content did not undergo any dimensional changes over a two–year period. In contrast, the amorphous samples with higher TMC content degraded and showed reduced dimensions. The hydrolysis rate *in vitro* was a function of the CL content in the co-polymer. Higher hydrolysis rates and subsequently higher water uptake and mass loss were detected in the polymers with higher ester content. Even in the *in vivo* study, the TMC–CL co-polymers degraded slower when compared to the TMC–DLLA co-polymers.[82] In contrast, the TMC–DLLA co-polymer degraded completely in 52 weeks, and the semicrystalline TMC–CL co-polymers suffered a mass loss below 7%. Apart from material characterization, the adhesion and proliferation of human Schwann cells on these co-polymers has been studied to determine their suitability as artificial nerve guides.[85,86]

Human Schwann cells (HSCs) were seeded on PTMC and poly(TMC–co–CL) co-polymers coated with fibronectin to evaluate the suitability of these elastomers as nerve guide materials. These materials are ideal for fabrication of nerve guides because of their long degradation rates, which are well suited to the long regeneration time of neural tissue. The number of primary HSCs which attached to the coated polymers was similar to the number of cells seeded on a control gelatin film.[85] In addition, *in vivo* studies have shown that poly(TMC–co–CL) co-polymers can be effective nerve guides in the regeneration of autonomous neural tissue.[93]

3.4 MEMS PRINCIPLES IN TISSUE ENGINEERING

In the past decade, microscale technologies have emerged as a powerful tool for biological and biomedical applications.[94] MEMS research and development has remained intense to solve complex problems at the cellular and molecular level.[2,95] Biological or Biomedical MEMS, BioMEMS, can be defined as the application of micro– and nanotechnology to develop devices or systems that are used for the processing, delivery, manipulation, analysis, or construction of biological and chemical modalities.[2,95] The advancement of BioMEMS technologies has progressed, and will have a broad and significant impact in the fields of biology and medicine if fully realized.[96]

Few other engineering techniques are able to closely match the micro to millimeter size dimension of tissues in the human body with the precision and accuracy of BioMEMS techniques.[95] Due to these advantages, BioMEMS holds great promise in addressing the challenges found in many disciplines such as diagnostic, therapeutic, sensing, detection, and tissue engineering applications.[2,97,98] The potential to mimic complex tissue architecture and *in vivo* conditions makes BioMEMS a powerful tool for tissue engineering.

3.5 MEMS APPLICATIONS IN TISSUE ENGINEERING

Although BioMEMS based tissue engineering is a rapidly advancing field, research involving the use of biodegradable elastomers coupled with microfabrication processes is new and fairly limited. Discussed in the following section are BioMEMS based techniques involving hydrogels and biodegradable elastomers to construct 3D structures, control cell adhesion, control cell morphology, and create microvasculature for 3D constructs.

The recent progress of MEMS based technologies has lead to new approaches to study *in vitro* cell culture environments. Many of these new techniques utilize a soft lithography approach to rapidly produce 3D microstructures. Leclerc *et al.* used a photosensitive caprolactone and lactide based polymer to fabricate biodegradable polymer microstructures down to 50 μm for tissue engineered liver constructs.[99]

As seen in Fig. 3.7, Leclerc *et al.* successfully created various single and multistepwise microstructures using a soft lithographic technique. In addition, the single stepwise microstructures supported the attachment, spreading, and growth of a variety of mammalian cell types. Other groups have also successfully created complex 3D polymer constructs for hepatic tissue engineering. In 2007, Tsang *et al.* created PEGDA hydrogel constructs for hepatic cell encapsulation. By combining a PEG based hydrogel with a multilayer fabrication method, Tsang and co-workers were able to fabricate highly cell–encapsulated scaffolds with architecture to facilitate nutrient delivery through convective flow.[100]

Figure 3.7. SEM photographs of fabricated microstructures. (A) microchambers and microchannels on pCLLA; (B) a microchannel network on pCLLA; (C) channels fabricated with direct UV exposure on pCLLA; (D) a single stepwise microstructure on pCLH fabricated by stamping; (E) a multistepwise microstructure on pCLLA fabricated by stamping; and (F) a multistepwise microstructure on pCLH fabricated by stamping. Reprinted from Biomaterials, 25(19), Leclerc E. *et al.*, Fabrication of microstructures in photosensitive biodegradable polymers for tissue engineering applications, 4683–4690, 2004, with permission from Elsevier.

In addition to creating 3D constructs, many research groups have incorporated micro scale technologies to promote and discourage cell adhesion. Mizutani *et al.* showed the ability to control cell adhesion on PLA films using photocured co-polymers.[101] Coating a PLA surface with a low molecular weight alcohol based co-polymer promoted endothelial cell adhesion, whereas the PLA surface coated with PEG–based co-polymer did not support cell adhesion. The different co-polymers coated on the PLA films were able to change the hydrophobicity of the surface to either encourage or deter endothelial cell adhesion.[101]

Another research group successfully proved to control tissue organization by immobilizing non-adhesive domains onto a surface. The group of Liu *et al.* used a photolithographic technique to immobilize a PEO–terminated triblock polymer onto various surfaces to deter cell adhesion for up to 4 weeks *in vitro*.[102] Expanding upon previous research by Neff *et al.*, the hydrophobic core of the polymer was modified with adhesive peptides to create non-adhesive domains.[103] This cell avoidance phenomenon can be explained by the polymer's ability to also deter proteins, which are necessary for cell attachment.[102]

The ability to control cell and protein behavior using mechanical cues in addition to chemical cues is critical in understanding tissue development.[102] While these mechanisms of cell behavior are not yet fully understood, research has shown that the extracellular matrix proteins of cells possess a 3D surface topography of sub-micron length scales.[104] The ability to control cellular structure and function by culturing cells on substrates modified with micron and sub-micron features is a field termed contact guidance.[105] Contact guidance has been shown to induce cellular responses in various cell types such as epithelial cells, fibroblasts, oligodendrocytes, and astrocytes.[106]

The use of poly (dimethyl siloxane) (PDMS) has been a major limitation of previous research involving contact guidance. PDMS, although elastic, is not biodegradable and has limited biocompatibility, which limits its use in tissue engineering applications. To overcome this limitation, Bettinger *et al.* is the first group to successfully use BioMEMS to introduce rounded sub-micron features onto an elastic, biodegradable substrate for contact guidance applications (Fig. 3.8). Using PGS, a novel biodegradable elastomer, the research group developed a photolithographic method to fabricate substrates with rounded features as small as 500 nm in scale.

Bovine aortic endothelial cells cultured on the microstructures exhibited a rounded and spindle–shaped morphology when compared to cells cultured on a flat substrate, which had a random orientation of cell projections and a flattened appearance. Thus, their results showed that filipodia of cells are able to detect regional gradients in substrate topography, which results in preferential cell adherence through cytoskeletal rearrangement.[106]

In addition to guiding the cell morphology, BioMEMS techniques have also been applied toward creating microvasculature for tissue engineered constructs. Creating tissue constructs on 3D scaffolds has been a heavily researched area.[18] However, creating constructs that provide adequate nutrient and oxygen transport

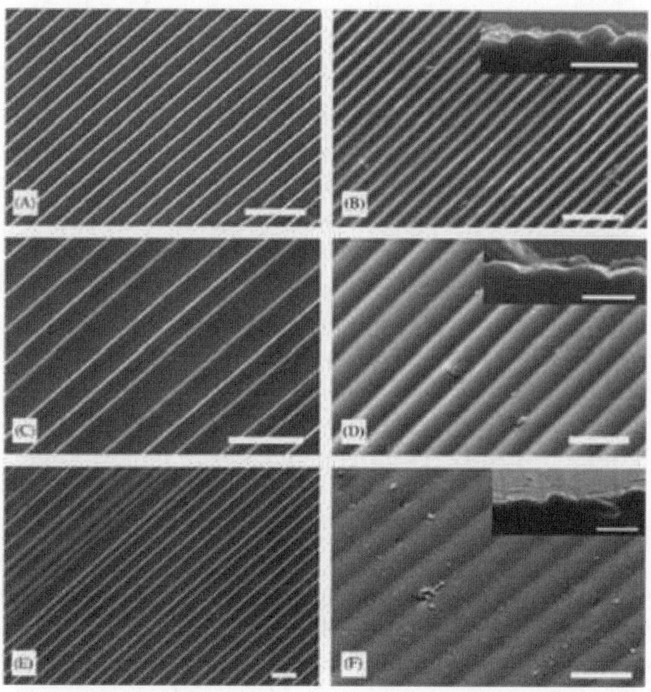

Figure 3.8. SEM photographs of silicon masters with cross–sections and their resulting PGS substrates. Reprinted from Biomaterials, 27(12), Bettinger CJ *et al.*, Microfabrication of poly (glycerol–sebacate) for contact guidance applications, 2558–2565, 2006, with permission from Elsevier.

to cells deeply embedded in the substrate has proven to be a formidable task.[100] The development of an established vasculature system to provide oxygen, nutrients, and waste removal is critical in the survival of tissue engineered organs.[107] From this limitation, current engineered tissue is limited to 150–200 micron thicknesses due to oxygen diffusion limitations.[108]

Fidkowski *et al.* have used BioMEMS to build capillary networks onto synthetic substrates. Using standard soft photolithography techniques, the research group patterned intricate capillary networks 10 microns in size onto PGS using silicon wafers as molds. Human umbilical vein endothelial cells (HUVECs) were seeded onto the PGS substrates and perfused under flow conditions to create confluent endothelialized two–dimensional cell layers. The HUVECs could be lifted from the PGS substrate and incorporated into other devices. Thus, this study showed the potential for using PGS in combination with BioMEMS techniques to create microvasculature *in vitro* towards the fabrication of vascularized organs.[107]

3.6 OUTLOOK

Many of the tissues in the body are soft and elastic. Much attention has been paid in using biodegradable soft and elastic scaffolds for tissue engineering soft tissues such as skin, blood vessel, tendon, ligament, cartilage, bladder etc. The roles of biodegradable elastomeric materials in tissue engineering have been increasingly emphasized as the evolving progress in understanding the cell/materials/host interactions. Soft and elastic scaffolds made of biodegradable elastomeric scaffolds not only provide a substrate for cells to adhere and proliferation, but also minimize the compliance mismatch with surround tissues and provide cues and signals to promote tissue development and functional integration with the host.

The design and synthesis of biodegradable elastomers will continuously evolve owing to the more stringent material requirements in personalized tissue regeneration. Despite the recognized importance of the mechanical properties of tissue engineering scaffolds on the tissue development, there has been a dearth on fundamental understanding on how the soft and elastic scaffolds affect the inflammatory response of the host and the tissue/graft integration.

The application of BioMEMS in tissue engineering has resulted in more understanding on how cells respond to micro/nano structure created by BioMEMS. Constructing vasculature with the aid of BioMEMS on biodegradable elastomeric scaffolds for tissue engineering is still in its infancy. The current studies lie on fabricating channels on two-dimensional films, and then stacking them into 3D channels on elastomers, mostly on PDMS. More studies should be focused on using biodegradable elastomeric substrates. More importantly, the vasculature should be built up within 3D porous scaffolds instead of just in between two-dimensional solid films. Our recent studies have resulted in 3D scaffolds with vasculature-like channels built using our recently developed CUPE polymers via the scaffold-sheet tissue engineering strategy combined with BioMEMS technology.

References

1 Webb A. R., Yang J. and Ameer G. A., Biodegradable polyester elastomers in tissue engineering. *Expert Opin. Biol. Ther.*, **4**(6), 801–812 (2004).

2 Bashir R., BioMEMS: State-of-the-art in detection, opportunities and prospects. *Adv. Drug. Deliv. Rev.*, **56**(11), 1565–1586, (2004).

3 Pomfret E. A., Sung R. S., Allan J., Kinkhabwala M., Melancon J. K. and Roberts J. P., Solving the organ shortage crisis: The 7th annual American Society of Transplant Surgeons' State-of-the-Art Winter Symposium. *Am J. Transplant*, **8**(4), 745–752 (2008).

4 Kulig K. M. and Vacanti J. P., Hepatic tissue engineering. *Transpl. Immunol.*, **12**(3–4), 303–310 (2004).

5 Chan C., Berthiaume F., Nath B. D., Tilles A. W., Toner M. and Yarmush M. L., Hepatic tissue engineering for adjunct and temporary liver support: Critical technologies. *Liver Transpl.*, **10**(11), 1331–1342 (2004).

6 Ikada Y., Challenges in tissue engineering. *J. R. Soc. Interface*, **3**(10), 589–601 (2006).

7 Evans G. R., Approaches to tissue engineered peripheral nerve. *Clin. Plast. Surg.* **30**(4), 559–563, viii (2003).

8 Selvam S., Thomas P. B. and Yiu S. C., Tissue engineering: Current and future approaches to ocular surface reconstruction. *Ocul. Surf.*, **4**(3), 120–136 (2006).

9 Metcalfe A. D. and Ferguson M. W., Tissue engineering of replacement skin: The crossroads of biomaterials, wound healing, embryonic development, stem cells and regeneration. *J. R. Soc. Interface*, **4**(14), 413–437 (2007).

10 Nahmias Y., Berthiaume F. and Yarmush M. L., Integration of technologies for hepatic tissue engineering. *Adv. Biochem. Eng. Biotechnol.*, **103**, 309–329 (2007).

11 Beck J., Angus R., Madsen B., Britt D., Vernon B, Nguyen K. T., Islet encapsulation: Strategies to enhance islet cell functions. *Tissue Eng.*, **13**(3), 589–599 (2007).

12 Lee S. H. and Shin H., Matrices and scaffolds for delivery of bioactive molecules in bone and cartilage tissue engineering. *Adv. Drug. Deliv. Rev.*, **59**(4–5), 339–359 (2007).

13 Stern-Straeter J., Riedel F., Bran G., Hormann K. and Goessler U. R., Advances in skeletal muscle tissue engineering. *In Vivo* **21**(3), 435–444 (2007).

14 Sievert K. D., Amend B. and Stenzl A., Tissue engineering for the lower urinary tract: A review of a state of the art approach. *Eur. Urol.*, **52**(6), 1580–1589 (2007).

15 Zhang W. J., Liu W., Cui L. and Cao Y., Tissue engineering of blood vessel. *J. Cell Mol. Med.*, **11**(5), 945–957 (2007).

16 Hiles M. and Hodde J., Tissue engineering a clinically useful extracellular matrix biomaterial. *Int. Urogynecol. J. Pelvic Floor Dysfunct*, **Suppl 1**, S39–43 (2006).

17 Rabkin E. and Schoen F. J., Cardiovascular tissue engineering. *Cardiovasc. Pathol.*, **11**(6), 305–317 (2002).

18 Vacanti J. P. and Langer R., Tissue engineering: The design and fabrication of living replacement devices for surgical reconstruction and transplantation. *Lancet* 1999 Jul; 354 Suppl 1:SI32–34.

19 Lee J., Cuddihy M. J. and Kotov N. A., Three-dimensional cell culture matrices: State of the art. *Tissue Engineering Part B: Reviews*, **14**(1), 61–86 (2008).

20 Yang J., Webb A. R. and Ameer G. A., Novel citric acid-based biodegradable elastomers for tissue engineering. *Adv. Mater* **16**(6), 511–516 (2004).

21 Ratner B. D., Hoffman A. S., Schoen F. J. and Lemmons J. E., *Biomaterial Science: An Introduction to Materials in Medicine*. San Diego: Elsevier Academic Press, (2004).

22 Levental I., Georges P. C. and Janmey P. A., Soft biological materials and their impact on cell function. *Soft Matter,* **3**, 299–306 (2007).

23 Hodde J., Naturally occurring scaffolds for soft tissue repair and regeneration. Tissue Eng., **8**(2), 295–308 (2002).

24 Wang Y., Ameer G. A., Sheppard B. J. and Langer R. A tough biodegradable elastomer. *Nat. Biotechnol.*, **20**(6), 602–606 (2002).

25 Younes H. M., Bravo-Grimaldo E. and Amsden B. G., Synthesis, characterization and *in vitro* degradation of a biodegradable elastomer. *Biomaterials,* **25**(22), 5261–5269 (2004).

26 Stegemann J. P. and Nerem R. M., Phenotype modulation in vascular tissue engineering using biochemical and mechanical stimulation. *Ann. Biomed. Eng.*, **31**(4), 391–402 (2003).

27 Niklason L. E., Gao J., Abbott W. M., Hirschi K. K., Houser S., Marini R. and Langer R., Functional arteries grown *in vitro*. *Science,* **284**(5413), 489–493 (1999).

28 Lee J. H., Kisiday J. and Grodzinsky A. J,. Tissue-engineered versus native cartilage: Linkage between cellular mechano-transduction and biomechanical properties. *Novartis Found. Symp.*, **249**, 52–64, discussion 64–59, 170–174, 239–141.

29 Sarkar D., Yang J. C. and Lopina S. T. Structure-property relationship of L-tyrosine-based polyurethanes for biomedical applications. *Journal of Applied Polymer Science,* **108**, 2345–2355 (2008).

30 Wright E. R. and Conticello V. P., Self-assembly of block copolymers derived from elastin-mimetic polypeptide sequences. *Adv. Drug Deliv. Rev. Oct.*, **54(8)**, 1057–1073 (2002).

31 Wang Y.-Q. and Cai J.-Y. Enhanced cell affinity of poly(L-lactic acid) modified by base hydrolysis: Wettability and surface roughness at nanometer scale. *Current Applied Physics*, **7S1**, e108–e111 (2007).

32 Wang S, Cui W, Bei J. Bulk and surface modifications of polylactide. *Anal. Bioanal. Chem.*, **381**(3), 547–556 (2005).

33 Amsden B., Curable, biodegradable elastomers: emerging biomaterials for drug delivery and tissue engineering. *Soft Matter,* **3**, 1335–1348 (2007).

34 Gallocher S. L., Aguirre A. F., Kasyanov V., Pinchuk L., Schoephoerster R. T., A novel polymer for potential use in a trileaflet heart valve. *Inc. J. Biomed. Mater. Res. Part B: Appl. Biomater.*, **79B**, 325–334 (2006).

35 Kunioka M., Ninomiya F. and Funabashi M., Biodegradation of poly(lactic acid) powders proposed as the reference test materials for the international standard of biodegradation evaluation methods. *Polymer Degradation and Stability,* **91**, 1919–1928 (2006).

36 Nijst C. L., Bruggeman J. P. Karp J. M., Ferreira L., Zumbuehl A., Bettinger C. J. and Langer R., Synthesis and characterization of photocurable elastomers from poly(glycerol-co-sebacate). *Biomacromolecules,* **8**(10), 3067–3073 (2007).

37 Yoda R., Elastomers for biomedical applications. *J. Biomater. Sci. Polym. Ed.*, **9**(6), 561–626 (1998).

38 Gunatillake P. A. and Adhikari R. Biodegradable synthetic polymers for tissue engineering. *European Cells & Materials* **5**, 1–16; discussion 16 (2003).

39 Nair L. and Laurencin C., Biodegradable polymers as biomaterials. *Progress in Polymer Science*, **32**, 762–798 (2007).

40 Lee S., Bacterial Polybhydroxyalkanoates. *Biotechnology and Bioengineering,* **49**, 1–14 (1996).

41 Poirier Y., Nawrath C. and Somerville C., Production of polyhydroxyalkanoates, a family of biodegradable plastics and elastomers, in bacteria and plants. Bio/technology (Nature Publishing Company), **13**(2), 142–150 (1995).

42 Hocking P. and Marchessault R., Syndiotactic poly[(R, S)-b-hydroxybutyrate] isolated from methylaluminoxane-catalyzed polymerization. *Polym Bull*, **30**, 163–170 (1993).

43 Kemnitzer J., McCarthy S. and Gross R., Preparation of predominantly syndiotactic poly(b-hydroxybutyrate) by the tributyltin methoxide catalyzed ring-opening polymerization of racemic b-butyrolactone. *Macromolecules*, **26**, 1221–1229 (1993).

44 Kemnitzer J., McCarthy S. and Gross R., Syndiospecific ringopening polymerization of b-butyrolactone to form predominantly syndiotactic poly(b-hydroxybutyrate) using tin(IV) catalysts. *Macromolecules*, **16**, 6143–6150 (1993).

45 Hori Y. and Hagiwara T., Ring-opening polymerisation of beta-butyrolactone catalysed by distannoxane complexes: study of the mechanism. *International Journal of Biological Macromolecules*, **25**(1-3), 237–245 (1999).

46 Williams S., Martin P., Horowitz D. and Peoples O. PHA applications: Addressing the price performance issue. *International Journal of Biological Macromolecules*, **25**, 111–121 (1999).

47 Akhtar S., Physiochemical properties of bacterial P(HB-HV) polyesters and their uses in drug delivery.: University of Bath; 1990.

48 Stock U. A., Nagashima M., Khalil P. N., Nollert G. D., Herden T., Sperling J. S., Moran A., Lien J., Martin D. P., Schoen F. J., Vacanti J. P. and Mayer J. E., Jr., Tissue-engineered valved conduits in the pulmonary circulation. *The Journal of Thoracic and Cardiovascular Surgery*, **119**(4 Pt 1), 732–740 (2000).

49 Stock U. A. and Mayer J. E., Jr., Tissue engineering of cardiac valves on the basis of PGA/PLA co-polymers. *Journal of Long-term Effects of Medical Implants*, **11**(3–4), 249–260 (2001).

50 Nagata M., Kiyotsukuri T., Ibuki H., Tsutsumi N. and Sakai W., Synthesis and enzymatic degradation of regular network aliphatic polyesters. *Reactive Functional Polymers*, **30**, 165–171 (1996).

51 Sundback C. A., Shyu J. Y., Wang Y., Faquin W. C., Langer R. S., Vacanti J. P. and Hadlock T. A., Biocompatibility analysis of poly(glycerol sebacate) as a nerve guide material. *Biomaterials*, **26**(27), 5454–5464 (2005).

52 Chen Q. Z., Bismarck A., Hansen U., Junaid S., Tran M. Q., Harding S. E., Ali N. N. and Boccaccini A. R., Characterisation of a soft elastomer poly(glycerol sebacate) designed to match the mechanical properties of myocardial tissue. *Biomaterials*, **29**(1), 47–57 (2008).

53 Ding T., Liu Q., Shi R., Tian M., Yan J. and Zhang L. Synthesis, characterization and in vitro degradation study of a novel and rapidly degradable elastomer. *Polym. Degrad. Stab.*, **91**, 733–739 (2006).

54 Yang J., Webb A. R., Pickerill S. J., Hageman G. and Ameer G. A., Synthesis and evaluation of poly(diol citrate) biodegradable elastomers. *Biomaterials*, **27**(9), 1889–1898 (2006).

55 Yang J., Webb A. R., Hageman G., Ameer G. A., Biodegradable elastomeric polymer for tissue engineering. In: Mallapragada S. NB, editor. Biodegradable Polymer Materials and their Application: *American Scientific Publishers*, pp. 191–232 (2006).

56 Lei L., Ding T., Shi R., Liu Q., Zhang L., Chen D. and Tian W., Synthesis, characterization and in vitro degradation of a novel degradable poly((1,2-propanediol-sebacate)-citrate) bioelastomer. *Polymer Degradation and Stability*, **92**, 389–396 (2007).

57 Wan Y., Feng G., Shen F. H., Balian G., Laurencin C. T. and Li X., Novel biodegradable poly(1,8-octanediol malate) for annulus fibrosus regeneration. *Macromol. Biosci.*, **7**(11), 1217–1224 (2007).

58 Yang J., Gywali D. and Richard Tran M.-T., Composition and method of biodegradable elastic polymers, patents (2008).

59 Yang J. and Dey J., Biopolymer and scaffold-sheet method for tissue engineering. Patents USA, (2007).

60 Dey J., Xu H. Shen J., Gondi S. R., Nguyen K. T., Sumerlin B. S., Tang L. and Yang J., Development of biodegradable crosslinked urethane-doped polyester elastomers. *Submitted*, (2008).

61 Boretos J. and Pierce W., Segmented polyurethane: A new elastomer for biomedical applications. *Science*, **158**(3807), 1481 (1967).

62 Guelcher S. A., Biodegradable polyurethanes: Synthesis and applications in regenerative medicine. *Tissue Engineering: Part B*, **14**(1), 3–16 (2008).

63 Zdrahala R. and Zdrahala I., Biomedical applications of polyurethanes: A review of past promises, present realities, and a vibrant future. *Journal of Biomaterials Applications*, **14**(1), 67–90 (1999).

64 Skarja G. and Woodhouse K., Synthesis and characterization of degradable polyurethane elastomers containing and amino-acid based chain extender. *Journal of Biomaterial Science, Polymer Edition*, **9**(3), 271–295 (1998).

65 Skarja G. and Woodhouse K., Structure-property relationships of degradable polyurethane elastomers containing and amino acid based chain extender. *Journal of Applied Polymer Science*, **75**, 1522–1534 (2000).

66 Skarja G. and Woodhouse K., *In vitro* degradation and erosion of degradable, segmented polyurethanes containing and amino acid based chain extender. *Journal of Biomaterial Science, Polymer Edition*, **12**(8), 851–873 (2001).

67 Guan J. and Wagner W. R., Synthesis, characterization and cytocompatibility of polyurethaneurea elastomers with designed elastase sensitivity. *Biomacromolecules*, **6**, 2833–2842 (2005).

68 Guan J., Sacks M., Beckman E. and Wagner W. R., Synthesis, characterization and cytocompatibility of elastomeric, biodegradable poly(ester-urethane)ureas based on poly(caprolactone) and putrescine. *Journal of Biomedical Materials Research*, **61**, 493 (2002).

69 Kylma J. and Seppala J. V., Synthesis and characterization of a biodegradable thermoplastic poly(ester-urethane) elastomer. *Macromolecules*, **30**, 2876–2882 (1997).

70 Tatai L., Moore T. G., Adhikari R., Malherbe F., Jayasekara R., Griffiths I. and Gunatillake P. A., Thermoplastic biodegradable polyurethanes: The effect of chain extender structure on properties and *in-vitro* degradation. *Biomaterials*, **28**(36), 5407–5417 (2007).

71 Ciardelli G., Rechichi A., Cerrai P., Tricoli M., Barbani N. and Giusti P., Segmented polyurethanes for medical applications: Synthesis, characterization and *in vitro* enzymatic degradation studies. *Macromolecular Symposia*, **218**, 261–271 (2004).

72 Sarkar D., Yang J. C., Gupta A. S. and Lopina S. T., Synthesis and characterization of L-tyrosine based polyurethanes for biomaterial applications. *J. Biomed. Mater. Res., A* May 21 (2008).

73 Hirt T. D., Neuenschwander P. and Suter U. W., Synthesis of degradable, biocompatible, and tough block-copolyesterurethanes. *Macromol. Chem. Phys.*, **197**, 4253–4268 (1996).

74 Cohn D., Stern T., Gonzalez M. F. and Epstein J., Biodegradable poly(ethylene oxide)/poly(caprolactone) multiblock co-polymers. *Journal of Biomedical Materials Research*, **59**, 273–281 (2002).

75 Guan J., Sacks M., Beckman E. and Wagner W. R., Biodegradable poly(ether ester urethane)urea elastomers based on poly(ether ester) triblock copolymers and putrescine: Synthesis, characterization and cytocompatibility. *Biomaterials*, **25** 85 (2004).

76 Abraham G. A., Marcos-Fernandez A. and Roman J. S., Bioresorbable poly(ester-ether urethane)s from L-lysine diisocyanate and triblock copolymers with different hydrophilic character. *J. Biomed. Mater. Res., A* **76**(4), 729–736 (2006).

[77] Guan J., Fujimoto K. L., Sacks M. S. and Wagner W. R., Preparation and characterization of highly porous, biodegradable polyurethane scaffolds for soft tissue applications. *Biomaterials,* **26**(18), 3961–3971 (2005).

[78] Fujimoto K. L., Guan J., Oshima H., Sakai T. and Wagner W. R., *In vivo* evaluation of a porous, elastic, biodegradable patch for reconstructive cardiac procedures. *Ann. Thorac. Surg.,* **83**(2), 648–654 (2007).

[79] Nieponice A., Soletti L., Guan J., Deasy B. M., Huard J., Wagner W. R. and Vorp D. A., Development of a tissue-engineered vascular graft combining a biodegradable scaffold, muscle-derived stem cells and a rotational vacuum seeding technique. *Biomaterials,* **29**(7), 825–833 (2008).

[80] Borkenhagen M. , Stoll R. C., Neuenschwander P., Suter U. W. and Aebischer P., *In vivo* performance of a new biodegradable polyester urethane system used as a nerve guidance channel. *Biomaterials,* **19**(23), 2155–2165 (1998).

[81] Zhu K. J., Hendren R. W., Jensen K. and Pitt C. G., Synthesis, properties and biodegradation of poly(1,3-trimethylene carbonate). *Macromolecules,* **24**, 1736–1740 (1991).

[82] Pego A. P., Van Luyn M. J., Brouwer L. A., van Wachem P. B., Poot A. A., Grijpma D. W. and Feijen J., *In vivo* behavior of poly(1,3-trimethylene carbonate) and copolymers of 1,3-trimethylene carbonate with D,L-lactide or epsilon-caprolactone: Degradation and tissue response. *J. Biomed Mater. Res., A* **67**(3), 1044–1054 (2003).

[83] Pego A. P., Poot A. A., Grijpma D. W. and Feijen J., Copolymers of trimethylene carbonate and epsilon-caprolactone for porous nerve guides: synthesis and properties. *J. Biomater. Sci. Polym. Ed.,* **12**(1), 35–53 (2001).

[84] Pego A. P, Poot A. A., Grijpma D. W. and Feijen J., Physical properties of high molecular weight 1,3-trimethylene carbonate and D,L-lactide copolymers. *J. Mater. Sci. Mater. Med.,* **14**(9), 767–773 (2003).

[85] Pego A. P., Vleggeert-Lankamp C. L., Deenen M., Lakke E. A., Grijpma D. W., Poot A. A., Marani E. and Feijen J., Adhesion and growth of human Schwann cells on trimethylene carbonate (co)polymers. *J. Biomed. Mater. Res., A* **67**(3), 876–885 (2003).

[86] Pego A. P., Poot A. A., Grijpma D. W. and Feijen J., Biodegradable elastomeric scaffolds for soft tissue engineering. *J. Control Release,* **87**(1–3), 69–79 (2003).

[87] Pego A. P., Siebum B., Van Luyn M. J., Gallego Y., Van Seijen X. J., Poot A. A., Grijpma D. W. and Feijen J., Preparation of degradable porous structures based on 1,3-trimethylene carbonate and D,L-lactide (co)polymers for heart tissue engineering. *Tissue Eng.,* **9**(5), 981–994 (2003).

[88] Pego A. P., Grijpma D. W. and Feijen J. Enhanced mechanical properties of 1,3-trimethylene carbonate polymers and networks. *Polymer,* **44**, 6495–6504 (2003).

[89] Jie C. and Zhu K. J., Preparation, characterization and biodegradable characteristics of poly(D, L-lactide-co-1,3-trimethylene carbonate). *Polymer International,* **42**, 373–379 (1997).

[90] Buchholz B., Analysis and characterization of resorbable DL-lactide-trimethylene carbonate copolyesters. *Journal of Materials Science: Materials in Medicine,* **4**, 381–388 (1993).

[91] Pego A. P, Poot A. A, Grijpma D. W. and Feijen J., *In Vitro* degradation of trimethylene carbonate based (co)polymers. *Macromolecular Bioscience,* **2**, 411–419 (2002).

[92] Chen Q., Harding S. E., Ali N. N., Lyon A. R. and Boccaccini A. R., Biomaterials in cardiac tissue engineering: Ten years of research survey. *Materials Science and Engineering R,* **59**, 1–37 (2008).

[93] Fabre T., Schappacher M., Bareille R., Dupuy B., Soum A., Bertrand-Barat J. and Baquey C., Study of a (trimethylenecarbonate-co-epsilon-caprolactone) polymer–part 2: in vitro cytocompatibility analysis and *in vivo* ED1 cell response of a new nerve guide. *Biomaterials,* **22**(22), 2951–2958 (2001).

94 Khademhosseini A., Langer R., Borenstein J. and Vacanti J. P., Microscale technologies for tissue engineering and biology. *Proc. Natl. Acad Sci., USA* **103**(8), 2480–2487 (2006).

95 Desai T. A., Micro- and nanoscale structures for tissue engineering constructs. *Med. Eng. Phys.*, **22**(9), 595–606 (2000).

96 Chatni M. R., The onset of a revolution in drug discovery. *Trends Biotechnol.*, **25**(4), 142–144 (2007).

97 Puleo C. M., Yeh H. C. and Wang T. H., Applications of MEMS technologies in tissue engineering. *Tissue Eng.*, **13**(12), 2839–2854 (2007).

98 Peppas N. A., Hilt J. Z., Khademhosseini A. and Langer R., Hydrogels in biology and medicine: From molecular principles to bionanotechnology. *Advanced Materials*, **18**(11), 1345–1360 (2006).

99 Leclerc E., Furukawa K. S., Miyata F., Sakai Y., Ushida T. and Fujii T., Fabrication of microstructures in photosensitive biodegradable polymers for tissue engineering applications. *Biomaterials*, **25**(19):4683–4690 (2004).

100 Tsang V. L. and Bhatia S. N., Fabrication of three-dimensional tissues. *Adv. Biochem. Eng. Biotechnol*, **103**, 189–205 (2007).

101 Mizutani M., Arnold S. C. and Matsuda T., Liquid, phenylazide-end-capped copolymers of epsilon-caprolactone and trimethylene carbonate: Preparation, photocuring characteristics, and surface layering. *Biomacromolecules*, **3**(4), 668–675 (2002).

102 Liu V. A., Jastromb W. E. and Bhatia S. N., Engineering protein and cell adhesivity using PEO-terminated triblock polymers. *J. Biomed Mater. Res.*, **60**(1), 126–134 (2002).

103 Neff J. A., Tresco P. A. and Caldwell K. D., Surface modification for controlled studies of cell-ligand interactions. *Biomaterials*, **20**(23–24), 2377–2393 (1999).

104 Abrams G. A., Schaus S. S., Goodman S. L., Nealey P. F. and Murphy C. J., Nanoscale topography of the corneal epithelial basement membrane and Descemet's membrane of the human. *Cornea*, **19**(1), 57–64 (2000).

105 Weiss P., Experiments on cell and axon orientation *in vitro*: The role of colloidal exudates in tissue organization. *Journal of Experimental Zoology*, **100**(3), 353–386 (1945).

106 Bettinger C. J., Orrick B., Misra A., Langer R. and Borenstein J. T., Microfabrication of poly (glycerol-sebacate) for contact guidance applications. *Biomaterials*, **27**(12), 2558–2565 (2006).

107 Fidkowski C., Kaazempur-Mofrad M. R., Borenstein J., Vacanti J. P., Langer R. and Wang Y., Endothelialized microvasculature based on a biodegradable elastomer. *Tissue Eng.* **11**(1–2), 302–309 (2005).

108 Colton C. K., Implantable biohybrid artificial organs. *Cell Transplant* **4**(4), 415–436 (1995).

Chapter Four

MEMS in the Nervous System

Sharon Norman* and Ravi Bellamkonda†

**Bioengineering Program, School of Electrical and Computer Engineering, Atlanta GA 30332, USA. E-mail: sbyers3@gatech.edu*
†Wallace H Coulter Department of Biomedical Engineering at Georgia Institute of Technology and Emory School of Medicine, Atlanta GA 30332-0535, USA

Seamless interfacing with the human nervous system is likely to have significant implications for enhanced understanding of its function, as well as exciting therapeutic potential. However, it is important to recognize the biological landscape setting the design criteria for any engineering interface to the nervous system. The nervous system of mammalian species is incredibly complex, but can be divided into two basic parts, central and peripheral. The central nervous system (CNS) is composed of the brain and spinal cord, while the peripheral system (PNS) consists of sensory, motor, and pain fibers that connect the body's extremities to the spinal cord and the brain. Therefore there exists a neuroanatomical map that is intricate as well as a neurochemical map that has great, micron level topographical specificity that any sensor would have to accommodate. Essentially, the brain has a high cell density and particular regions of the cortex are responsible for specific functions.

The cortex is a striated structure, and each layer contains different types of cells that code for different functions. The simplest unit of the nervous system is the neuron, which has a cell body, or soma, and two types of processes, called dendrites and axons. The primary mode of "communication" between neurons involves electrical signals, also known as action potentials, that propagate down axons to connect with the processes or somas of other neurons. These action potentials are the main means of communication within the nervous system and are commonly initiated by the release of neurotransmitters from surrounding neurons. Therefore from an analytical perspective, electrical and chemical signals are the information carriers in the nervous system and they are dispersed with great regional specificity throughout the X, Y, Z and temporal space of the nervous system. Any sensor/actuator system that interfaces to the nervous system, then, is judged by how efficiently it is able to monitor and influence this spatio–temporally organized information processing system.

Biomaterials for MEMS, Edited by M. Chiao and J.-C. Chiao

www.panstanford.com
978-981-4241-46-5

In this chapter, we examine the case for microelectromechanical systems being valuable tools for interfacing with the nervous system as effective diagnostic and therapeutic tools by facilitating both sensation and actuation in the brain. This chapter will focus on both *in vitro* and *in vivo* applications. There are several important advantages of MEMs devices when working with the nervous system, notably size, electrical functionality, chemical functionality, reproducibility, model system simplification, and inexpensive fabrication and some of these advantages are discussed below.

The small size of MEMs devices makes them both minimally invasive for *in vivo* studies and allows them to interface with neurons at a scale comparable to cells. Minimizing invasion and the resulting damage to neuronal tissue is important because neural communication can be partially or completely disrupted after trauma and functional deficits can result; the less tissue damage initiated when trying to achieve recordings or stimulation, the better the functional outcome for the animal, human or otherwise. The neuronal tissue reaction to implants can be challenging to alleviate as well, so smaller devices may trigger a less severe reaction. In addition, neuronal tissues tend to be cell–dense and have a high regional specificity; the brain, for example, is organized into functional areas at different places on the cortex, and each of the six cortical layers contains different combinations of cell types carrying different kinds of information. The ability to tightly control the geometry of MEMs devices allows for exploration of specific functional areas, including cells at a certain cortical depth and in a particular functional region, such as the visual or auditory cortex. MEMs devices usually use small volumes of reagents, so resources such as supplements and media can be conserved when performing *in vitro* studies. Also, fewer cells are needed to cover or inhabit a MEMs device, so significant results can be achieved using fewer cells.

Because of the nervous system's electrical nature, microwires and microelectrodes have been used for years to measure electrical signals in the brain. The direct interface of electrical MEMs with the electrical nervous system allows signals to be sent and received to and from the nervous system fairly reliably. Another advantage of microscale electrical devices is that spatial resolution for both recording and stimulation is typically higher than for larger devices; individual action potentials and signals can be measured and analyzed, and the small contact area of a microelectrode allows for stimulation in a precise region of the nervous system, whether in the brain, spinal cord, or peripheral nerve. This spatial specificity gives greater control over the functional outcome of stimulation.

In addition to electrical communication, chemical communication is vital for normal neural function. The ability to sense and monitor the chemical environment of nervous tissue *in vivo* and control portions of it by applying drugs or other compounds is important for answering questions about the effects of drugs, bioactive molecules, chemicals, neuromodulation, and external stimulation on the nervous system. Localized drug delivery using microfluidic channels could lead to effective new therapies, perhaps for managing pain or reducing fibrous encapsulation of the device itself. Microchannels capable of sampling neurotransmitters and chemicals present at synapses after external stimulation could answer questions about what signaling cascades are initiated in response to certain stimuli and how the chemicals present affect nervous system transduction. The small volumes used by these chemical interfaces are advantageous because they do not displace a large volume of fluid in nervous system tissue and so such procedures should be minimally damaging.

To answer complex questions about the nervous system, it is often advantageous to have a simple model system that is easily controlled. Multielectrode arrays and microfluidic devices allow the investigator to control the environment of cells or tissues *in vitro*. The development of cells in culture can be modified by a multitude of stimuli, including but not limited to, chemical gradients of growth factors or adhesion proteins, supplements present or absent in culture medium, fluid flow rates, and electrical stimulation. *In vitro* cultures allow a great deal of control over the environment of cells or tissues, which can be helpful when determining which variables are critically important in a study.

MEMs devices are capable of being batch fabricated, which means many can be made at one time and this reduces the cost per device. This is especially applicable to silicon devices, where more than one device can be patterned and processed on a single wafer. Batch fabrication can also reduce the variability between devices, as many are made using the same solutions at the same time in the same environment. Because devices made at the same time are similar, this may help reduce the variability of the *in vivo* biological response to implants. In the case of polydimethylsiloxane (PDMS) or glass devices, fabrication is often inexpensive because the materials used are inexpensive. In addition, such devices may not require a cleanroom and can, in theory, be made in the laboratory.

4.1 *IN VITRO* DEVICES

In vitro devices allow an investigator to study neuronal tissues outside the body, where the environment and external factors can be easily controlled. The *in vitro* devices discussed here are microelectrode arrays, microperfusion devices, and microfluidic devices.

4.1.1 *Microelectrode Arrays*

Microelectrode Arrays (MEAs), also called multielectrode arrays, are valuable tools for studying network activity between neurons. An array is typically made by patterning a conductive material such as gold onto an insulating substrate like glass. Substances like platinum black can be used to reduce electrode impedance.[1] The recording site for each electrode is exposed to cells plated on the MEA surface, and leads transmitting this information to an external connection are insulated from the neurons above it (see Fig. 4.1).In some experiments, dissociated cultures of cells are plated onto the MEA surface and allowed to attach.[2] Other groups have had success using electrically active whole tissues, such as retina.[3,4] Signals

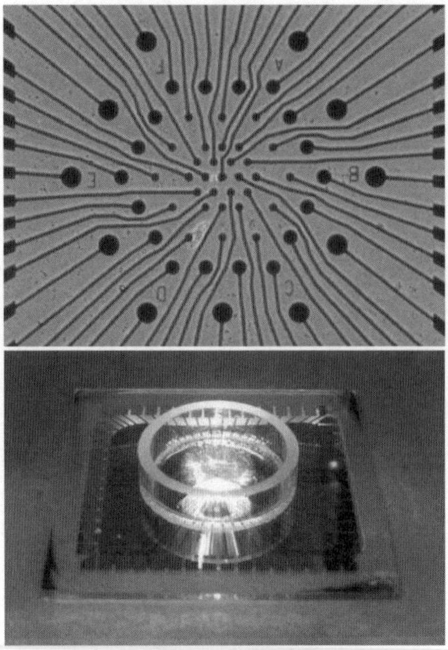

Figure 4.1. One example of a commercially available microelectrode array (Multichannel Systems Hexagonal Array). Neurons or neural tissues are plated on the array surface, top. The circular electrodes can be stimulated and recorded from, but the electrical traces are insulated from neurons plated above. The electrodes are found in the center of the well, which contains culture media to provide cells with nutrients, bottom. For color reference, see page 258.

can be recorded from the somas or neurite extensions of these neurons. In addition to recording from neurons, electrodes in MEAs can be used for other purposes, such as examining neurotransmitter release from activated cells using cyclic voltammetry.[5]

While dissociated cortical cultures do contain the cells found in brain tissue, this model system does not completely recapitulate the structural arrangement of the brain itself. Slice cultures are a model choice for experiments that require the structural components of the brain to be intact. Like dissociated cultures, brain slice cultures do not represent exactly what happens in a whole organism, because systemic effects are not present, but they are still an effective *in vitro* model for studies requiring preserved neural spatial arrangement. Slice cultures contain the appropriate combination of cell types in the appropriate organization of the brain, and are more physiologically relevant structurally than dissociated cortex.

Like dissociated cell cultures, slice cultures can be placed directly on traditional planar MEAs;[6] however, three dimensional MEAs are also used. Three dimensional MEAs contain raised pillars for electrodes, which form a three-dimensional surface, in contrast to the planar arrays discussed above. They are fabricated from a variety of materials and have a variety of geometries. For example, one group etched relatively short 60 micrometer pillars from glass substrates; the top portion of each pillar is used as an electrode.[7] A different silicon MEA model is anisotropically etched in potassium hydroxide after each tip area is masked by patterned silicon dioxide, which forms the pillars; a conductive platinum layer is then deposited on top of silicon nitride and silicon dioxide insulation layers and covered with another silicon nitride layer, after which the tips are shaped with a plasma treatment and etched out of the silicon nitride using deep reactive ion etching (DRIE).[8] The advantage of three-dimensional electrodes for brain or spinal cord slices is the ability to record or stimulate multiple neurons located in interior regions of the tissue, but researchers have found that recordings from dissociated cultures on three-dimensional MEAs tend to have approximately the same signal quality as dissociated cultures on planar electrodes.[9]

One problem with dissociated cells grown in two-dimensional culture is that planar arrangements do not represent the structural complexity found within normal brain because cells are not grown in three dimensions. Spatial cues can dictate how cells grow and develop, and these cues are lacking in planar MEAs. Three dimensional MEMs scaffolds offer the unique advantage of built-in perfusion ability and can be augmented with electrodes for recording and stimulation. These scaffolds are fabricated from silicon using normal microfabrication processing and then arranged in the three dimensional structure after fabrication of the individual two-dimensional parts is complete (see Fig. 4.2).[10]

4.1.2 Microperfusion Devices

Traditional slice cultures have complications, two of which are that they cannot be cut more than several hundred microns thick and they thin out over time,[11] thick slices can become necrotic in the center, probably because nutrients from media and oxygen from the surrounding air do not diffuse entirely through the slice.[12] Also, if a slice is laying flat on a planar MEA or other surface, nutrients must diffuse all the way through the slice to maintain it, which may not happen

Figure 4.2. Three dimensional scaffold for growing neuronal cultures. Rowe *et al.*, 2007, Lab on a Chip, Vol 4, pp 475-482.[10] Reproduced by permission of The Royal Society of Chemistry. http://dx.doi.org/10.1039/b700795g.

effectively. Some three-dimensional MEAs attempt to solve this problem by delivering media to the center of a slice. One type of 3D MEA for cultures contains 500 μm tall towers fabricated out of SU-8 and has functional electrodes and fluid-delivering capabilities.[13]

One way to prevent the center of slice cultures from becoming necrotic is to force media though the slice, but this does not necessarily have to be done through towers that traverse the slice; tissue can also be contained in a chamber and fresh media perfused through. This setup is advantageous because a relatively large amount of nutrients and fresh media can continually be put in contact with the brain slice. There are several varieties of microperfusion devices. Some employ SU-8 towers on glass substrates, and brain slices sit atop the towers while fresh media is perfused above and below the slice.[14] Similar microtower PDMS perfusion devices also exist,[15] and some contain multiple pore sites for enhanced exchange of media and nutrients.[16]

4.1.3 Microfluidic Devices

Microfluidic devices allow the researcher to study a particular cell line, collection of cells, or small organism in a controlled environment while using resources efficiently. Microfluidic devices that house cells are effective for drug screening, and can also be used to see how cells develop under different chemical and physical environmental conditions. Microfluidic devices with separate chambers have been used to study specific neurons involved in the movement and chemosensation of *Caenorhabditis elegans*.[17] Neural stem cells and neuron-like cell lines have been grown in microfluidic chambers made from polyimide, glass, and PDMS.[18] Some microfluidic devices are capable of creating concentration gradients; this is particularly important when studying cell development, as gradients supply neurons and their growth cones with guidance cues and development instructions.[19] Other devices have employed electroosmosis to expose neuronal cell lines to signaling molecules.[20] In addition, microfluidic systems capable of exposing cells to temperature gradients have been developed to explore the activity of temperature–sensitive neurons.[21]

An interesting modification of the planar MEA is the compartmental MEA. These devices are initially fabricated like traditional MEAs, with deposition of metal for electrodes and leads, and then a covering of an insulating material, but compartments are added to separate the processes of neurons from their somas.[22,23] PDMS microfluidic chambers have also been fabricated to create compartments when placed over MEAs.[24] This compartmentalization allows investigators to monitor the effects of different cellular environments on neuronal cultures.

4.2 *IN VIVO* DEVICES

Before the widespread microfabrication of silicon electrodes, electrophysiologists used microwires and glass micropipettes to record signals from brains, spinal cords, peripheral nerves, invertebrate ganglia and other electrically active tissues. Microwire arrays and glass pipette electrodes are still widely used for electrophysiological recording and stimulation. Groups have performed long term studies in monkeys using microwire arrays, and they perform well several months after implantation.[25]

With the refinement of CMOS and MEMs processes and the development of new techniques and equipment, silicon electrodes have become less expensive to manufacture and more popular than before; while many groups do create their own custom electrodes for stimulation and recording, the best known silicon devices used in mammalian brains are the Utah Array and the Michigan Probe electrodes.

4.2.1 The Utah Electrode Array

The original Utah array is composed of 100 pillars arranged in a 1×1 mm square base (see Fig. 4.3). It was first fabricated in the lab of Richard Normann at the University of Utah in the late 1970s. The electrodes are electrically isolated from one another, and signals can be individually measured from each one. Electrodes are approximately 90 micrometers wide and 1000 to 1500 micrometers long.[26,27] These arrays have been used in primate studies and have been implanted into humans.[28] Arrays with differing numbers of pillars can also be created; for a study of the cat sciatic nerve, a Utah array of 25 electrodes (5 by 5 configuration) was manufactured instead of the typical 100.[29]

This particular procedure is described in Campbell *et al.*,[27] though alternative methods have been explored since. The device is fabricated from a 1.7 mm thick n-type <100> silicon wafer. Aluminum is evaporated onto the surface and patterned into a 10 by 10 grid of 325 μm by 325 μm squares. The wafer is then heated on

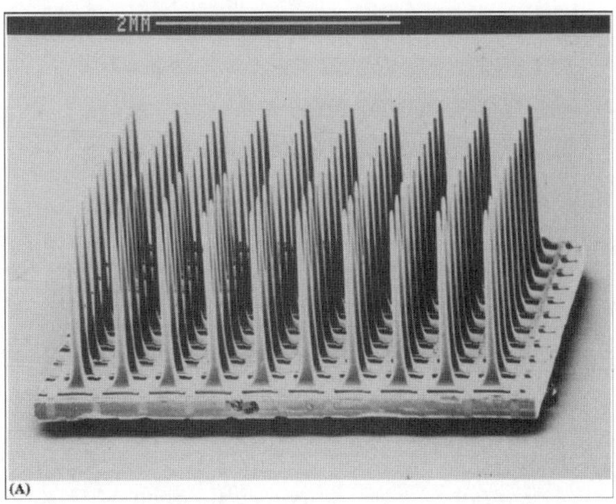

Figure 4.3. The Utah Electrode Array.[26] Reprinted from *Journal of Neuroscience Methods*, Rousche and Normann, 1998, Vol 82, Issue 1, pp 1–15, Copyright 1998, with permission from Elsevier.

one side and a silicon–aluminum compound traverses the wafer, essentially p-type doping columns in the silicon with aluminum. The wafer is polished and electrical contacts placed where the aluminum squares were previously. The n-type areas between the p-doped squares are ground down with a dicing saw, creating the pillars. The pn junctions formed electrically isolate each electrode from the next. The electrodes are then sharpened with a chemical etching process in a mixture of hydrofluoric and nitric acids. First, the electrode tips are put into an etch bath that is agitated with a stir bar. Afterward, the electrodes are etched in a bath with no stirring, cleaned, and coated with gold and platinum. Polyamide is then used to insulate the entire electrode, except for the recording tips, and the electrode is complete.[27] Newer electrodes are fabricated slightly differently; some use glass around the base of each electrode, instead of the pn junction, for electrical insulation, and the insulating coatings on the electrode shafts are different as well.[30]

A pneumatic insertion device is used to place the Utah array into the cortex.[26] Utah arrays have shown incredible longevity in the clinical setting–primate experiments using the Cyberkinetics 100-electrode array have documented recordings for over 1.5 years.[31]

4.2.2 Michigan Probes

The Michigan probes are a planar array of batch-fabricated silicon electrodes. Unlike the Utah array, which has recording sites at the end of each probe, the Michigan array has multiple electrodes on each shank (Fig. 4.4). Today there are many custom-fabricated varieties of the electrode, but two of the most common are a four-shank, four electrodes per shank configuration[32,33] (shown in Fig. 4.4), and a one-shank, 16 electrodes per shank configuration.[34] The Michigan array

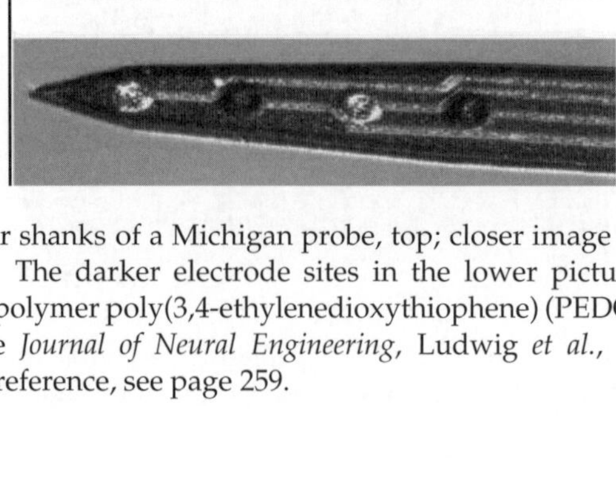

Figure 4.4. The four shanks of a Michigan probe, top; closer image of the four electrodes per shank, bottom. The darker electrode sites in the lower picture have been treated with the conductive polymer poly(3,4-ethylenedioxythiophene) (PEDOT).[32] Reprinted with permission from the *Journal of Neural Engineering*, Ludwig *et al.*, 2006, Vol 3, Issue 1, pp. 59–70. For color reference, see page 259.

has undergone design iterations over time, shifting from single shank to multi-shank, and utilizing different electrode configurations.[35–38] It has since been commercialized and adapted by other investigators.

The details of the basic fabrication process presented here can be found in Kipke *et al.*,[39] but certain applications may merit a deviation from this procedure. The process starts with a boron diffusion through silicon to create a p^{++}etch stop. Silicon dioxide and silicon nitride are then deposited as insulation layers, and conductive polysilicon is patterned on top. The insulation layers are again deposited on top of the polysilicon and etching reveals the interconnect material in specific places. Irridium electrodes and gold bond pads are patterned and, following a dry etch, the structures are released from the bulk silicon. A flexible silicon cable attached to the electrode carries signals to a commercial microconnector.[40] This connector transfers the neuronal signals to the outside world.

Performance of the Michigan electrode is encouraging. Studies of implants into rat barrel cortex have shown recordings for 28 weeks and one rodent had recordings for more than 1 year.[39] Three-dimensional arrangements of the classic two-dimensional probe have also been fabricated,[41] presumably for use in applications similar to the Utah Electrode Array.

4.2.3 Custom Electrodes and Combination Devices

While the Utah and Michigan arrays are probably some of the best known microelectrodes, they are not the only ones. Many labs have fabricated custom cortical electrodes. Polyamide and parylene are both used as substrate materials, and both are more flexible than silicon, which makes them desirable for some applications. One method to create polyamide electrodes uses a three layer process; first, polyamide is deposited on a pretreated metal surface, titanium and platinum are deposited and patterned using reactive ion etching, these steps are repeated and finally a polyamide layer covers the top.[42]

Some interesting new devices have emerged which combine microelectrode technology with other types of sensors. One such device is a "state" detector that has been tested on rats. The detector is essentially a microelectrode array combined with accelerometers. The microelectrode records the electroencephalogram (EEG), while the accelerometers record movements of the rodent's head. With this information, one can determine the state of wakefulness or sleep an animal is in Ref. 43.

Another important modification of the cortical probe is the inclusion of microfluidic channels. These channels are capable of delivering drugs to tissue next to the electrode and others can sample the local electrode environment for neurotransmitters and other factors. Microelectrodes with channels are typically fabricated similar to standard polymer electrodes, but a few extra steps are involved. The channel can be formed by several methods, two of which are discussed here. One method entails depositing parylene, depositing a placeholder sacrificial photoresist layer where the channel will be, and then depositing more parylene to complete the probe; the photoresist is then dissolved in a solvent and the hollow channel remains.[44] Another method of parylene channel formation involves depositing parylene on a glass substrate and bonding this parylene to parylene deposited in a silicon mold lined with aluminum; while the process is slightly more complicated, the seal between bonded parylene layers appears to make a reliable channel.[45] Because parylene is softer than silicon, it is more susceptible to bending. To prevent buckling, one group filled the microfluidic electrode with polyethylene glycol (PEG) before insertion, then dissolved the PEG using saline.[44] Polyamide channels have also been fabricated with a technique where partially cured polyamide is bonded onto another polyamide sheet to create a seal; electrodes are first patterned on the polyamide pieces.[46] Silicon microfluidic channels have been made to deliver bio-active molecules or drugs to the area around insertion.[47]

The ability to sample the brain environment in response to behavioral triggers or after trauma is an exciting prospect to researchers. One group has used capillary liquid chromatography tandem mass spectrometry to monitor the neural environment of rat striatum using small dialysis probes.[48] Using the microdialysis technique and a slightly different analysis system, they have been able to identify glutamate, aspartate, and serine in fluid taken from mouse striatum.[49] Push-pull methods for drug delivery and fluid sampling have also been developed, in which fluid is pulled from the region of interest and an equal volume of fluid, such as saline or drug, is pushed into its place.[50] This assures that there is no volumetric displacement in the test area, as this could cause excess trauma to the tissue involved.

4.2.4 Deep Brain Stimulation Electrodes

Deep brain stimulation (DBS) is now approved as a therapy for advanced Parkinson's disease, and has also been used to treat the symptoms of epilepsy[51] and depression.[52] Electrodes used for DBS are approximately 1 mm in diameter and usually have four stimulation electrodes (not all of which are used); they are commercially available from Medtronic, and the Medtronic models used in clinical practice include 3387 and 3389.[53] They are inserted into an awake patient, and electrical recordings can be taken before insertion to guide the physician to proper placement.[54] To alleviate Parkinson's symptoms, including tremor, the electrode is usually placed in one of three brain regions: the internal segment of the globus pallidus (iGP), the subthalamic nucleus (STN) or the motor thalamus.[55]

Movement disorders associated with Parkinson's disease, such as tremor and rigidity, appear to respond well to stimulation treatments. The clinical standard in the past was to lesion areas in the subthalamic nucleus or globus pallidus, but lesioning too much of the brain can have adverse consequences. Stimulation appears to have a similar effect to lesioning, and can more readily be done bilaterally. Patients that receive DBS electrodes have fewer motor deficiencies and more time per day when they are not affected by motor deficiencies.[53] Five years after implantation, patients still report improved motor function.[56]

In addition to effects on movement disorders, DBS appears able to elicit strong emotional responses as well. In one patient being treated for Parkinson's related movement disorders, stimulation of a certain electrode elicited a case of severe depression, which stopped upon cessation of stimulation.[57] In other patients, stimulation of Broadmann's area 25 (subgenual cingulated region) alleviated depression in patients unresponsive to traditional pharmaceutical and behavioral therapies.[52] DBS used in the nucleus acumbens has been suggested as a treatment for patients with obsessive compulsive disorder that does not respond to conventional treatments.[58] There is evidence that DBS can also help patients who have had a traumatic brain injury recover some functionality.[59]

4.2.5 Peripheral Prosthetic Devices

One of the most common reasons for interfacing MEMs with the human body are prosthetic devices. Prosthetics range from artificial limbs to retinal and cochlear implants. They span a wide range of applications, from supplying sensory feedback to recording and interpreting motor commands. Artificial limbs with integrated microelectrodes may one day be capable of supplying a patient with bio-mimetic closed-loop control, effectively mimicking natural function of an artificial limb. Sensors in the prosthetic device could monitor information, perhaps about pressure and temperature, and relay that back to the person so they could module their actions accordingly. This next generation of prosthetic devices could function essentially just like the lost limb.

Peripheral nerve electrodes are used to monitor and stimulate nerves in the PNS. There are several varieties of peripheral nerve electrodes. Cuff electrodes wrap around a peripheral nerve and stimulate it to generate a muscular response. Polyamide substrates with platinum electrodes have been used to elicit activity from rat muscle fibers.[60] Such electrodes could be used to evoke muscular activity in a patient where communication from the spinal cord to the muscle is disconnected. A flexible substrate material is highly desirable in this situation, as it allows the electrode to be curved around the nerve of interest quite easily. Sieve electrodes contain holes and are placed between regenerating neurons to allow fibers to grow through the holes. Electrodes around or near the holes allow for electrical stimulation and recording.[61,62] Severed nerves are capable of physically regenerating through silicon sieve electrodes[62] and have shown electrophysiological signals for over one year.[63] Flat interface nerve electrodes (FINE) consist of a silicone tube with electrical contacts in it; the FINE electrode flattens peripheral nerve to make the central portion of it more accessible for stimulation and recording.[64] This is not an exhaustive list of peripheral electrodes but rather just a sampling of what is currently used.

4.2.6 Visual Prosthetics

Visual prosthetics are devices have the potential to help the visually impaired regain sight or some aspects of it. Blindness can occur at several levels in the visual system, so different methods have been devised to help patients with different impairments. If the photoreceptive cells, i.e. the rods and cones which are respectively sensitive to grayscale and colored light, are damaged, then an electronic array to mimic their function can be placed over the optic nerve, forming an artificial retina. An artificial silicon retina (ASR) implanted into the subretinal space, between the retina and the optic nerve, of patients with retinal pigmentosa improved patients' general sensitivity to light, and some patients noticed an improvement in the size of their perceived visual field.[65] The ASR in this study was a 2 mm wide, 25 micrometer thick disk with an array of approximately 5000 microphotodiodes, which were 9 μm by 9 μm squares of iridium oxide isolated from each other and bonded to the substrate. Flexible polyamide planar arrays with oxidized iridium electrodes have also been implanted into the eye on top of the retina; these epiretinal arrays deliver monopolar stimulation, and experimenters showed that shorter stimulation pulses require less charge density to elicit a response than longer pulses.[66] Parylene electrode arrays have been fabricated for use atop the retina as well; the parylene is patterned on top of a photoresist release layer, and platinum electrodes are patterned onto it.[67] While retinal implants are promising and can allow patients to distinguish between high-contrast objects and attain a wider visual field, the scenes created with a retinal prosthesis are not as refined as those seen with a natural functioning eye, and improvements must be made to create a device that adequately transposes a scene from the external environment to the brain of a blind patient.

Another approach to visual restoration is the stimulation of the visual cortex (V1) directly. V1 is located in the occipital cortex, at the most posterior portion of the cerebrum. Studies have shown that blind patients, even those blind for years, can perceive phosphenes, or small areas of light, when areas of V1 are stimulated with an electrode array.[68] The stimulation arrays used in this study were 8 by 8 arrays of 1 mm^2 platinum disk electrodes in Teflon ribbon cables, and penetrating cortical arrays such at the Utah arrangement have been used for recording studies of the visual cortex.[69] Microwire arrays, made from gold and iridium wires coated in parylene, have also been used in V1 stimulation experiments; these experiments show that less current is needed from penetrating electrodes than from surface electrodes to generate phosphenes in a patient.[70] V1 stimulation is another promising technology for repairing sight, but the techniques are not perfect and refinements are still needed.

4.2.7 Auditory Prosthetics

Auditory prosthetic devices are another MEMs application. If the auditory nerve is intact and functional, a cochlear prosthetic can be used to return some hearing to a patient. The main parts of a cochlear implant include a microphone for collecting sounds, an external and internal processor, and a stimulation electrode array.[71] Cochlear electrode arrays contain sites for stimulation. The electrode array is inserted into the cochlea and placed in close proximity to the auditory nerve. When

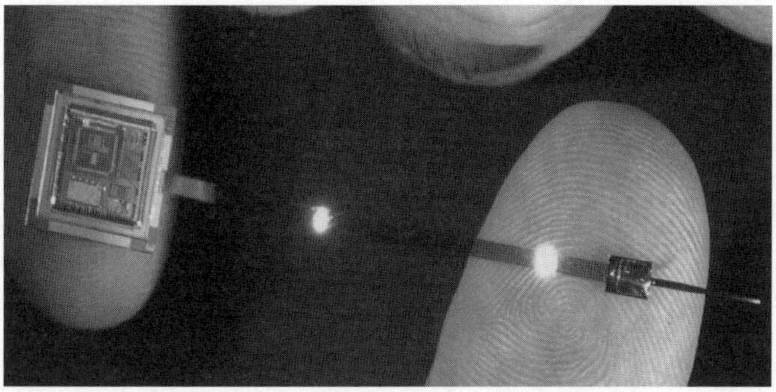

Figure 4.5. A cochlear implant, complete with CMOS circuitry. Image from Bhatti and Wise, *IEEE Journal of Solid State Circuits*, Vol 41, Iss 12, pp 2965-2973. ©2005 IEEE

the ear is functioning properly, the basilar membrane vibrates. Each location on the membrane corresponds to a certain frequency. The auditory nerve interprets these vibrations of a specific area of the membrane as specific frequencies of sound. This frequency-location relationship of the membrane is called tonotopic mapping. Because of this tonotopic mapping, stimulation at different points in the cochlea is interpreted by the auditory cortex as sound of different frequencies.[72] The University of Michigan has created a 2.5 μm thick boron-doped silicon cochlear microelectrode that is 8 mm long and has 32 iridium electrodes on it; this electrode also possesses on-chip circuitry to stimulate and record inside the cochlea (see Fig. 4.5).[73]

A cochlear implant only works when the cochlea is intact and the auditory nerve from it is functional. If this is not the case, a cochlear implant will not restore hearing, so one must move further down the auditory processing pathway to restore hearing. Another area of interest for auditory prosthetics is the auditory cortex. Early experiments exploring the cat auditory cortex were performed using single tungsten electrodes.[74] Experimenters have shown that two-shank 16-electrode Michigan probes implanted in the inferior colliculus central nucleus can elicit activity in the auditory cortex (A1) when stimulated.[75] Brainstem implants can be placed directly into the auditory cochlear nucleus, which also has a tonotopic mapping.[76] Experiments have shown that rats trained to respond to different auditory signals can respond to microelectrode stimulation in the auditory cortex,[77] and a human experiment using a polyurethane hybrid density electrode, a recording device with both high impedance Teflon-coated platinum-iridium microwires and low impedance electroencephalogram (EEG) electrodes,[78] has shown that human auditory cortex shows a tonotopic mapping as well.[79]

4.2.8 Spinal Cord Electrodes

Spinal cord injury (SCI) affects a staggering number of otherwise healthy young individuals each year. Patients with SCI are often left with permanent motor and sensory deficits. A spinal cord lesion interrupts the signal transduction from the brain to distal portions of the body. Electrical stimulation of the spinal cord distal to the injury may restore function or sensation to patients with SCI.

When the spinal cord, nerves, or muscles are stimulated directly, they can elicit a desired response. In one study, cats were implanted with either microwire arrays in the lumbar spinal cord or microwires in the hindlimb muscles, and stimulation was sufficient for standing and balance.[80] A review on intraspinal microstimulation states that the technology looks promising for a variety of applications, including the reestablishment of egestive and sexual function to SCI patients.[81] Intraspinal stimulation has been shown to elicit force from and contraction of the hindlimbs of anesthetized and decerebrate cats;[82] the ability to control such muscular activation is of primary interest for stimulation devices designed to help SCI patients maintain balance and possibly walk.

Microelectrodes for spinal stimulation are often made from silicon or microwires. Smaller electrodes presumably result in less damage to neural tissue, so micromachined devices are important for use in the spinal cord. Investigators using silicon probes manufactured at the University of Michigan and iridium microwire arrays implanted into the sacral spinal cord of cats have shown them to be adequate for modulating bladder pressure.[83]

In addition to reestablishing lost function, spinal cord stimulation has also been used to reduce discomfort experienced by chronic pain patients. In an implanted spinal cord stimulation study, more than half the patients reported at least a 50% reduction in pain and some were even able to return to work.[84] Continual refinement of microfabricated spinal cord devices will likely improve the number of chronic pain patients that achieve relief using spinal cord stimulators.

4.2.9 Brain Computer Interfaces

While a visual or auditory prosthesis translates signals from the external environment into electrical stimulation that patients can interpret, brain computer interfaces (BCIs) are the reciprocal devices; a BCI uses a patient's brain activity to interact with the external environment in some way. Patients with extreme physical trauma or debilitating disease can undergo a major loss of function, such that they are rendered unable to move and sometimes even unable even to blink; such patients may have completely intact thought and mental functioning, but they are unable to interact with the world around them, and these patients are said to be "locked-in" to their internal environment.[85] BCIs offer locked-in patients clinical hope, as they may one day be able to control mobility devices, computers, and household appliances simply thinking by about them.

At the highest level, a brain computer interface records signals from the patient while they perform a task and then gives them feedback (Fig. 4.6). Neuronal signals for BCIs can be acquired from the patient in a number of ways: classical electroencephalogram (EEG) has been shown in the literature, as well as the more invasive electrocorticogram (ECoG), local field potentials (LFPs) and single neuron action potentials (single unit recordings).[86] Generally, more invasive recording

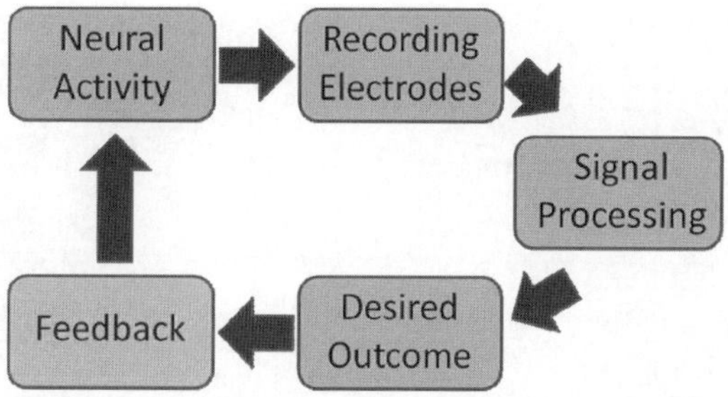

Figure 4.6. General diagram for a closed-loop brain computer interface. Neural signals are recorded by an electrode or array of electrodes, processed, and translated into a desired action. Feedback is usually visual in BCIs, such as a cursor on a computer screen, but it does not have to be. Feedback can influence neural activity such that a more precise desired outcome is achieved.

methods have greater spatial resolution than less invasive methods, but complications like infection are higher for the patient.

Surface EEG recordings have been used as brain computer interfaces with much success. Studies performed by humans have shown that scalp EEG signals are adequate for the performance of simple tasks, such as moving computer screen cursors to desired targets.[87]

Implanted arrays are also used, and one patient with an implanted BCI device was able to move a mouse cursor on a computer screen and spell his name and the names of the study investigators.[88] The microelectrode arrays discussed earlier can also be used for recording signals for a BCI. Monkeys implanted with microelectrode arrays were able to feed themselves with a robotic arm using only cortical signals.[89] Cyberkinetics, the commercial manufacturer of the Utah Electrode Array, has sponsored the BrainGate program to further develop BCI technology for use in humans; the initial patient, a complete tetraplegic known as MN, showed that he was capable of moving a robotic arm as well as a computer screen cursor with his implanted Utah Electrode Array.[28] An electrode array such as this one, which was placed over the motor cortex and measured signals related to motor movements, is known as a neuromotor prosthesis (NMP); such devices may one day be refined enough to allow paralyzed patients to walk.

Another device used to acquire signals from the brain is the ECoG microelectrode. This microelectrode sits atop the cortex, instead of penetrating into it (Fig. 4.7). While still invasive to the skull, the electrode does not penetrate into and damage the brain tissue it is measuring signals from, which may help improve signal longevity. Since the array is inside the skull and therefore closer to the neurons it is recording from, the spatial resolution is presumably better than recordings from EEG. Patients with grids of either 48 or 64 2 mm electrodes were able to perform a target tracing task using an ECoG device placed over the left frontal-parietal-temporal cortex.[90]

Once signals from the brain are acquired, they must be processed to translate into something meaningful. There are many signal processing schemes available,

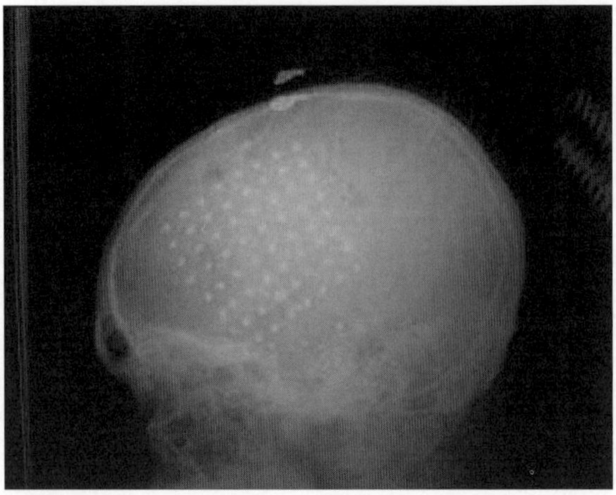

Figure 4.7. An ECoG electrode array located on top of the cortex but inside the skull.[91] Reprinted with permission from the *Journal of Neural Engineering*, Leuthanrdt *et al.*, 2004, Vol 1, Issue 2, pp 63–71. For color reference, see page 259.

and they are constantly being improved to increase accuracy and decrease the amount of training time needed. It has been shown in monkeys with Utah arrays implanted over the arm representation in the premotor cortex that predictive models about arm movement speed up processing and could allow BCI users to perform functions more quickly than they are able to with traditional BCI algorithms.[91] Signal processing algorithms are dependent on the method of signal acquisition to some degree; programs to process single neuron unit data will behave differently from programs to interpret EEG scalp recordings, which typically consist of the summation of thousands of neurons near the cortical surface.

BCIs have a great deal of clinical promise, but they also need refinement before they can be used in widespread clinical trials. Once developed and refined, BCIs have the potential to allow paralyzed and locked-in patients to interact with the world around them and improve their quality of life.

4.3 DEVICE CONCERNS AND TISSUE RESPONSE

As with any biological implant, the materials used for a MEMs neural interface must be biologically compatible. The device should not leach toxic materials into the organism and should not be detrimental to the health of the animal; it also should not deteriorate over time, unless biodegradation is the goal. Most implant materials, even if biocompatible, trigger an inflammatory response when left in the body for extended periods of time, and the nervous system is no exception.

The initial insertion of microelectrodes into brain or spinal cord is likely to be harsh on the tissue. It is speculated that when a probe is inserted, it tears cells and the surrounding tissue, and this damage triggers a cascade of events that will manifest themselves as the inflammatory response.[92] The inflammatory response is divided into acute and chronic phases; during the acute phase, microglia become activated and respond to the foreign electrode.[92] It is suspected that immune cells circulating through the blood may also enter the brain,[93] possibly via breached blood vessels. Microglia can be seen around an electrode as soon as 24 hours after the implant.[93] The acute phase lasts for approximately 1–3 weeks after insertion, after which the chronic phase begins.[92]

One of the attributes of the chronic phase of inflammation is an astrocytic sheath formed around the implant (Fig. 4.8).[92] Resting astrocytes form the blood brain barrier, buffer excess potassium and neurotransmitters, and may supply neurons with nutrients.[94] When astrocytes become reactive, they can secrete compounds inhibitory to neural growth, such as chondroitin sulfate proteoglycans (CSPGs).[95,96] The sheath formed around the electrode grows in size for about 6 weeks and then becomes thinner and denser.[93] Microglia are located on the inside of this astroglial sheath;[97] activated microglia have been shown to release inflammatory cytokines and reactive oxygen species (see Block and Hong for review[98]), and this constant release of compounds from activated microglia can be neurotoxic.[99] It has been shown that a region devoid of neurons appears next to an implanted electrode.[97] Thus, the neurons from which the probe is most likely to record signals may be killed because of this chronic foreign body response. The dense glial encapsulation may also interfere with electrical recording by forming a high-impedance layer around the electrode,[100] so several strategies have been employed to lessen the severity of the glial response. Groups have experimented with modifying electrode cross section, geometry, and tip shape, but none of these parameters appears to mitigate the chronic glial response.[93] Electrode coatings have been developed to enhance neuronal attachment,[101] but it is not clear if this

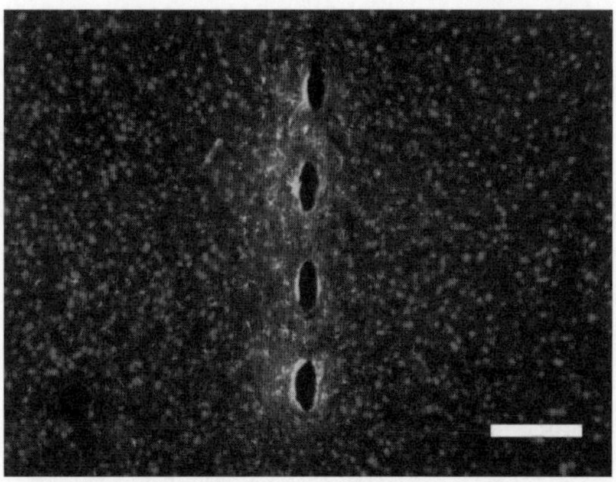

Figure 4.8. The response of rat cortical tissue 8 weeks after insertion. The green stain marks glial fibrillary acidic protein (GFAP), a marker for astrocytes, and the red NeuN stain marks neuron nuclei. The scale bar is 100 micrometers. Notice the astrocytes apparent at the electrode interface, and that the neurons appear further away. Image courtesy of George McConnell, Bellamkonda Lab, Georgia Institute of Technology. For color reference, see page 260.

will alleviate the tissue reaction. Surface modification of electrodes with anti-inflammatory compounds[102] appears to lessen the glial encapsulation, but some gliosis is still present, so research in electrode surface modification remains an important topic for the realization of successful chronic neural electrodes.

The insertion of probes into the brain can be done quickly or slowly. Faster insertion speeds and sharper needles result in less strain to the surrounding tissue.[103] This reduced strain may minimize tissue damage due to insertion, because processes and cells near the electrode remain relatively undisturbed while those in the electrode's path are completely severed.[92] However, some investigators insist that slower insertions lead to greater longevity in chronic recordings.[104] This may be because slow insertions allow cells and their processes time to be displaced and moved, rather than impaled, by microelectrodes.[92] Insertion technique is far from standardized in the neural interfacing fields, and no one technique may be best for all MEMs devices, so insertion method will probably continue to be chosen on a device-by-device basis.

4.4 CONCLUDING REMARKS

MEMs devices hold immense promise for future neuroprosthetics and investigative biological studies. *In vitro* devices allow the observation of individual cell types and their reactions to environmental stimuli; the small volumes used, inexpensive materials, and ability to batch fabricate devices reduces the cost to perform such studies. *In vivo* microdevices are less damaging to neural tissues than larger devices, some can be batch fabricated, and most are electrically or chemically active. Such devices may one day allow patients who have lost normal neural function to regain it.

Neuronal interfacing technology is far from completely optimized, but new developments are always refining existing processes and opening new frontiers. The implications for improved understanding of the nervous system and the ability to improve patient quality of life are some of the strongest factors driving the field today, and because the promise is evident, we can expect to see important strides being made towards seamless integration of sensor and actuator technology with the human nervous system in the future.

References

1 Nisch, W., *et al.*, A thin-film microelectrode array for monitoring extracellular neuronal-activity *in vitro*. *Biosensors & Bioelectronics*, 1994, **9**(9–10): 737–741.

2 Wagenaar, D. A., J. Pine, and S. M. Potter, An extremely rich repertoire of bursting patterns during the development of cortical cultures. *BMC Neuroscience*, 2006, **7**: p.17.

3 Hammerle, H., *et al.*, Extracellular recording in neuronal networks with substrate integrated microelectrode arrays. *Biosensors & Bioelectronics*, 1994, **9**(9–10): 691–696.

4 Meister, M., J. Pine, and D. A. Baylor, Multi-neuronal signals from the retina: Acquisition and analysis. *Journal of Neuroscience Methods*, 1994. **51**: 95–106.

5 Chen, Y., *et al.*, Compact microelectrode array system: Tool for *in situ* monitoring of drug effects on neurotransmitter release from neural cells. *Analytical Chemistry*, 2008. **80**(4): 1133–1140.

6 Jimbo, Y. and H. P. C. Robinson, Propagation of spontaneous synchronized activity in cortical slice cultures recorded by planar electrode arrays. *Bioelectrochemistry*, 2000. **51**(2): 107–115.

7 Heuschkel, M. O., *et al.*, A three-dimensional multi-electrode array for multi-site stimulation and recording in acute brain slices. *Journal of Neuroscience Methods*, 2002. **114**(2): 135–148.

8 Thiebaud, P., *et al.*, An array of Pt-tip microelectrodes for extracellular monitoring of activity of brain slices. *Biosensors & Bioelectronics*, 1999. **14**(1): 61–65.

9 Nam, Y., B. C. Wheeler, and M. O. Heuschkel, Neural recording and stimulation of dissociated hippocampal cultures using microfabricated three-dimensional tip electrode array. *Journal of Neuroscience Methods*, 2006. **155**(2): 296–299.

10 Rowe, L., *et al.*, Active 3–D microscaffold system with fluid perfusion for culturing in vitro neuronal networks. *Lab on a Chip*, 2007. **7**(4): 475–482.

11 Xiang, Z. M., *et al.*, Long-term maintenance of mature hippocampal slices *in vitro*. *Journal of Neuroscience Methods*, 2000. **98**(2): 145–154.

12 Noraberg, J., B. W. Kristensen, and J. Zimmer, Markers for neuronal degeneration in organotypic slice cultures. *Brain Research Protocols*, 1999. **3**(3): 278–290.

13 Rajaraman, S., *et al.*, Microfabrication technologies for a coupled three-dimensional microelectrode, microfluidic array. *Journal of Micromechanics and Microengineering*, 2007. **17**(1): 163–171.

14 Passeraub, P. A., A. C. Almeida, and N. V. Thakor, Design, microfabrication and analysis of a microfluidic chamber for the perfusion of brain tissue slices. *Biomedical Microdevices*, 2003. **5**(2): 147–155.

15 Blake, A. J., *et al.*, Multilayer PDMS microfluidic chamber for controlling brain slice microenvironment. *Lab on a Chip*, 2007. **7**(7): 842–849.

16 Choi, Y., *et al.*, MEMS–based fabrication and microfluidic analysis of three-dimensional perfusion systems. *Biomedical Microdevices*, 2008. **10**(3): 437–446.

17 Chronis, N., M. Zimmer, and C. I. Bargmann, Microfluidics for *in vivo* imaging of neuronal and behavioral activity in Caenorhabditis elegans. *Nature Methods*, 2007. **4**(9): 727–731.

18 Mahoney, M. J., *et al.*, The influence of microchannels on neurite growth and architecture. *Biomaterials*, 2005. **26**(7): 771–778.

19 Chung, B. G., *et al.*, Human neural stem cell growth and differentiation in a gradient-generating microfluidic device. *Lab on a Chip*, 2005. **5**(4): 401–406.

20 Peterman, M. C., *et al.*, Localized chemical release from an artificial synapse chip. *Proceedings of the National Academy of Sciences of the United States of America*, 2004. **101**(27): 9951–9954.

21 Pearce, T. M., *et al.*, Integrated microelectrode array and microfluidics for temperature clamp of sensory neurons in culture. *Lab on a Chip*, 2005. **5**(1): 97–101.

22 Ravula, S. K., *et al.*, A multielectrode microcompartment culture platform for studying signal transduction in the nervous system. *Lab on a Chip*, 2006. **6**(12): 1530–1536.

23 Taylor, A. M., *et al.*, Microfluidic multicompartment device for neuroscience research. *Langmuir*, 2003. **19**(5): 1551–1556.

24 Morin, F., *et al.*, Constraining the connectivity of neuronal networks cultured on microelectrode arrays with microfluidic techniques: A step towards neuron–based functional chips. *Biosensors & Bioelectronics*, 2006. **21**(7): 1093–1100.

[25] Donoghue, J. P., *et al.*, Neural discharge and local field potential oscillations in primate motor cortex during voluntary movements. *Journal of Neurophysiology*, 1998. **79**(1): 159–173.

[26] Rousche, P. J. and R. A. Normann, Chronic recording capability of the Utah Intracortical Electrode Array in cat sensory cortex. *Journal of Neuroscience Methods*, 1998. **82**(1): 1–15.

[27] Campbell, P. K., *et al.*, A silicon-based, three-dimensional neural interface: Manufacturing processes for an intracortical electrode array. *IEEE Transactions on Biomedical Engineering*, 1991. **38**(8): 758–768.

[28] Hochberg, L. R., *et al.*, Neuronal ensemble control of prosthetic devices by a human with tetraplegia. *Nature*, 2006. **442**(7099): 164–171.

[29] Branner, A. and R. A. Normann, A multielectrode array for intrafasicular recording and stimulation in sciatic nerve cats. *Brain Research Bulletin*, 2000. **51**(4): 293–306.

[30] Hillman, T., *et al.*, Cochlear nerve stimulation with a 3-dimensional penetrating electrode array. *Otology & Neurotology*, 2003. **24**(5): 764–768.

[31] Suner, S., *et al.*, Reliability of signals from a chronically implanted silicon–based electrode array in non-human primate primary motor cortex. *IEEE Transactions on Neural Systems and Rehabilitation Engineering*, 2005. **13**(4): 524–541.

[32] Ludwig, K. A., *et al.*, Chronic neural recordings using silicon microelectrode arrays electrochemically deposited with a poly(3,4-ethylenedioxythiophene) (PEDOT) film. *Journal of Neural Engineering*, 2006. **3**(1): 59–70.

[33] McConnell, G. C., *et al.*, Extraction force and cortical tissue reaction of silicon microelectrode arrays implanted in the rat brain. *IEEE Transactions on Biomedical Engineering*, 2007. **54**(6): 1097–1107.

[34] Martinez–Santiesteban, F. M., *et al.*, Magnetic field perturbation of neural recording and stimulating microelectrodes. *Physics in Medicine and Biology*, 2007. **52**(8): 2073–2088.

[35] Drake, K. L., *et al.*, Performance of planar multisite microprobes in recording extracellular single-unit intracortical activity. *IEEE Transactions on Biomedical Engineering*, 1988. **35**(9): 719–732.

[36] Anderson, D. J., *et al.*, *Batch-fabricated thin-film electrodes for stimulation of the central auditory-system. IEEE Transactions on biomedical Engineering*, 1989. **36**(7): 693–704.

[37] Weiland, J. D. and D. J. Anderson, Chronic neural stimulation with thin-film, iridium oxide electrodes. *IEEE Transactions on Biomedical Engineering*, 2000. **47**(7): 911–918.

[38] Snellings, A., D. J. Anderson, and J. W. Aldridge, Improved signal and reduced noise in neural recordings from close-spaced electrode arrays using independent component analysis as a preprocessor. *Journal of Neuroscience Methods*, 2006. **150**(2): 254–264.

[39] Kipke, D. R., *et al.*, Silicon-substrate intracortical microelectrode arrays for long-term recording of neuronal spike activity in cerebral cortex. *IEEE Transactions on Neural Systems and Rehabilitation Engineering*, 2003. **11**(2): 151–155.

[40] Hetke, J. F., *et al.*, Silicon ribbon cables for chronically implantable microelectrode arrays. *IEEE Transactions on Biomedical Engineering*, 1994. **41**(4): 314–321.

[41] Bai, Q., K. D. Wise, and D. J. Anderson, A high-yield microassembly structure for three-dimensional microelectrode arrays. *IEEE Transactions on Biomedical Engineering*, 2000. **47**(3): 281–289.

[42] Cheung, K. C., *et al.*, Flexible polyimide microelectrode array for in vivo recordings and current source density analysis. *Biosensors & Bioelectronics*, 2007. **22**(8): 1783–1790.

[43] Sunderam, S., *et al.*, Improved sleep–wake and behavior discrimination using MEMs accelerometers. *Journal of Neuroscience Methods*, 2007. **163**: 373–383.

[44] Takeuchi, S., *et al.*, Parylene flexible neural probes integrated with microfluidic channels. *Lab on a Chip*, 2005. **5**(5): 519–523.

45 Ziegler, D., T. Suzuki, and S. Takeuchi, Fabrication of flexible neural probes with built-in microfluidic channels by thermal bonding of Parylene. *Journal of Microelectromechanical Systems*, 2006. **15**(6): 1477–1482.

46 Metz, S., *et al.*, Flexible polyimide probes with microelectrodes and embedded microfluidic channels for simultaneous drug delivery and multi-channel monitoring of bioelectric activity. *Biosensors & Bioelectronics*, 2004. **19**(10): 1309–1318.

47 Retterer, S. T., *et al.*, Model neural prostheses with integrated microfluidics: A potential intervention strategy for controlling reactive cell and tissue responses. *IEEE Transactions on Biomedical Engineering*, 2004. **51**(11): 2063–2073.

48 Haskins, W. E., *et al.*, Discovery and neurochemical screening of peptides in brain extracellular fluid by chemical analysis of in vivo microdialysis samples. *Analytical Chemistry*, 2004. **76**(18): 5523–5533.

49 Presti, M.F., *et al.*, Behavior–related alterations of striatal neurochemistry in a mouse model of stereotyped movement disorder. *Pharmacology Biochemistry and Behavior*, 2004. **77**(3): 501–507.

50 Kottegoda, S., I. Shaik, and S. A. Shippy, Demonstration of low flow push-pull perfusion. *Journal of Neuroscience Methods*, 2002. **121**(1): 93–101.

51 Hodaie, M., *et al.*, Chronic anterior thalamus stimulation for intractable epilepsy. *Epilepsia*, 2002. **43**(6): 603–608.

52 Mayberg, H. S., *et al.*, Deep brain stimulation for treatment–resistant depression. *Neuron*, 2005. **45**(5): 651–660.

53 Obeso, J. A., *et al.*, Deep-brain stimulation of the subthalamic nucleus or the pars interna of the globus pallidus in Parkinson's disease. *New England Journal of Medicine*, 2001. **345**(13): 956–963.

54 Breit, S., J. B. Schulz, and A. L. Benabid, Deep brain stimulation. *Cell and Tissue Research*, 2004. **318**(1): 275–288.

55 Dostrovsky, J. O., *et al.*, Microstimulation–induced inhibition of neuronal firing in human globus pallidus. *Journal of Neurophysiology*, 2000. **84**(1): 570–574.

56 Krack, P., *et al.*, Five-year follow-up of bilateral stimulation of the subthalamic nucleus in advanced Parkinson's disease. *New England Journal of Medicine*, 2003. **349**(20): 1925–1934.

57 Bejjani, B. P., *et al.*, Transient acute depression induced by high-frequency deep-brain stimulation. *New England Journal of Medicine*, 1999. **340**(19): 1476–1480.

58 Tass, P. A., *et al.*, Obsessive–compulsive disorder: Development of demand controlled deep brain stimulation with methods from Stochastic phase resetting. *Neuropsychopharmacology*, 2003. **28**: S27–S34.

59 Schiff, N. D., *et al.*, Behavioural improvements with thalamic stimulation after severe traumatic brain injury. *Nature*, 2007. **448**(7153): 600–U10.

60 Rodriguez, F. J., *et al.*, Polyimide cuff electrodes for peripheral nerve stimulation. *Journal of Neuroscience Methods*, 2000. **98**(2): 105–118.

61 Bradley, R. M., *et al.*, Long term chronic recordings from peripheral sensory fibers using a sieve electrode array. *Journal of Neuroscience Methods*, 1997. **73**(2): 177–186.

62 Akin, T., *et al.*, A micromachined silicon sieve electrode for nerve regeneration applications. *IEEE Transactions on Biomedical Engineering*, 1994. **41**(4): 305–313.

63 Kovacs, G. T. A., C. W. Storment, and J. M. Rosen, Regeneration microelectrode array for peripheral-nerve recording and stimulation. *IEEE Transactions on Biomedical Engineering*, 1992. **39**(9): 893–902.

[64] Tyler, D. J. and D. M. Durand, Functionally selective peripheral nerve stimulation with a flat interface nerve electrode. *IEEE Transactions on Neural Systems and Rehabilitation Engineering*, 2002. **10**(4): 294–303.

[65] Chow, A. Y., *et al.*, The artificial silicon retina microchip for the treatment of vision loss from retinitis pigmentosa. *Archives of Ophthalmology*, 2004. **122**(4): 460–469.

[66] Rizzo, J. F., *et al.*, Methods and perceptual thresholds for short-term electrical stimulation of human retina with microelectrode arrays. *Investigative Ophthalmology & Visual Science*, 2003. **44**(12): 5355–5361.

[67] Rodger, D. C., *et al.*, Flexible parylene-based multielectrode array technology for high-density neural stimulation and recording. *Sensors and Actuators B–Chemical*, 2008. **132**(2): 449–460.

[68] Dobelle, W. H., M. G. Mladejov and J. P. Girvin, Artificial vision for blind–electrical-stimulation of visual-cortex offers hope for a functional prosthesis. *Science*, 1974. **183**(4123): 440–444.

[69] Normann, R. A., *et al.*, High–resolution spatio–temporal mapping of visual pathways using multi-electrode arrays. *Vision Research*, 2001. **41**(10–11): 1261–1275.

[70] Schmidt, E. M., *et al.*, Feasibility of a visual prosthesis for the blind based on intracortical microstimulation of the visual cortex. *Brain*, 1996. **119**: 507–522.

[71] Spelman, F. A., Cochlear electrode arrays: Past, present and future. *Audiology and Neuro-Otology*, 2006. **11**(2): 77–85.

[72] Middlebrooks, J. C., J. A. Bierer, and R. L. Snyder, Cochlear implants: The view from the brain. *Current Opinion in Neurobiology*, 2005. **15**(4): 488–493.

[73] Bhatti, P. T. and K. D. Wise, A 32–site 4–channel high–density electrode array for a cochlear prosthesis. *IEEE Journal of Solid–State Circuits*, 2006. **41**(12): 2965–2973.

[74] Hubel, D. H., *et al.*, Attention units in the auditory cortex. *Science*, 1959. **129**(3358): 1279–1280.

[75] Lim, H. H. and D. J. Anderson, Auditory cortical responses to electrical stimulation of the inferior colliculus: Implications for an auditory midbrain implant. *Journal of Neurophysiology*, 2006. **96**(3): 975–988.

[76] Rauschecker, J. P. and R. V. Shannon, Sending Sound to the Brain. *Science*, 2002. **295**(5557): 1025.

[77] Otto, K. J., P. J. Rousche, and D. R. Kipke, Microstimulation in auditory cortex provides a substrate for detailed behaviors. *Hearing Research*, 2005. **210**(1–2): 112–117.

[78] Howard, M. A., *et al.*, A hybrid clinical-research depth electrode for acute and chronic in vivo microelectrode recording of human brain neurons – Technical note. *Journal of Neurosurgery*, 1996. **84**(1): 129–132.

[79] Howard, M. A., *et al.*, A chronic microelectrode investigation of the tonotopic organization of human auditory cortex. *Brain Research*, 1996. **724**(2): 260–264.

[80] Lau, B., L. Guevremont, and V. K. Mushahwar, Strategies for generating prolonged functional standing using intramuscular stimulation or entraspinal microstimulation. *IEEE Transactions on Neural Systems and Rehabilitation Engineering*, 2007. **15**(2): 273–285.

[81] Pikov, V., Clinical applications of intraspinal microstimulation. *Proceedings of the IEEE*, 2008. **96**(7): 1120–1128.

[82] Lemay, M. A. and W. M. Grill, Modularity of motor output evoked by intraspinal microstimulation in cats. *Journal of Neurophysiology*, 2004. **91**(1): 502–514.

[83] McCreery, D., *et al.*, Arrays for chronic functional micro stimulation of the lumbosacral spinal cord. *IEEE Transactions on Neural Systems and Rehabilitation Engineering*, 2004. **12**(2): 195–207.

[84] North, R. B., *et al.*, Spinal-cord stimulation for chronic, intractable pain-experience over two decades. *Neurosurgery*, 1993. **32**(3): 384–395.

[85] Schnakers, C., *et al.*, Cognitive function in the locked–in syndrome. *Journal of Neurology*, 2008. **255**(3): 323–330.

[86] Schwartz, A. B., *et al.*, Brain-controlled interfaces: Movement restoration with neural prosthetics. *Neuron*, 2006. **52**(1): 205–220.

[87] McFarland, D. J., *et al.*, Emulation of computer mouse control with a noninvasive brain-computer interface. *Journal of Neural Engineering*, 2008. **5**(2): 101–110.

[88] Kennedy, P. R., *et al.*, Direct control of a computer from the human central nervous system. *IEEE Transactions on Rehabilitation Engineering*, 2000. **8**(2): 198–202.

[89] Velliste, M., *et al.*, Cortical control of a prosthetic arm for self-feeding. *Nature*, 2008. **453**(7198): 1098–1101.

[90] Leuthardt, E.C., *et al.*, A brain-computer interface using electrocorticographic signals in humans. *Journal of Neural Engineering*, 2004. **1**(2): 63–71.

[91] Santhanam, G., *et al.*, A high-performance brain-computer interface. *Nature*, 2006. **442**(7099): 195–198.

[92] Polikov, V. S., P. A. Tresco, and W. M. Reichert, Response of brain tissue to chronically implanted neural electrodes. *Journal of Neuroscience Methods*, 2005. **148**(1): 1–18.

[93] Szarowski, D. H., *et al.*, Brain responses to micro-machined silicon devices. *Brain Research*, 2003. **983**(1–2): 23–35.

[94] Kandel, E. R., J. H. Schwartz, and T. M. Jessell, eds. *Principles of Neural Science*. Fourth ed. 2000, McGraw–Hill: New York.

[95] Lemons, M. L., D. R. Howland, and D. K. Anderson, Chondroitin sulfate proteoglycan immunoreactivity increases following spinal cord injury and transplantation. *Experimental Neurology*, 1999. **160**(1): 51–65.

[96] Smith, G. M. and C. Strunz, Growth factor and cytokine regulation of chondroitin sulfate proteoglycans by astrocytes. *Glia*, 2005. **52**(3): 209–218.

[97] Biran, R., D. C. Martin, and P. A. Tresco, Neuronal cell loss accompanies the brain tissue response to chronically implanted silicon microelectrode arrays. *Experimental Neurology*, 2005. **195**(1): 115–126.

[98] Block, M. L. and J. S. Hong, Microglia and inflammation-mediated neurodegeneration: Multiple triggers with a common mechanism. *Progress in Neurobiology*, 2005. **76**(2): 77–98.

[99] Giulian, D., K. Vaca, and M. Corpuz, Brain glia release factors with opposing actions upon neuronal survival. *Journal of Neuroscience*, 1993. **13**(1): 29–37.

[100] Turner, J. N., *et al.*, Cerebral astrocyte response to micromachined silicon implants. *Experimental Neurology*, 1999. **156**(1): 33–49.

[101] He, W. and R. V. Bellamkonda, Nanoscale neuro-integrative coatings for neural implants. *Biomaterials*, 2005. **26**(16): 2983–2990.

[102] Zhong, Y. H. and R. V. Bellamkonda, Dexamethasone-coated neural probes elicit attenuated inflammatory response and neuronal loss compared to uncoated neural probes. *Brain Research*, 2007. **1148**: 15–27.

[103] Bjornsson, C. S., *et al.*, Effects of insertion conditions on tissue strain and vascular damage during neuroprosthetic device insertion. *Journal of Neural Engineering*, 2006. **3**(3): 196–207.

[104] Nicolelis, M. A. L., *et al.*, Chronic, multisite, multielectrode recordings in macaque monkeys. *Proceedings of the National Academy of Sciences of the United States of America*, 2003. **100**(19): 11041–11046.

Chapter Five

Hydrogel-Based Microfluidic Cell Culture

Michael C. W. Chen and Karen C. Cheung*

Department of Electrical & Computer Engineering, University of British Columbia, 2332 Main Mall, Vancouver, BC, V6T 1Z4, Canada
**E-mail: kcheung@ece.ubc.ca*

Microfluidic technology provides the tools for creating more *in vivo*-like environments for cell culture by permitting precise control of chemical concentration gradients, shear rates, and geometry, among other parameters. Hydrogels, which create a three-dimensional environment, are now increasingly used to encapsulate cells for on-chip incubation. As porous polymer networks, hydrogels allow the transport of nutrients and waste away from embedded cells, and the gel network can also include specific adhesive properties for cell attachment. The integration of cell culture and characterization onto a miniaturized platform would permit automated, hands-free assays. In particular, the ability to manipulate individual cells and small cell populations represents a large improvement over traditional methods involving large cell numbers.

5.1 INTRODUCTION

Today, microsystem technology for biomedical applications spans a wide range of biosensors, implantable microdevices, and lab-on-a-chip systems. By reducing reagent volumes and costs, decreasing analysis times, and increasing throughput, micro total analysis systems (microTAS) promise to have large impact on healthcare.

Microfluidics involves the study of flow phenomena and the manipulation of fluid volumes at the microscale. Microfabrication processes easily permit production of structures with dimensions of microns or tens of microns, thus defining fluid volumes in the range of picoliters (1 pL = cube with sides of 10 microns) or below. Since typical mammalian cells are in the range of ten microns in diameter, miniaturization facilitates manipulation of individual biological cells.

Biomaterials for MEMS, Edited by M. Chiao and J.-C. Chiao

www.panstanford.com
978-981-4241-46-5

Microfluidic devices also give advantages in fluid handling, mixing, separation, and detection.[1]

5.1.1 Traditional Cell Culture Methods

In traditional, well-established cell culture techniques, Petri dishes, culture well plates, and culture flasks are used to maintain either a two-dimensional monolayer of adherent cells below a layer of liquid culture medium, or a suspension of non-adherent cells within the culture medium. These bulk methods may require cell densities of hundreds of thousands to millions of cells per milliliter.

This two-dimensional petri dish-based method of cell culture has long been used in molecular and cell biology. Recently, there has been increased interest in three-dimensional cell culture, in which cells are embedded in a material that mimics the extracellular matrix.[2]

5.1.2 Two-dimensional Versus Three-dimensional Culture Methods

In traditional two-dimensional cultures, cells can be incubated until the monolayer reaches confluence, or maximum cell density. The culture environment can have a dramatic influence on cell behavior, with different responses observed between cells incubated in 2-D versus 3-D environments, such as modified morphology, gene expression, and proliferation.[2] A large part of tissues is composed of the extracellular matrix (ECM). The ECM comprises a network of proteins and polysaccharides that are mainly excreted by fibroblast cells. ECM proteins include collagen, fibronectin, elastin, and laminin. Other major components of ECM include glycosaminoglycans. The ECM mechanically stabilizes the tissue, and the fibrous network gives both physical and biochemical signals to the cells (Fig. 5.1). The ECM interacts with cells to influence their shape, function, and survival.[3]

Mammalian cells communicate with each other through several mechanisms, including cell-cell contact as well as cell-ECM interaction.[4] Cells are anchored to the ECM through integrins, which are proteins on the cell surface. Integrins link the cytoskeleton to the ECM. Specifically, fibronectin is the key ECM protein for cell attachment. The Arg-Gly-Asp (RGD) tripeptide sequence in fibronectin has been identified as the central component which binds to integrins on cell surfaces.

One major factor affecting cell phenotype, function, proliferation, and migration is cell shape.[5] Each cell type has a different context, or microenvironment (Fig. 5.2). The 3-D environment is a major factor affecting the differentiation of stem cells. Different cell signaling pathways are activated in 2-D and 3-D cell culture, and these pathways are mediated in part by cell adhesion matrices.[6] The 2-D and 3-D adhesion sites differ in chemical composition and spatial organization, with different integrin complexes formed depending on the culture geometry.[7]

In cell-based drug screening, the different cellular responses exhibited in 2-D versus 3-D culture will have a crucial impact in the pharmacological response to drugs, which may differ between cells in 2-D and 3-D culture.[8,9] For cell-based testing to give predictive data during further in vivo testing, the cells in culture must mimic as closely as possible the *in vivo* behavior. Thus, drugs developed using a monolayer screening system may not have the same activity against solid tumors.[10]

Currently, several three-dimensional cell culture systems are already used in tumor and normal cell research.[9] Multicellular spheroids, in which spherical aggregates are grown in suspension, are used to investigate tumor cell metabolism. Cellular multilayers, in which layers of cells are cultured on porous membranes, are used in studies of drug transport. Established 3-D culture methods used to study drug penetration in tumors include multicellular spheroids, which are cell aggregates typically suspended in rotating flasks, or spinner culture.[11] In collagen cultures, a solid tumor specimen is embedded in a rat tail collagen gel, in order to maintain the three dimensional structure. Multilayered cell cultures (MCCs) feature thick layers of cells grown on permeable support membranes.[12] In one configuration, the MCC separates two reservoirs which contain stirred media; the drug is added to one reservoir and drug penetration through the thick ($\sim 200\,\mu$m) layer of cells is measured by sampling from the second reservoir.[13]

5.1.3 Microscale Cell Culture Using Hydrogels

Current trends aim to culture cells in a microscale three-dimensional environment that more closely mimics the natural extracellular environment.[15] Miniaturization and spatial control are two goals addressed in microfluidic cell culture systems.

First, miniaturization permits more precise control of the microenvironment, such as gradients in soluble factors. Gradients in biomolecules are essential — they guide cell growth, migration, and differentiation.[16] In Petri dishes, secreted molecules are quickly dispersed over the entire bulk volume, and this loss of

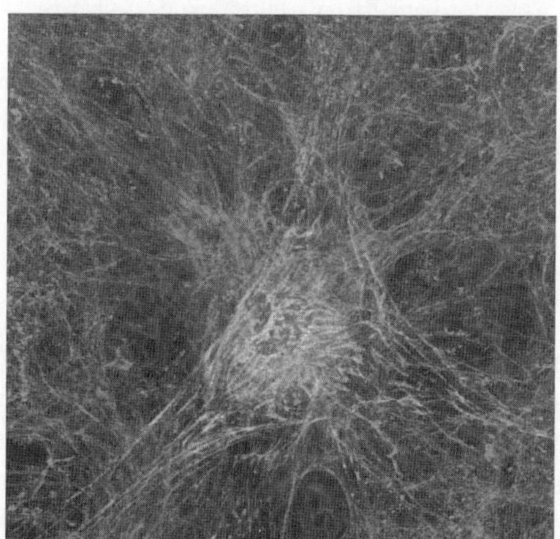

Figure 5.1. A cell in a three-dimensional culture forming links with the scaffold.[2] For color reference, see page 260.

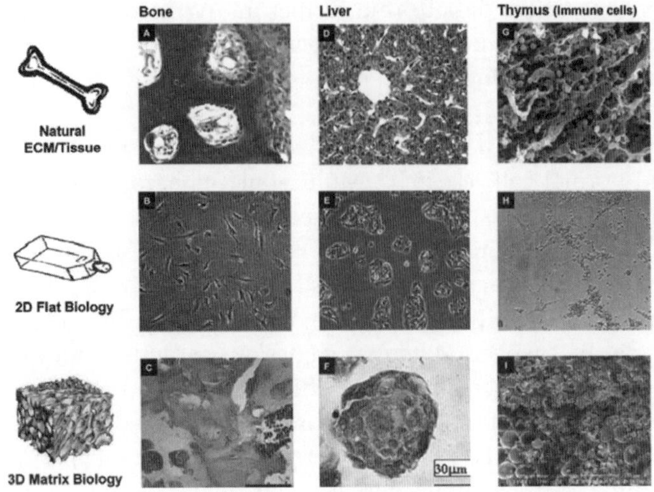

Figure 5.2. Comparison of natural cell and tissue morphology cultured on 2-D and 3-D substrates.[14]

spatial variability can interfere in signaling between cells.[17] The small dimensions of microfluidic devices offer short diffusion lengths and better control of the microenvironment.

Second, by providing a three-dimensional scaffold for cell organization, hydrogels offer attachment points and mechanical support to cells seeded inside. At the same time, the porous gel matrix allows the delivery of nutrients and removal of waste products from the cells through diffusion.

Miniaturizing to the microscale offers advantages in scaling.[18,19] For example, the low Reynolds numbers imply laminar flow, where diffusive effects dominate. Enzyme analysis, immunoassays, and DNA analysis or PCR can be conducted in much smaller timescales compared to macroscale systems. Microfluidic systems which integrate cell culture and characterization will contribute to improved culture conditions which will more closely resemble the physiological environment, and by permitting tracking of small cell numbers or individual cells, advance work in areas such as stem cell research or drug discovery.

5.2 HYDROGELS

A variety of three-dimensional cell culture matrices, or scaffolds, have been developed for tissue engineering applications. Tissue engineering uses living cells to restore or maintain the function of tissue and organs, and involves cell biology, materials science, and medicine.[20] Scaffolds have been used as implant materials as well as *in vitro* cell culture materials.[14] Implantable materials should be biodegradable and should match the properties of the surrounding tissue, whether it be bone or skin. Cell culture scaffolds should approximate the ECM as closely as possible.

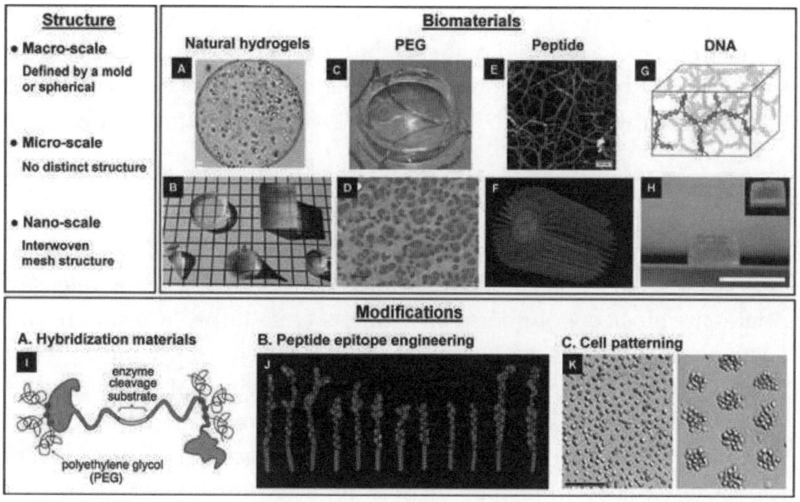

Figure 5.3. Overview of hydrogels used for 3-D cell entrapment.[14] For color reference, see page 261.

Tissue scaffold materials include ceramics, polymers, and composites, but the majority of matrices are polymeric (Fig. 5.3). Synthetic polymers include polyglycolic acid (PGA), polylactic acid (PLA), polyethylene glycol (PEG), polyvinyl alcohol (PVA), and many others. Natural polymers include collagen, gelatin, chitosan, hyaluronic acid, and alginate. Hydrogels are hydrophilic polymer networks which swell with water.[21] The polymer chains may be crosslinked via covalent chemical bonds, ionic interactions, or physical bonds. Collagen, agarose, alginate, PVA, and PEG are all examples of hydrogels.

Hydrogels that undergo gentle gelation processes, which will minimally affect the entrapped cells, are desired for cell immobilization.[22] Gelation can occur through chemical crosslinking, ionic crosslinking, or thermal phase change.

5.2.1 Naturally Derived Hydrogels

In tissue engineering, the most commonly used natural hydrogels include collagen, Matrigel, and alginate.[23]

Matrigel is a mixture of proteins extracted from mouse tumor cells.[2] Matrigel solidifies above 4°C. The chilled liquid is often coated onto plastic tissue culture dishes, and incubated at 37°C to form a thin gel film. Cells can be cultured on the gel surface, or their migration behavior into thick gel films can be studied. Alternatively, cells can be directly mixed into the chilled liquid, which forms a gel with embedded cells upon incubation at warm temperature. Matrigel has been widely used in angiogenesis assays, in which endothelial cells are cultured on the gel.

Matrigel contains many proteins, including laminin, and collagen IV, as well as heparan sulfate proteoglycans, many other proteins in low concentrations, and growth factors. Not all of its components have been identified. For many applications, the use of a scaffold material that more closely mimics the natural cell environment would be more appropriate. In addition, the use of animal-derived materials may be controversial in some tissue engineering applications.

5.2.2 *Alginate*

Alginates are naturally derived polysaccharide polymers isolated from seaweed. They have been used as food stabilizers and thickeners, and as drug release materials. Alginates are block copolymers consisting of (1-4)-linked β-D-mannuronic acid (M units) and α-L-guluronic acid (G units) (Fig. 5.4). The mechanical properties of the hydrogel depend on the ratio of the M to G units. Divalent cations such as Ca^{2+} or Ba^{2+} bind between the G-blocks of adjacent alginate chains, giving ionic cross-linking.[24] A higher M-unit content gives the polymer chains more flexibility in solution. The gel can be stiffened by increasing the G-unit content or by increasing the length of the G-unit blocks.[25]

Alginate hydrogels have been extensively used in cell encapsulation. Islet transplantation has been investigated as an approach for treatment of insulin-dependent diabetes,[26] and hepatocyte encapsulation has been investigated in the development of bioartificial livers.[27] For islet transplantation, the alginate microcapsule provides immunoisolation for the transplanted xenogeneic (cross-species) cells. Calcium alginate capsules have already demonstrated long term biocompatibility in animal studies and clinical trials[28] (Fig. 5.5). Since sodium ions in the physiological environment will exchange with the divalent cations in the alginate, the gels can weaken mechanically over time after implantation. One approach used to increase the mechanical strength of cell capsules has been to coat the capsules with a layer of poly-L-lysine. Alginate gel beads can be produced by using a syringe pump to push droplets of an alginate-cell suspension into a bath of calcium chloride. In addition, coaxial air jets and electrostatic bead generators are now commercially available.

Alginates can be modified with cell adhesion ligands to improve cell attachment.[22] Covalent modification of the carboxylic acid functional groups

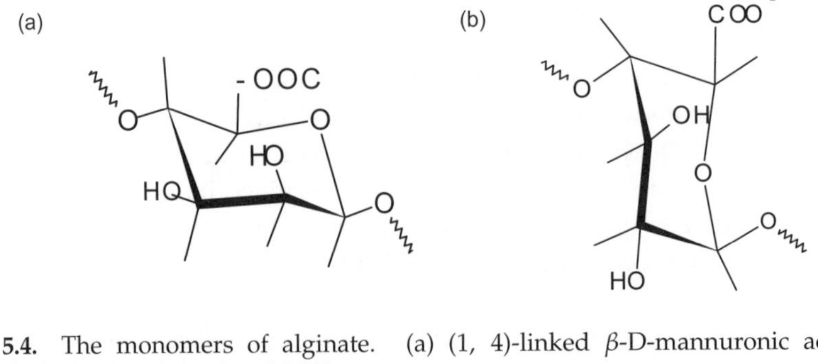

Figure 5.4. The monomers of alginate. (a) (1, 4)-linked β-D-mannuronic acid (M). (b) (1, 4)-linked α-L-guluronic acid (G).

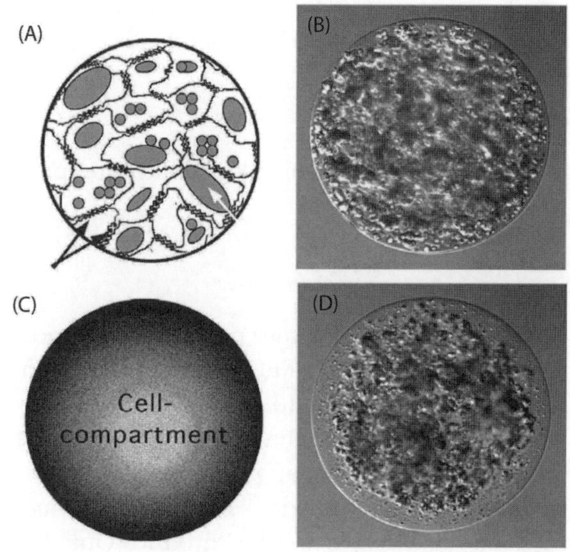

Figure 5.5. Alginate microcapsules. (A) Schematic model of alginate microcapsules: the "egg-box" model. G-units (black arrows) are linked by cations, forming the porous network that surrounds the producer cells (white arrow). (B) Light microscopic image of a homogeneous alginate bead and producer cells (650 µm in diameter). (C) Schematic model of an inhomogeneous alginate bead comprising an outer rim of alginate and an inner cell compartment. (D) Light microscopic image of an inhomogeneous bead. Original magnification $\times 100$.[28]

on the alginate chain can include addition of RGD peptides to encourage cell attachment.[24] Cell viability can be increased by decreasing the molecular weight of the alginate, which decreases the low shear viscosity of pre-gelled solutions.[29] High viscosity of the pre-gel solution may cause higher shear forces during cell suspension. The molecular weight of the gels can be decreased by irradiation.

5.2.3 Agarose

Agar gels are also polysaccharides derived from seaweed or algae. Agar gels have been used as food thickeners and as a vegetarian gelatin substitute. Agar gels have been used to coat petri dishes in bacterial cell culture.

Agarose is one of the main polysaccharide components of agar. Agarose gels have been used in electrophoresis for DNA separation.[3] Dilute solutions of agarose form gels which permit the separation of large DNA molecules through the pores of the gel under an applied electric field.

Agarose is a copolymer consisting of (1,3)-linked β-D-galactose and (1,4)-linked (3-6)-anhydro-α-L-galactose. Agarose forms a thermally reversible gel in a range of temperatures depending on the molecular weight and concentration of the polymer. Agarose melts around 85°C and solidifies around 35°C, although a low gel temperature agarose will melt at 70°C. Agarose can also be formed into microbeads. Although agarose has also been investigated as a cell encapsulation material for transplantation,[30] it has not demonstrated immunoisolation.[31]

5.2.4 *Synthetic Hydrogels*

Synthetic polymers can offer more controllable chemical composition and physical properties compared to natural polymers.

5.2.5 *Pluronic*

The thermally reversible triblock copolymer poly(ethylene oxide)-poly(propylene oxide)-poly(ethylene oxide) (PEO-PPO-PEO) are known as Poloxamers or Pluronic. Historically, Pluronic has been added to mammalian cell culture medium in low concentrations, from 0.05-0.1%. This additive lowers the surface tension of the medium to improve transport of metabolites to cells, and also protects the cells against shear-induced mechanical damage.[32] During experimental and industrial manufacturing processes, gas bubbles may be introduced into the bulk liquid. Cell damage occurs when the bubbles break at the liquid surface, and bubble rupture is a major cause of cell lysis. Cell lysis can also occur due to the shear force in draining liquid films if cells are trapped in a foam layer. The Pluronic protects cells by preventing cell attachment to air bubbles.[33]

The polymer self-assembles into micelles due to the hydrophilic PEO and hydrophobic PPO blocks. At low temperatures the aqueous polymer solution is a liquid. The polymer undergoes thermoreversible gelation to form a close-packed arrangement of micelles. Pluronic F127 (molecular weight = 120 000 g/mol, PEO_{99}-PPO_{65}-PPO_{99}) has been approved by the Food and Drug Administration (FDA) for controlled drug delivery applications. Pluronic F127 has also been used as a tissue engineering scaffold material for chondrocytes, epithelial cells, and fibroblasts.[34] Pluronic solutions in the range of 10–15% have an LCST at physiological temperatures. Since non-modified Pluronic capsules require an external membrane to prevent bead dissolution, Pluronic has also been modified to add covalent chemical crosslinking in the polymer for better stability in cell encapsulation.[35]

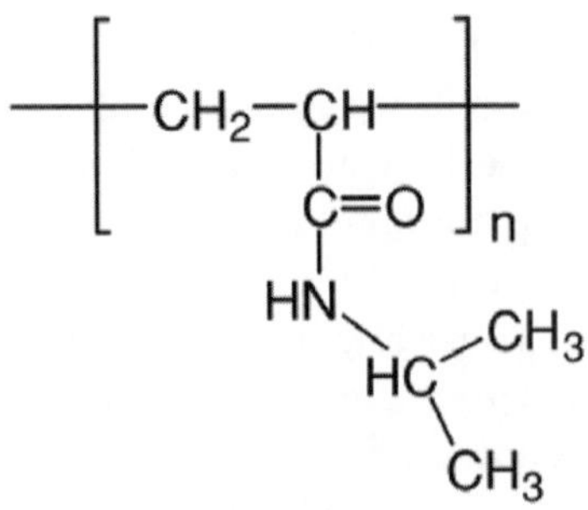

Figure 5.6. *N*-isopropylacrylamide (NiPAAm).[36]

5.2.6 N-isopropylacrylamide Polymers (NiPAAm)

N-isopropylacrylamide (NiPAAm) homopolymers (Fig. 5.6) and copolymers are thermoreversible polymers which have been used for drug delivery and tissue engineering. At low temperature, the polymer dissolves is a liquid solution since hydrogen bonds form between polar groups in the polymer and water. The polymer chains form extended coils surrounded by ordered water molecules. The gel is observed at high temperature. There, the polymer precipitates out of solution because hydrophobic interactions dominate, and the polymer chains transition to a globular state.[36] NiPAAm homopolymer in pure water has a lower critical solution temperature (LCST) of 32–34°C, where the transition occurs between liquid and gel. NiPAAm copolymers which contain hydrophilic groups such as polyethylene glycol (PEG) will increase the LCST. Reversible gelation occurs due to polymer chain entanglements and the increased hydration.

In contrast to other thermally reversible systems such as Pluronics, NiPAAm copolymers when gelled do not revert to the liquid state upon dilution. The NiPAAm phase transition also occurs over a narrower temperature range, permitting more precise control for on-chip gelation.

Copolymers of NiPAAm have already been used for 3-D culture of chondrocytes[37] and pancreatic islets.[38] Acrylic acid or polyethylene glycol (PEG) copolymers are used to increase the hydration of the gel, since NiPAAm homopolymer gel exhibits high water loss. This change in hydration with temperature has been used to fabricate cell sheets. When cells are cultured on surfaces treated with NiPAAm at 37°C, the cells adhere to the slightly hydrophobic surface. When the temperature is brought below the LCST, the gel swells and becomes more hydrated, and the cells spontaneously detach. This effect has been used to create confluent sheets of myocardial cells which can be layered to build 3-D electrically communicative structures.[39]

5.3 MICROFABRICATION

Microfabrication processes are the basis for integrated circuit (IC) manufacture. They involve thin film deposition, patterning, and etching. In addition to fabrication of planar silicon-based ICs, microfabrication processes have enabled the development of micro-electro-mechanical systems (MEMS), as well as an increasing variety of glass- and polymer-based lab-on-a-chip and implantable devices for biomedical applications.

More recently, the technique of soft lithography has been developed to create microstructures using printing, molding, and embossing processes.[40] This technique most commonly uses poly(dimethylsiloxane) (PDMS) as an elastomeric mold material. Elastomeric PDMS structures are replicated from a mold master, which is commonly either a silicon or glass substrate patterned with photolithography (Fig. 5.7). If a high-aspect ratio photoresist such as SU-8 is used, the mold master can be produced using only one photolithography step. Alternatively, the

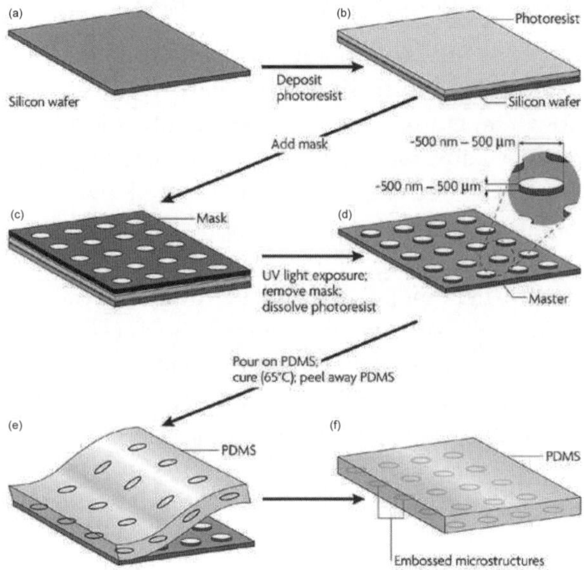

Figure 5.7. Microfabrication of PDMS structures. (a–b) Photoresist is spin-coated on a silicon wafer. (c) A mask is placed in contact with the layer of photoresist. (d) The photoresist is illuminated with ultraviolet (UV) light through the mask. An organic solvent dissolves and removes photoresist that is not crosslinked. The master consists of a silicon wafer with features of photoresist in bas-relief. An expanded view of one of the microfabricated structures with its characteristic critical dimensions is shown. (e) PDMS is poured on the master, cured thermally and peeled away. (f) The resulting layer of PDMS has microstructures embossed in its surface.[47] For color reference, see page 261.

mold can be produced through a photolithography step followed by reactive ion etching of the silicon substrate. PDMS is poured onto the mold master, cured, and then removed. The PDMS structure can be used as a stamp for microcontact printing of proteins or self-assembled monolayers. This PDMS structure can also form closed microfluidic channels when bonded to another substrate.

PDMS has many advantages for use in biomedical applications: flexibility, transparency, biocompatibility, and gas permeability. Prototype PDMS structures can be made rapidly and at relatively low cost. Elastomeric structures made in this method have been used in microstamping, microfluidic patterning, and stencil patterning to create well-defined areas of cellular adhesives, polymers, self-assembled monolayers (SAMs), and biomolecules to study cellular interactions on the microscale.[41] PDMS microchambers have been used as stamps for patterning cell attachment on surfaces,[42] chambers for DNA amplification via the polymerase chain reaction (PCR),[43–45] and as chambers for bacterial[46] (Fig. 5.7) and mammalian cell culture.[47]

PDMS will swell and absorb liquid media, which can be a major consideration for cell culture applications involving small volumes. The amount of liquid absorbed depends on the ratio of siloxane oligomer precursors used to make the PDMS.[48]

The biocompatibility of PDMS can be increased by serially washing the PDMS to extract uncrosslinked oligomers, solvents, and catalyst which can leach out, reducing cell viability in cell culture devices.[49] Further work is needed to investigate whether PDMS has a significant absorption for the drugs or other elements from the culture media. The uptake of small molecules into PDMS microstructures or release of uncured oligomers[50–52] can lead to adverse results for cell culture, drug screening and discovery, or cytotoxicity assays. In this case, the culture chambers may need to be coated in order to reduce uptake of small molecules into PDMS microstructures and leaching of uncured oligomers.

PDMS has been used to make valves, both as an elastomeric diaphragm integrated into glass or silicon devices, and also as membranes in multilayer PDMS structures fabricated using soft lithography.[53] In these structures, one layer of the device contains the control channels and the other layer contains the flow channels. The membrane where the control channel and flow channel intersect is deflected into the flow channel, often pneumatically, to effect valving (Fig. 5.8). Peristaltic pumps, created using three such membrane valves in a row, can be used to achieve pumping rates as high as 2 cm/s at 100 Hz. Microfluidic valves can also be used to create mixers by putting peristaltic pumps in a closed loop.

In cases where controlling liquid loss through evaporation is critical, the PDMS can be coated with parylene to reduce gas and moisture permeability. Parylene is a biocompatible polymer that is conformally deposited inside the channel. The deposition is done from the vapor phase at room temperature to give a thin-film coating that has low residual stress. Parylene films are pinhole-free even at low thickness and have low moisture permeability. Parylene coatings have already been used to reduce the sample evaporation and protein adsorption in polymerase chain reaction (PCR) chambers,[45] as well as control moisture loss in microfluidic cell culture devices.[54] Alternatives to PDMS as a structural material include glass, polystyrene, and polyimide. Other solvent-resistant materials include microfluidic channels molded from thiolene or structures molded in photocurable teflon. Of these materials, however, the production of prototypes and test devices with multiple layers/heights and valving functionality using PDMS is still the most cost effective and time efficient in most cases.

5.4 HYDROGEL-BASED MICROFLUIDIC CELL CULTURE

Rather than continuing to miniaturize conventional cell culture systems, such as moving from 96-well to 384- or even 1536-well plates, moving to microfluidic systems can offer advantages such as fluid handling and transport. For systems with integrated sensors, scaling down the physical dimensions can lead to improved sensor signal/noise ratio and response time.[55] Microfabricated systems can offer lower power consumption, lower fabrication costs, and smaller footprint than conventional systems. Controlled spatial and temporal gradients are easily produced using microfabricated systems[56] (Fig. 5.9).

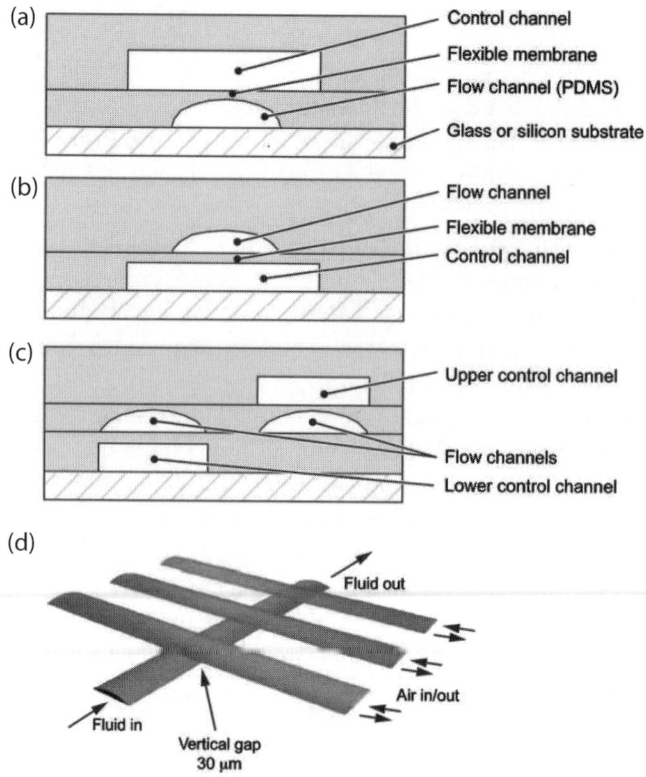

Figure 5.8. (a) A two-layer polydimethylsiloxane (PDMS) push-down microfluidic valve. An elastomeric membrane is formed where the flow channel is positioned orthogonal to the control channel directly above. Fluid flow is out of the page. (b) A two-layer PDMS push-up microfluidic valve where a control channel lies orthogonal to and below the flow channel. (c) A three-layer device with both push-up and push-down valves. (d) Schematic of a linear peristaltic pump using three membrane valves in a series.[53] For color reference, see page 262.

Shear stress in the culture environment can also affect cell differentiation.[57] Although a wide range of fluid velocities and shear stresses are seen in different areas of the body, shear stress in perfusion culture, in which cells adhere to the bottom of a microfluidic culture channel, may not be desirable for all cell types. Shear stress has been found to induce tumor cell cycle arrest, thus affecting tumor cell sensitivity to anticancer agents.[58] Even low perfusion rate in microfluidic cell culture has been found to affect growth kinetics and morphology.[59] By reducing the flow so that shear stress is on the order of 0.01 Pa, which approximates many *in vivo* conditions, a continuous flow environment can be used to maintain cell viability for long-term culture.[60,61]

Encapsulation within a gel may be one way to protect cells from shear stress due to the perfusion of culture medium at higher flowrates. Perfusion culture can be combined with 3-D culture matrices within microfluidic devices.[62] In this case,

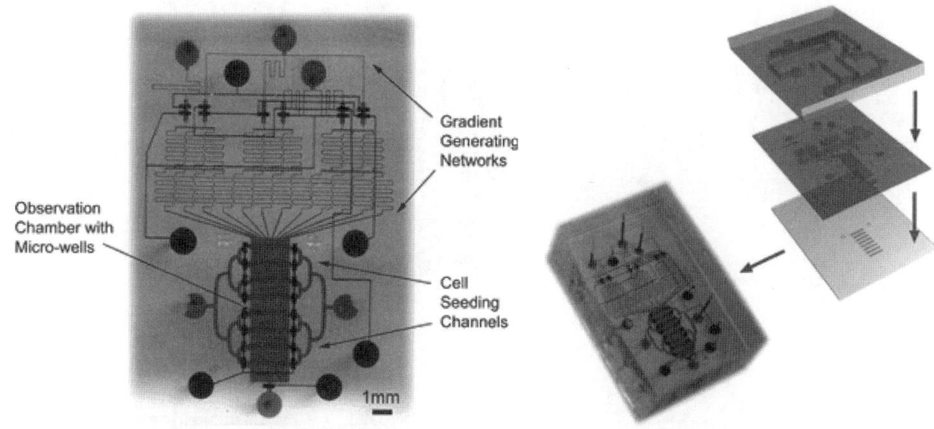

(a) (b)

Figure 5.9. A composite-gradient generator composed of two parts: a custom microfabricated PDMS chip and a glass coverslip. (a) The PDMS chip has two layers: the bottom fluidic layer (red) contains a gradient generating network, a symmetric cell seeding network and a chemotaxis observation chamber; the top layer (blue) contains pneumatic control valves which can turn on or off the flow in the specified fluidic channels on the bottom layer. The glass coverslip is wet-etched with an array of micro-wells (darker red). (b) The three-dimensional relationship between all the layers is schematically demonstrated on the right panel. The actual device is filled with food dyes to facilitate visualization.[56] For color reference, see page 262.

the flow can be used to precisely define a variable chemical microenvironment, which is not achievable using current static culture techniques.

5.4.1 On-chip Alginate Cell Encapsulation

Alginate-based cell entrapment and release in microfluidic chips has been demonstrated with yeast cells.[63,64] The hydrogel is formed when laminar flows of alginic acid solution and calcium ion solution are brought into contact. The calcium stream diffuses into the second stream and alginate hydrogel forms at the boundary (Fig. 5.11). The alginate hydrogel can be dissolved in EDTA (ethylene diaminetetraacetic acid), which is a calcium complexing agent.

Microfluidic patterning of alginate hydrogels has been demonstrated with stacked gel layers having widths as low as 10 μm.[65] The gel layer width is varied through control of the alginate solution flowrate. Each gel layer can be used to entrap a different cell type. Such fine control of cell positioning may be used to create models of the natural structure of tissues. Cells can be released from the gel by applying a flow of EDTA into the channel, which dissolves the alginate within seconds. Since cell release from the hydrogel can be effected in a short time period, the exposure to EDTA is minimized, giving high viability after release. This system has been used for co-culture by seeding different cell types into separate adjacent gel layers.[66] A laminar alginate-based system has been used to encapsulate breast

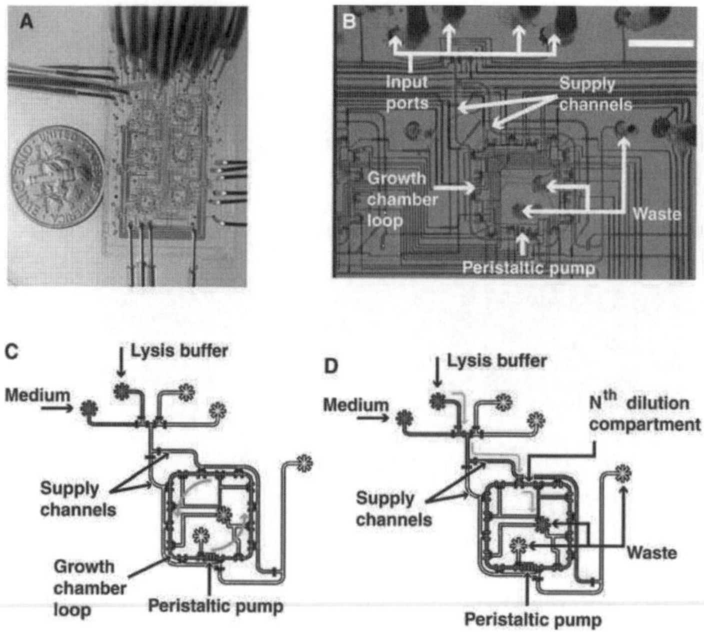

Figure 5.10. (A) Optical micrograph showing six microchemostats that operate in parallel on a single chip. Various inputs have been loaded with food dyes to visualize channels and sub-elements of the microchemostats. The coin is 18 mm in diameter. (B) Optical micrograph showing a single microchemostat and its main components. Scale bar, 2 mm. (C) Schematic diagram of a microchemostat in continuous circulation mode. Elements such as the growth loop with individually addressable connected segments, the peristaltic pump, supply channels, and input/output ports are labeled. (D) Isolation of a segment from the rest of the growth chamber during cleaning and dilution mode. A lysis buffer (indicated in red) is introduced into the chip through the lysis buffer port. Integrated microvalves direct the buffer through the segment, flushing out cells, including those adhering to chamber walls. The segment is then rinsed with fresh sterile medium and reunited with the rest of the growth chamber.[46] For color reference, see page 263.

tumor cells, permitting tumor spheroid formation and anticancer drug testing.[67]

5.4.2 *Microfluidic Agarose Cell Culture*

In an alternative configuration, the microfluidic channel itself can be made of hydrogel in which the cells have been seeded. Ling *et al.* have molded microfluidic channels made of agarose seeded with cells[68] (Fig. 5.12). The mold master was made using SU-8 patterned on a silicon wafer. Low gel-temperature agarose was mixed with phosphate buffered saline and cooled to 70°C, when it was loaded with AML-12 hepatocytes. The mixture was poured onto the mold master and gelled. A flat slab agarose was sealed onto the de-molded structure to created closed microchannels. Culture medium was supplied using a syringe pump through the microchannels. Although they were able to achieve uniform cell seeding, and good

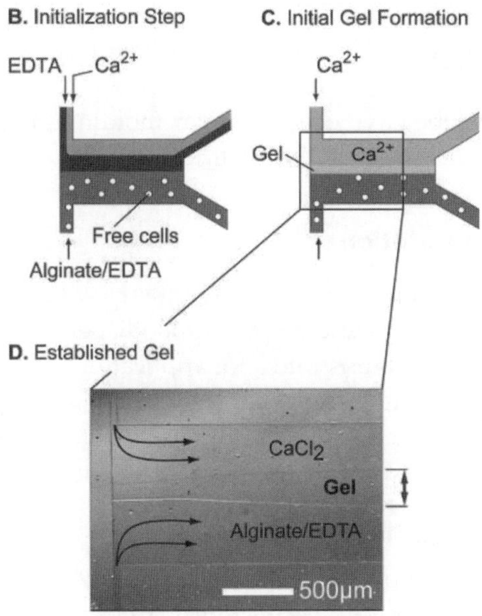

Figure 5.11. Microchannel layout for on-chip formation of alginate layers with embedded cells using laminar flows of alginate and Ca^{2+}.[63]

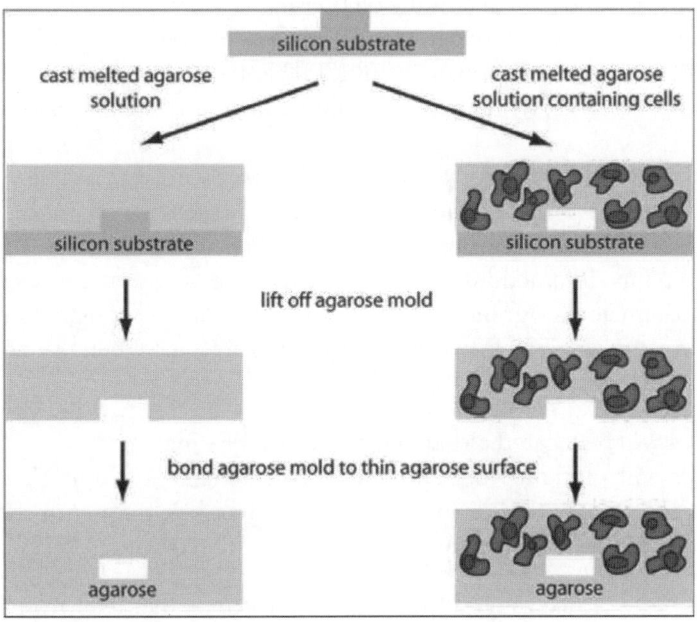

Figure 5.12. Schematic of the fabrication of agarose microfluidic devices with (right) and without (left) embedded cells.[68] For color reference, see page 263.

initial cell viability after device construction, the cell viability dropped after 3 days except in the vicinity of the microfluidic channels through which culture medium was supplied.

Paguirigan and Beebe have used a similar molding process to create gelatin-based microfluidic devices seeded with cells.[69]

5.4.3 Droplet Encapsulation

Cells can be encapsulated on-chip within gel spheres with diameters ranging from tens to hundreds of microns. Generation of monodisperse droplets in microfluidic systems has already been demonstrated, for applications including micromixing,[70] reverse transcriptase PCR, molecular synthesis, and cell encapsulation.[71]

Droplets can be generated using two immiscible flows, in which the hydrogel is the dispersed aqueous phase and a solvent is the continuous phase. In the flow-focusing design, droplets are generated through shear at the interface between two parallel streams, where the flows pass through a constriction. The continuous phase places viscous stress on the immiscible dispersed phase, which is balanced by the interfacial surface tension. Viscous shear stress tends to extend the interface, while the competing surface tension effect tends to reduce the interfacial area. In this situation, droplets are created above a critical stress. The Capillary number $Ca = \mu v/\gamma$ is a dimensionless parameter. It is the ratio of viscous forces to surface tension, where μ is the viscosity of the oil phase, v is the velocity of the droplet, and γ is the interfacial tension between the two phases. Droplet size is a function of the fluid viscosities, surface tension, microfluidic channel geometry, and flowrates.[72,73] Droplet size decreases with increasing flow rates, and increases with channel diameter.

Recent work has demonstrated monodisperse bead and single-cell loading within aqueous drops using the flow-focusing geometry.[75] A high-aspect-ratio channel was used to achieve single-particle loading within aqueous drops due to the hydrodynamic interactions. A high density suspension of particles, in which the particle diameter is significant compared to the channel constriction diameter, will become self-organized into one of two configurations when traveling rapidly through this channel:

(a) particles will be located along one side of the channel, or
(b) particles will alternate from one side of the channel to the other. The flowrates can be adjusted so that droplets form with the same frequency that particles arrive at the flow constriction.

The droplet volume ranged from 14–21 pL (droplet diameter from 30–34 μm), which encapsulated HL60 cells and 9.9 μm-diameter beads, respectively.

The flow-focusing method has been used to generate liquid NiPAAm droplets at room temperature, which then gelled further downstream in a heated region of the channel.[76,77] Embedded thin-film electrodes were used: as current was applied through the heater on the bottom of the channel, the local temperature

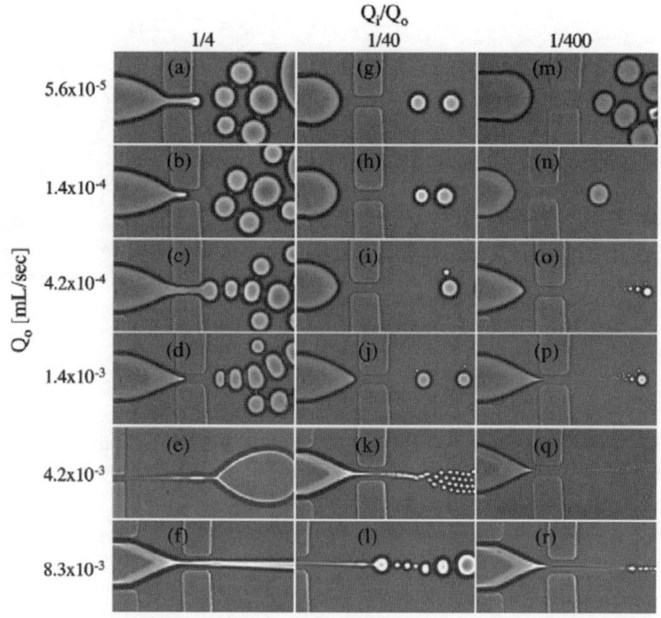

Figure 5.13. Droplet formation using the flow focusing design. Droplet size decreases with increasing oil (continuous phase) flow rate relative to water (dispersed phase) flow rate.[74]

in the channel increased above the gel formation temperature to 36°C. The gelled droplets were then collected in a simple grid to form a packed bed.

Although the majority of droplet-encapsulated cells have used macroscale extrusion, air-jet or electrostatic methods to generate droplets which gel subsequently in a large bath, microfluidic hydrogel droplets have also been made. Alginate gel droplets have been generated using the flow-focusing method.[78,79]

Alginate microspheres with encapsulated cells have also been generated using microfluidic devices. Choi *et al.* used alginate and $CaCl_2$ flows in hexadecane, and they reported blocking problems depending on the aqueous phase flowrate, due to gelation at the interface between alginate and calcium flows.[80] Hexadecane is highly immiscible with water and has low solubility (9.0×10^{-8} g/100 g water at 25°C) in the aqueous phase, allowing high cell viability in beads formed in hexadecane. However, for specific flowrates they did report successful on-chip crosslinking and encapsulation of yeast into alginate beads. Tan and Takeuchi used insoluble $CaCO^3$ particles interspersed in aqueous alginate solution, and droplets were created in corn oil which contained acetic acid.[81] Workman *et al.* have also used a flow-focusing geometry in which one phase contains sodium alginate mixed with $CaCO_3$ and cells and another phase contains acetic acid in oil. These two phases are separated by flows of plain oil, so that during laminar flow the H^+ does not diffuse into the alginate stream. After droplet formation, the H^+ then diffuses inside the alginate droplet to release Ca^{2+}, thus permitting gelation of the droplet. This method has been used to encapsulate the PC12, HEK293, and

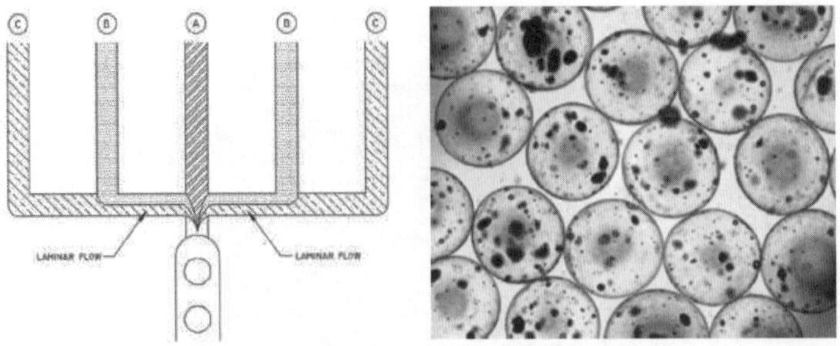

Figure 5.14. (left) The shielded junction employed to generate alginate microspheres. Aqueous sodium alginate mixed with CaCO3 and cells is introduced into the central channel (A). Sunflower oil mixed with acetic acid is supplied to the outermost channels (C). Sunflower oil is supplied to the intermediate channels (B) to act as a shield preventing the alginate solution from coming into contact with the acidified oil flow. Between channels B and A the two oils flow in a laminar fashion, with minimal diffusion of H+ into the protective sunflower oil. After droplet formation at the junction, H+ diffuses into the alginate droplet, thus liberating Ca2+ from CaCO3, which causes gelation of the alginate. (right) Light microscope image of encapsulated PC12 cells, showing small cell clumps present after encapsulation.[82]

U-2 OS cell lines.[82] A droplet-based microfluidic system has also been used to form alginate beads with entrapped breast tumor cells. The alginate environment permitted cell proliferation and the formation of multicellular spheroids. The dose-dependent response of the tumor spheroids to doxorubicin, and anticancer drug, showed multicellular resistance compared to conventional monolayer culture.[83]

5.4.4 Other Configurations

A photopolymerizable polyethylene glycol (PEG) was used in combination with dielectrophoresis (DEP) to trap cells in controlled, reproducible organizations.[84] The ability to pattern cell clusters with varying size and spacing showed that microscale organization and tissue architecture can affect cell behavior.

A combination of micropillars and gel encapsulation has been used in a 3-D perfusion system[85] by Toh *et al.* In this work, the cells were immobilized using the micropillar array, and then the cells were stabilized by a coacervated methylated collagen/HEMA-MMA-MAA terpolymer. Their work has also used polyethyleneimine-hydrazide as an inter-cellular linker, instead of the coacervated hydrogel, to created a "gel-free" system.[86]

Soft lithography and replica molding has been used to cast 3-D gel objects from collagen, Matrigel, and agarose.[87] The cell densities in the gel modules ranged from 10^8 – 10^9 cells/cm^3, which approaches the density in tissues. The gel objects had lateral dimensions from 40 to 1000 μm.

On-chip gelation of NiPAAm has been achieved using microheaters to encapsulate single cells which have been positioned using optical tweezers.[88] Since NiPAAm is a thermoreversible gel, the entrapped cell can be released by turning off the microheater.

Matrigel has been used on-chip to create cell-embedded regions within a microfluidic chamber.[89] The cell-gel suspension was injected into the device using a syringe. After the studies involving nanoparticle penetration into the gel were completed, the Matrigel was dissolved using the metalloprotease dispase at 37°C to recover the cells for subsequent flow cytometry analysis.

5.4.5 Transport Considerations

One of the key issues in cell encapsulation is ensuring adequate transport of nutrients (oxygen, electrolytes) to the cells and metabolites away from the cells. For example, stem cells are extremely sensitive to parameters in their microenvironment such as the oxygen concentration. High oxygen conditions can cause oxidative stress, which may alter protein expression and thus morphogenesis. For encapsulated islet cells, one persistent issue has been oxygen availability at the interior of the capsules, with cell necrosis at the capsule core. A gradient in oxygen is also seen in some tumors, and it has been found that some drugs are less active in hypoxic environments.[13]

The oxygen supply in PDMS devices may be insufficient without perfusion. Leclerc *et al.* have found that either culture medium exchange was required to maintain Hep G2 cell viability[90] or an additional oxygen supply chamber was necessary.[91]

An oxygen concentration gradient within a microfluidic cell culture system could result in non-uniform results. Oxygen diffusion through permeable PDMS structures is balanced by cellular uptake and convective transport by the flow of culture medium. The diffusivity of oxygen in PDMS has been found to be $D = 4.1 \times 10^{-5}$ cm^2/s, while its solubility is 0.18 cm^3 (STP)/cm^3 atm.[92] However, the permeability to oxygen can be significantly modified by protein adsorption or plasma surface modification. Oxygen concentration in PDMS microfluidic structures can be monitored using fluorescent dyes such as ruthenium tris(2,2'-dipyridyl) dichloride hexahydrate (RTDP), which are quenched by oxygen in a linearly dependent manner.

Recently, perfluorocarbons have been investigated as an oxygen carrier which would increase oxygen solubility by 15–20 times compared to oxygen solubility in water.[93] Perfluorocarbons are currently used as contrast markers in magnetic resonance imaging (MRI). HepG2 cells in alginate capsules with added perfluorocarbons showed improved cellular metabolic activity and cell viability.

In order to better control the distribution of solutes within scaffold materials, Choi *et al.* have fabricated microfluidic networks directly in the cell-seeded scaffold material to permit delivery of nutrients into the bulk (Fig. 5.15). They have defined a Krogh length, $\lambda_c \sim \sqrt{D_{s,g}c_0/R}$, where $D_{s,g}$ is the molecular diffusivity of the solute in the scaffold, R is the consumption rate, and c_0 is the concentration of the solute in the channel.[94] In order to minimize variations in concentrations of metabolites or waste products, the microfluidic scaffold should have a spacing

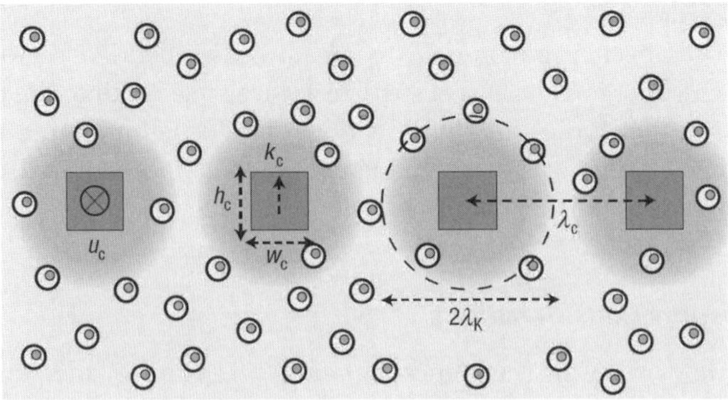

Figure 5.15. Schematic cross-sectional view of a cell-seeded microfluidic scaffold. The dispersed cells (circles) surround the microchannels (squares). The pink shading represents steady-state distributions of solutes. Here, a reactive solute is delivered via the channels and is consumed by the cells as it diffuses into the matrix. λ_K is the Krogh length, λ_c is the interchannel distance, w_c and hc are the microchannel width and height, k_c is the mass transfer coefficient of the flow in the microchannels, and u_c is the speed of the flow in the microchannels.[94] For color reference, see page 264.

$\lambda_c < 2\lambda_K$, where λ_c is the interchannel distance. By adjusting the flow rate through the microchannels, they can ensure high Peclet number ($Pe = u_c h_c / D_{s,c} >> L/h_c$, where $D_{s,c}$ is the diffusivity of a solute in the solution in the microchannels), to avoid depletion of solutes along the length of the microfluidic channels, and high Biot number ($Bi = k_c \lambda_K / D_{s,g}$ for reactive solutes and $Bi = k_c H / D_{s,g}$ for non-reactive solutes), to avoid having higher convective mass transfer than diffusive mass transfer in the gel. Here, λ_K is the Krogh length, λ_c is the interchannel distance, w_c and h_c are the microchannel width and height, k_c is the mass transfer coefficient of the flow in the microchannels, and u_c is the speed of the flow in the microchannels.

Using primary bovine chondrocytes embedded in calcium alginate structures, they have demonstrated the ability to control the spatial and temporal distribution of solutes within the alginate bulk.

5.5 APPLICATIONS

Currently, the majority of characterization techniques give information averaged over large cell populations. Although a large population of cells may be required to ensure statistically relevant data, the ability to measure each cell individually with high throughput analysis would provide both information over the cell population as well as the distribution or variability. For example, while flow cytometry does analyze each cell individually, the resulting data still reflects the aggregation of the cell population. Another example of single cell analysis is high-throughput screening using patch clamp. Highly integrated systems, which incorporate cell culture and characterization onto a single platform, will permit characterization

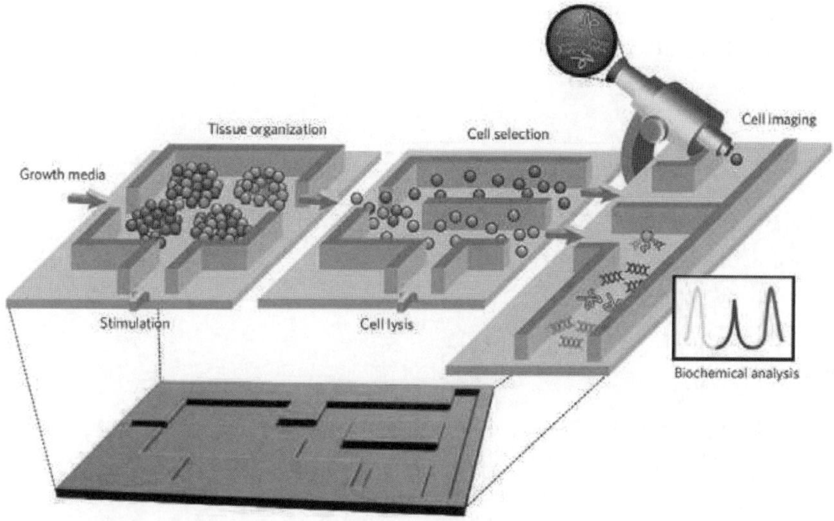

Figure 5.16. Tissue organization, culture and analysis in microsystems. Microsystems can incorporate 3D scaffolds to guide cell growth, microfluidic systems for nutrient transport, different techniques for biochemical analysis (such as image-based analysis), to give multiple functionalities on a single chip.[96] For color reference, see page 264.

of smaller numbers of cells or even individual cells. Integrated sensors include the microphysiometer to monitor extracellular pH, which indicates functional responses from cells.[95]

Highly integrated systems will also contribute to the development of portable point-of-care devices[96] (Fig. 5.16). Such systems can also be used to study changes in intracellular processes due to external variables.[55] This technology also promises to streamline the process of drug screening by offering simple, controlled techniques for repeatable assays of *in situ* cell response. Cancer cells respond differently to drugs when cultured in Petri dishes as compared to three-dimensional culture. Hepatocytes, which are important in drug metabolism, may be used for toxicity studies if more in-vivo-like culture conditions can be developed. This would also reduce the requirement for animal testing. Three-dimensional microfluidic cell culture systems have been developed to examine drug cytotoxicity.[58,67,83,85]

Applications of microfluidic hydrogel-based cell culture include *in vitro* cell-based testing in drug discovery research, to screen potential drug candidates and identify promising molecules.

5.6 CONCLUSIONS AND OUTLOOK

The rate of publications in the area of microfluidic cell culture is rapidly increasing. Many aspects of cell culture — proliferation, growth, signaling — are still not well understood, and much further work will be required to realize the many systems currently imagined: miniature automated labs, portable analysis systems, new technologies drug discovery, and more.

ACKNOWLEDGMENTS

This work was supported by the Natural Sciences and Engineering Research Council of Canada (NSERC).

References

1 Weigl B. H., Bardell R. L., Cabrera C. R., Lab-on-a-chip for drug development. *Advanced Drug Delivery Reviews.* **55**(3), 349–77 (2003).

2 Abbott A., Cell culture: Biology's new dimension. *Natur.* **424**(6951), 870–2 (2003).

3 Alberts B., Bray D., Lewis J., Raff M., Roberts K., Watson J. D., editors. *Molecular Biology of the Cell*. 3rd ed. New York: Garland Publishing; (1994).

4 Rosso F., Giordano A., Barbarisi M., Barbarisi A., From Cell-ECM interactions to tissue engineering. *Journal of Cellular Physiology.* **199**(2), 174–80 (2004).

5 Liu W. F., Chen C. S., Cellular and multicellular form and function. *Advanced Drug Delivery Reviews.* **59**(13), 1319–28 (2007).

6 Asthagiri A. R., Lauffenburger D. A., Bioengineering models of cell signaling. *Annual Review of Biomedical Engineering.* **2**(1), 31–53 (2002).

7 Vogel V., Baneyx G., The tissue engineering puzzle: A molecular perspective. *Annual Review of Biomedical Engineering.* **5**(1), 441–63 (2003).

8 Bhadriraju K., Chen C. S., Engineering cellular microenvironments to improve cell-based drug testing. *Drug Discovery Today.* **7**(11), 612–20 (2002).

9 Kunz-Schughart L. A., Freyer J. P., Hofstaedter F., Ebner R., The Use of 3-D Cultures for High-Throughput Screening: The Multicellular Spheroid Model. *Journal of Biomolecular Screening.* **9**(4), 273–85 (2004).

10 Padrón J. M., van der Wilt C. L., Smid K., Smitskamp-Wilms E., Backus H. H. J., Pizao P. E., *et al.*, The multilayered postconfluent cell culture as a model for drug screening. *Critical Reviews in Oncology/Hematology.* **36**(2-3), 141–57 (2000).

11 McLimans W. F., Davis E. V., Glover F. L., Rake G. W., The submerged culture of mammalian cells: The spinner culture. *The Journal of Immunology.* **79**(5), 428–33 (1957).

12 Minchinton A. I., Wendt K. R., Clow K. A., Fryer K. H., Multilayers of cells growing on a permeable support: *An in Vitro Tumour Model.* Acta Oncologica., **36**(1), 13–6 (1997).

13 Minchinton A. I., Tannock I. F., Drug penetration in solid tumours. *Nature Reviews Cancer.* **6**(8), 583–92 (2006).

14 Lee J., Cuddihy M. J., Kotov N. A., Three-Dimensional Cell Culture Matrices: State of the Art. *Tissue Engineering Part B: Reviews.* **14**(1), 61–86 (2008).

15 Edelman D. B., Keefer E. W., A cultural renaissance: *In vitro* cell biology embraces three-dimensional context. *Experimental Neurology.* **192**(1), 1–6 (2005).

16 Keenan T. M., Folch A., Biomolecular gradients in cell culture systems. *Lab on a Chip.* **8**(1), 34–57 (2008).

17 Yu H., Alexander C. M., Beebe D. J., Understanding microchannel culture: Parameters involved in soluble factor signaling. *Lab on a Chip.* **7**(6), 726–30 (2007).

18 Beebe D. J., Mensing G. A., Walker G. M., Physics and applications of microfluidics in biology. *Annual Review of Biomedical Engineering.* **4**(1), 261–86 (2002).

19 Mitchell P., Microfluidics — downsizing large-scale biology. *Nature Biotechnology.* **19**(8), 717–21 (2001).

20 Sipe J. D., Tissue engineering and reparative medicine. *Annals of the New York Academy of Sciences.* **961**(1), 1–9 (2002).

21 Jen A. C., Wake M. C., Mikos A. G., Review: Hydrogels for cell immobilization. *Biotechnology and Bioengineering.* **50**(4), 357–64 (1996).

22 Bhatia S. R., Khattak S. F., Roberts S. C., Polyelectrolytes for cell encapsulation. *Current Opinion in Colloid & Interface Science.* **10**(1-2) 45–51 (2005).

23 Lavik E., Langer R., Tissue engineering: Current state and perspectives. *Applied Microbiology and Biotechnology.* **65**(1), 1–8 (2004).

24 Rowley J. A., Madlambayan G., Mooney D. J., Alginate hydrogels as synthetic extracellular matrix materials. *Biomaterials.* **20**(1), 45–53 (1999).

25 Amsden B., Turner N., Diffusion characteristics of calcium alginate gels. *Biotechnology and Bioengineering.* **65**(5), 605–10 (1999).

26 Peirone M., Ross C. J. D., Hortelano G., Brash J. L., Chang P. L., Encapsulation of various recombinant mammalian cell types in different alginate microcapsules. *Journal of Biomedical Materials Research.* **42**(4), 587–96 (1998).

27 Maguire T., Davidovich A. E., Wallenstein E. J., Novik E., Sharma N., Pedersen H., *et al.*, Control of hepatic differentiation via cellular aggregation in an alginate microenvironment. *Biotechnology and Bioengineering.* **98**(3), 631–44 (2007).

28 Visted T., Bjerkvig R., Enger P. O., Cell encapsulation technology as a therapeutic strategy for CNS malignancies. *Neuro-oncology.* **3**(3), 201–10 (2001).

29 Kong H. J., Smith M. K., Mooney D. J., Designing alginate hydrogels to maintain viability of immobilized cells. *Biomaterials.* **24**(22), 4023–9 (2003).

30 Uludag H., De Vos P., Tresco P. A., Technology of mammalian cell encapsulation. *Advanced Drug Delivery Reviews.* **42**(1-2), 29–64 (2000).

31 Shoichet M. S., Li R. H., White M. L., Winn S. R., Stability of hydrogels used in cell encapsulation: An *in vitro* comparison of alginate and agarose. *Biotechnology and Bioengineering.* **50**(4), 374–81 (1996).

32 Mizrahi A., Pluronic polyols in human lymphocyte cell line cultures. *Journal of Clinical Microbiology.* **2**(1), 11–3 (1975).

33 Ma N., Chalmers J. J., Aunins J. G., Zhou W., Xie L., Quantitative Studies of Cell-Bubble Interactions and Cell Damage at Different Pluronic F-68 and Cell Concentrations. *Biotechnology Progress.* **20**(4), 1183–91 (2004).

34 Matthew J. E., Nazario Y. L., Roberts S. C., Bhatia S. R., Effect of mammalian cell culture medium on the gelation properties of Pluronic® F127. *Biomaterials.* **23**(23), 4615–9 (2002).

35 Cellesi F., Tirelli N., Hubbell J. A., Materials for cell encapsulation via a new tandem approach combining reverse thermal gelation and covalent crosslinking. *Macromolecular Chemistry and Physics.* **203**(10-11), 1466–72 (2002).

36 Jeong B., Kim S. W., Bae Y. H., Thermosensitive sol-gel reversible hydrogels. *Advanced Drug Delivery Reviews.* **54**(1), 37–51 (2002).

37 Jasionowski M., Krzyminski K., Chrisler W., Markille L. M., Morris J., Gutowska A., Thermally-reversible gel for 3-D cell culture of chondrocytes. *Journal of Materials Science: Materials in Medicine.* **15**(5), 575–82 (2004).

38 Lu H.-F., Targonsky E. D., Wheeler M. B., Cheng Y.-L., Thermally induced gelable polymer networks for living cell encapsulation.*Biotechnology and Bioengineering.* **96**(1), 146–55 (2007).

[39] Masuda S., Shimizu T., Yamato M., Okano T., Cell sheet engineering for heart tissue repair. *Advanced Drug Delivery Reviews.* **60**(2), 277–85 (2008).

[40] Whitesides G. M., Ostuni E., Takayama S., Jiang X., Ingber D. E., Soft Lithography in Biology and Biochemistry. *Annual Review of Biomedical Engineering.* **3**, 335–73 (2001).

[41] Folch A., Toner M., Microengineering of Cellular Interactions. *Annual Review of Biomedical Engineering.* **2**(1), 227–56 (2000).

[42] Mrksich M., Chen C. S., Xia Y., Dike L. E., Ingber D. E., Whitesides G. M., Controlling cell attachment on contoured surfaces with self-assembled monolayers of alkanethiolates on gold. *Proceedings of the National Academy of Science USA.* **93**(20), 10775–8 (1996).

[43] Zhang C., Xu J., Ma W., Zheng W., PCR microfluidic devices for DNA amplification. *Biotechnology Advances.* **24**(3), 243–84 (2006).

[44] Liu J., Enzelberger M., Quake S., A nanoliter rotary device for polymerase chain reaction. *Electrophoresis.* **23**(10), 1531–6 (2002).

[45] Shin Y. S., Cho K., Lim S. H., Chung S., Park S. J., Chung C., *et al.*, PDMS-based micro PCR chip with Parylene coating. *Journal of Micromechanics and Microengineering.* **5**, 768 (2003).

[46] Balagadde F. K., You L., Hansen C. L., Arnold F. H., Quake S. R., Long-Term Monitoring of Bacteria Undergoing Programmed Population Control in a Microchemostat. *Science.* **309**(5731), 137–40 (2005).

[47] Weibel D. B., DiLuzio W. R., Whitesides G. M., Microfabrication meets microbiology. *Nature Reviews Microbiology.* **5**(3), 209–18 (2007).

[48] Chang W.-J., Akin D., Sedlak M., Ladisch M. R., Bashir R., Poly(dimethylsiloxane) (PDMS) and Silicon Hybrid Biochip for Bacterial Culture. *Biomedical Microdevices.* **5**(4), 281–90 (2003).

[49] Millet L. J., Stewart M. E., Sweedler J. V., Nuzzo R. G., Gillette M. U., Microfluidic devices for culturing primary mammalian neurons at low densities. *Lab on a Chip.* **7**(8), 987–94 (2007).

[50] Regehr K. J., Domenech M., Koepsel J. T., Carver K. C., Ellison-Zelski S. J., Murphy W. L., *et al.*, Biological implications of polydimethylsiloxane-based microfluidic cell culture. *Lab on a Chip.* **9**(15), 2132–2139 (2009).

[51] Toepke M. W., Beebe D. J., PDMS absorption of small molecules and consequences in microfluidic applications. *Lab on a Chip.* **6**(12), 484–486 (2006).

[52] Sun X., Kelly R. T., Tang K., Smith R. D., Ultrasensitive nanoelectrospray ionization-mass spectrometry using poly(dimethylsiloxane) microchips with monolithically integrated emitters. Analyst. 2010, in press.

[53] Melin J., Quake S. R., Microfluidic Large-Scale Integration: The Evolution of Design Rules for Biological Automation. *Annual Review of Biophysics and Biomolecular Structure.* **36**(1), 213–31 (2007).

[54] Heo Y. S., Cabrera L. M., Song J. W., Futai N., Tung Y. C., Smith G. D., *et al.*, Characterization and Resolution of Evaporation-Mediated Osmolality Shifts That Constrain Microfluidic Cell Culture in Poly(dimethylsiloxane) Devices. *Analytical Chemistry.* **79**(3), 1126–34 (2007).

[55] Prokop A., Prokop Z., Schaffer D., Kozlov E., Wikswo J., Cliffel D., *et al.*, NanoLiter-BioReactor: Long-Term Mammalian Cell Culture at Nanofabricated Scale. *Biomedical Microdevices.* **6**(4), 325–39 (2004).

[56] Wang C. J., Li X., Lin B., Shim S., Ming G.-L., Levchenko A., A microfluidics-based turning assay reveals complex growth cone responses to integrated gradients of substrate-bound ECM molecules and diffusible guidance cues. *Lab on a Chip.* **8**(2), 227–37 (2008).

[57] Meyvantsson I., Beebe D. J., Cell Culture Models in Microfluidic Systems. *Annual Review of Analytical Chemistry.* **1**(1), 423–49 (2008).

58 Sung J. H., Shuler M. L., A micro cell culture analog (μCCA) with 3-D hydrogel culture of multiple cell lines to assess metabolism-dependent cytotoxicity of anti-cancer drugs. *Lab on a Chip.* **9**(10), 1385–1394 (2009).

59 Korin N., Bransky A., Dinnar U., Levenberg S., A parametric study of human fibroblasts culture in a microchannel bioreactor. *Lab on a Chip.* **7**(5), 611–7 (2007).

60 Lee P. J., Hung P. J., Rao V. M., Lee L. P., Nanoliter scale microbioreactor array for quantitative cell biology. *Biotechnology and Bioengineering.* **94**(1), 5–14 (2006).

61 Hung P. J., Lee P. J., Sabounchi P., Lin R., Lee L. P., Continuous perfusion microfluidic cell culture array for high-throughput cell-based assays. *Biotechnology and Bioengineering.* **89**(1), 1–8 (2005).

62 Kim L., Toh Y.-C., Voldman J., Yu H., A practical guide to microfluidic perfusion culture of adherent mammalian cells. *Lab on a Chip.* **7**(6), 681–94 (2007).

63 Braschler T., Johann R., Heule M., Metref L., Renaud P., Gentle cell trapping and release on a microfluidic chip by *in situ* alginate hydrogel formation. *Lab on a Chip.* **5**(5), 553–9 (2005).

64 Johann R., Cell trapping in microfluidic chips. *Analytical and Bioanalytical Chemistry.* **385**(3), 408–12 (2006).

65 Johann R. M., Renaud P., Microfluidic patterning of alginate hydrogels. *Biointerphases.* **2**(2), 73–9 (2007).

66 Chen M. C. W., Gupta M., Cheung K. C., Hydrogel-based microfluidic systems for co-culture of cells. *30th Annual International Conference of the IEEE Engineering in Medicine and Biology Society (EMBC '08);* 2008; Vancouver, Canada; 2008.

67 Chen M. C. W., Gupta M., Cheung K. C., Alginate-based microfluidic system for tumor spheroid formation and anticancer agent screening. *Biomedical Microdevices.* **12**(4), 647–654 (2010).

68 Ling Y., Rubin J., Deng Y., Huang C., Demirci U., Karp J. M., *et al.*, A cell-laden microfluidic hydrogel. *Lab on a Chip.* **7**(6), 756–62 (2007).

69 Paguirigan A., Beebe D. J., Gelatin based microfluidic devices for cell culture. *Lab on a Chip.* **6**(3), 407–13 (2006).

70 Bringer M. R., Gerdts C. J., Song H., Tice J. D., Ismagilov R. F., Microfluidic systems for chemical kinetics that rely on chaotic mixing in droplets. *Philosophical Transactions of the Royal Society A: Mathematical, Physical and Engineering Sciences.* **362**(1818), 1087–104 (2004).

71 Teh S.-Y., Lin R., Hung L.-H., Lee A. P., Droplet microfluidics. *Lab on a Chip.* **8**(2), 198–220 (2008).

72 Cristini V., Tan Y.-C., Theory and numerical simulation of droplet dynamics in complex flows-a review. *Lab on a Chip.* **4**(4), 257–64 (2008).

73 Squires T. M, Quake S. R., Microfluidics: Fluid physics at the nanoliter scale. *Reviews of Modern Physics.* **77**, 977–1026 (2005).

74 Anna S. L, Bontoux N., Stone H. A., Formation of dispersions using "flow focusing" in microchannels. *Appl. Phys. Lett.* **82**(3), 364–6 (2003).

75 Edd J. F., Carlo D. D., Humphry K. J., Koster S., Irimia D., Weitz D. A., *et al.*, Controlled encapsulation of single-cells into monodisperse picolitre drops. *Lab on a Chip.* **8**(8), 1262–4 (2008).

76 Flueckiger J., Cheung K. C., Integrated microfluidic systems for cell culture and characterization. *14th IEEE International Mixed-Signals, Sensors, and Systems Test Workshop;* 2008 June 18-20; Vancouver, Canada; 2008. p. 1–6.

77 Flueckiger J., Cheung K. C., Thermally reversible hydrogel-based microfluidic systems for cell culture. *IEEE Transactions on Biomedical Circuits and Systems.* 2008: *in review.*

[78] Zhang H., Tumarkin E., Peerani R., Nie Z., Sullan R. M. A., Walker G. C., *et al.*, Microfluidic Production of Biopolymer Microcapsules with Controlled Morphology. *Journal of the American Chemical Society.* **128**(37), 12205–10 (2006).

[79] Zhang H., Tumarkin E., Sullan R. M. A., Walker G. C., Kumacheva E., Exploring Microfluidic Routes to Microgels of Biological Polymers. *Macromolecular Rapid Communications.* **28**(5), 527–38 (2007).

[80] Choi C.-H., Jung J.-H., Rhee Y., Kim D.-P., Shim S.-E., Lee C.-S., Generation of monodisperse alginate microbeads and in situ encapsulation of cell in microfluidic device. *Biomedical Microdevices.* **9**(6), 855–862 (2007).

[81] Tan W. H, Takeuchi S., Monodisperse alginate hydrogel microbeads for cell encapsulation. *Advanced Materials.* **19**(18), 2696–2701 (2007).

[82] Yu L., Chen M. C. W., Cheung K. C., Droplet-based microfluidic system for multicellular tumor spheroid formation and anticancer drug testing. *Lab on a Chip.* (2010), DOI:10.1039/c004590j.

[83] Workman V. L., Dunnett S. B., Kille P., Palmer D. D., On-Chip Alginate Microencapsulation of Functional Cells. *Macromolecular Rapid Communications.* **29**(2), 165–70 (2008).

[84] Albrecht D. R., Underhill G. H., Wassermann T. B., Sah R. L., Bhatia S. N., Probing the role of multicellular organization in three-dimensional microenvironments. *Nature Methods.* **3**(5), 369–75 (2006).

[85] Toh Y.-C., Zhang C., Zhang J., Khong Y. M., Chang S., Samper V. D., *et al.*, A novel 3D mammalian cell perfusion-culture system in microfluidic channels. *Lab on a Chip.* **7**(3), 302–9 (2007).

[86] Ong S.-M., Zhang C., Toh Y.-C., Kim S. H., Foo H. L., Tan C. H., *et al.*, A gel-free 3D microfluidic cell culture system. *Biomaterials.* **29**(22), 3237–44 (2008).

[87] McGuigan A. P., Bruzewicz D. A., Glavan A., Butte M., Whitesides G. M., Cell Encapsulation in Sub-mm Sized Gel Modules Using Replica Molding. PLoS ONE. **3**(5), e2258 (2008).

[88] Arai F., Ng C., Maruyama H., Ichikawa A., El-Shimy H., Fukuda T., On chip single-cell separation and immobilization using optical tweezers and thermosensitive hydrogel. *Lab on a Chip.* **5**(12), 1399–403 (2005).

[89] Ng C. P., Pun S. H., A perfusable 3D cell-matrix tissue culture chamber for in situ evaluation of nanoparticle vehicle penetration and transport. *Biotechnology and Bioengineering.* **99**(6), 1490–501 (2008).

[90] Leclerc E., Sakai Y., Fujii T., Cell Culture in 3-Dimensional Microfluidic Structure of PDMS (polydimethylsiloxane). *Biomedical Microdevices.* **5**(2), 109–14 (2003).

[91] Leclerc E., Sakai Y., Fujii T., Microfluidic PDMS (Polydimethylsiloxane) Bioreactor for Large-Scale Culture of Hepatocytes. *Biotechnology Progress.*, 20(3), 750–5 (2004).

[92] Mehta G., Mehta K., Sud D., Song J., Bersano-Begey T., Futai N., *et al.*, Quantitative measurement and control of oxygen levels in microfluidic poly(dimethylsiloxane) bioreactors during cell culture. *Biomedical Microdevices.* **9**(2), 123–34 (2007).

[93] Khattak S. F., Chin K.-S., Bhatia S. R., Roberts S. C., Enhancing oxygen tension and cellular function in alginate cell encapsulation devices through the use of perfluorocarbons. *Biotechnology and Bioengineering.* **96**(1), 156–66 (2007).

[94] Choi N. W., Cabodi M., Held B., Gleghorn J. P., Bonassar L. J., Stroock A. D., Microfluidic scaffolds for tissue engineering. *Nature Materials.* **6**(11), 908–15 (2007).

[95] Park T. H., Shuler M. L., Integration of Cell Culture and Microfabrication Technology. *Biotechnology Progress.* **19**(2), 243–53 (2003).

[96] El-Ali J., Sorger P. K., Jensen K. F., Cells on chips. *Nature.* **442**(7101), 403–11 (2006).

Chapter Six

Flow Control in Biomedical Microdevices using Thermally Responsive Fluids

Vahid Bazargan* and Boris Stoeber†

*Department of Mechanical Engineering, The University of British Columbia, Vancouver, BC V6T 1Z4, Canada
†Department of Electrical and Computer Engineering, The University of British Columbia, Vancouver, BC V6T 1Z4, Canada
E-mail: stoeber@mech.ubc.ca

Microfluidic lab-on-a-chip systems for biomedical applications require specialized flow control devices. While several those for microfluidic devices have been developed to date, this text focuses on flow control concepts based on thermally responsive polymer solutions. In particular, flow control concepts using the thermally triggered reversible phase change of aqueous solutions of the polymer Pluronic will be discussed. The three example microflow control concepts presented in this text include selective heating of small regions of microfluidic channels, which leads to localized gel formation in these channels and reversible channel blockage. Furthermore, a timed channel blockage due to viscous heating within the flow will be described. Finally, a concept for cross-channel transport of Particles such as cells is presented that relies on the continuous regeneration of a gel wall at the diffusive interface of two co-streaming fluids in a microfluidic channel.

6.1 INTRODUCTION

Transport of biological materials in microfluidic channels on-chip has gained increasing popularity over the past decade. Many efforts aim at building integrated microfluidic devices that combine several functions such as sample preparation, separation and detection on a single chip, to form a so-called "lab-on-a-chip". Highly integrated lab-on-a-chip devices are also known as micro total analysis systems (microTAS). These systems can manipulate sample volumes of less than picoliters, reducing the amount of required sample and analyte that are often difficult and expensive to acquire. By integrating conventional lab bench analysis methods on-chip, cost for laboratory equipment can be significantly reduced, and automatic sample handling between the different lab processes requires less

Biomaterials for MEMS, Edited by M. Chiao and J.-C. Chiao

www.panstanford.com
978-981-4241-46-5

human intervention and reduces the risk of sample loss. Another advantage of the compactness of mocro TAS is faster completion of a sequence of laboratory processes. In addition, the time required for many analytical steps and chemical reactions scale favorably at small length scales, as will be described in more detail further below. The small size and reduced power consumption of microTAS makes them attractive for portable analytical instruments for the biomedical field or for environmental monitoring and control.

The transport and precise control of small volumes of fluid on a microfluidic chip requires adequate flow control devices such as pumps, valves and mixers that provide the required performance and are easily integrated. In addition, these flow control devices should not compromise the biological samples through electrical or mechanical interventions; furthermore, the biocompatibility of the flow environment such as the channel walls is a concern, and correspondingly, biocompatibility of materials is an important field of research for the microTAS community. The following discussion will concentrate on flow control mechanisms for microfluidic devices with an emphasis on thermally responsive fluids for microflow manipulation.

6.2 TRANSPORT IN MICROFLUIDIC CHANNELS

Typical dimensions L of the height and width of microfluidic channels range from several microns to several 100s of microns. Aqueous liquids that are most commonly transported in microfluidic systems have a density $\rho = 1000\ \mathrm{kg/m^3}$ and a shear viscosity $\mu = 10^{-3}\ \mathrm{Pa \cdot s}$. With a typical flow velocity $U = 1\ \mathrm{mm/s}$, flow in microchannels is characterized by a very low Reynolds number $\mathrm{Re} = \rho\ L\ U\ /\ \mu \approx 10^{-2} << 1$. The Reynolds number relates inertial forces and viscous forces of a fluid. A low Reynolds number corresponds to high viscous forces relative to the inertia forces in a flow. Consequently, microfluidic flow is dominated by viscous forces, in most cases inertia can be neglected.[1,2] The flow regime in microfluidic channels can therefore be considered laminar in most cases[3]; while turbulence is not impossible in microfluidic devices, it is rather uncommon.[2,4,5]

The Péclet number, $\mathrm{Pe} = L\ U\ /\ D$, is a dimensionless number that describes the ratio of convection to diffusion, and it can be defined either for molecular diffusion Pe_m or for thermal diffusion Pe_T, using the molecular diffusion constant D or the thermal diffusivity α, respectively. The Peclet number can also be interpreted as the ratio of a characteristic diffusion time scale to a characteristic convection time scale. The molecular diffusion coefficient is low especially for liquids, such as $D = 6 \cdot 10^{-10}\ \mathrm{m^2/s}$ for molecular diffusion of sodium phosphate in water.[6] The associated Péclet number $\mathrm{Pe} \approx 100$ indicates high convective effects compared to molecular diffusion. However, in general there is no characteristic order of magnitude for the Péclet number in microfluidic devices.[7] In absence of turbulence in microfluidic systems as mentioned above, molecular diffusion is important for mixing, which can become rather difficult for long diffusion time scales.[8] Depending on the particular application purely diffusive mixing might be either desired, especially where controlled mixing or separation of the species is needed, or not sufficient, when rapid mixing is required. Several rapid mixing strategies have been proposed for microfluidic devices that are effective despite the diffusion process being relatively slow.[9–11]

For the thermal Péclet number heat transfer is analogous to mass transfer. With a much larger thermal diffusivity than molecular diffusivity, heat diffusion is often more significant than convection so that heat diffusion can be considered very effective over the typical short length scales of microfluidic devices. With a thermal diffusivity of water $\alpha = 1.41 \cdot 10^{-7}$ m^2 / s, a typical time scale $t \sim L^2 / \alpha = 0.7$ ms for heat diffusion can be achieved across a 10 μm deep channel.

6.3 FLOW CONTROL MECHANISMS

6.3.1 Microvalve Principles

As described in Section 6.2, fluids are expected to behave quite differently in microfluidic systems as compared to large scale systems because the different forces governing flow scale differently. Therefore it is difficult to transfer flow control concepts from macroscopic systems to microscale flow devices. This led to the development of new flow control strategies for small-scale systems as conventional flow control concepts cannot be simply miniaturized. At the same time, microfluidics allows taking advantage of forces and effects that are negligible in large scale systems, thus opening new opportunities for innovation. The category of microvalves that are closest to their large scale counterparts simply obstruct a fluid conduit by moving a solid into the flow path; some of the actuation methods including electrostatic, piezoelectric, pneumatic, thermopneumatic, and electromagnetic actuation are more effective at the microscale than in large-scale systems.[12–17] Other examples include, passive micro check valves[16] that use drag forces rather than inertial forces as well as microflow control strategies based on surface tension.[25] A common disadvantage of many of these designs is the requirement for complicated fabrication schemes, unacceptable leakage rates, complex external actuation, and the risk of irreversible blockage of the valve.

A different approach to valve actuation has been achieved using environmentally responsive polymers that undergo a volumetric or phase change in response to changes in temperature or concentrations. These concepts lead to fast actuator response taking advantage of the relatively fast propagation of temperature or concentration changes on the microscale that was discussed in Section 6.2. In one of the first examples of this approach, Fréchet and co-workers[20,21] grafted a temperature-sensitive polymer (poly N-isopropylacrylamide) to the pore surfaces of a porous polymer monolith in a microchannel. As the device temperature transitioned through the lower critical solution temperature, the grafted polymer chain would undergo reversible swelling and collapse, leading to closing or opening of the pores of the polymer monolith, thus opening or closing the channel to flow. Mastrangelo and co-workers[23,24] included paraffin in a sealed reservoir; upon heating the high volumetric expansion of the paraffin was used to actuate a membrane that then obstructed a flow channel, acting as a valve. This method separated the actuator from the fluid to be controlled, allowing a wide range of applications. A different category of environmentally responsive flow control methods uses hydrogels, which are either anchored inside microfluidic channels, or are transported with the flow.

6.3.2 Hydrogel-based Microvalve Principles

As mentioned in the previous section, a wide variety of environmentally responsive fluids has been employed for flow control purposes in microfluidic devices. The actuation of the fluids occurs via a change in temperature,[26–37,101,115] exposure to light,[38] a change in pH or ionic strength of the fluid[39–51] and leads directly to a change in viscosity, volume, or a phase transition. Most of these hydrogel microvalve principles take advantage of the favorable scaling of their underlying physical principles of operation at small length scales.

So-called environmentally responsive fluids or smart hydrogels were first reported to exhibit a significant expansion as a function of the pH value of the surrounding fluid by Kuhn *et al.* in 1948.[48] This work was followed by work demonstrating the reversible volume change of these materials as a function of salt concentration in large scale systems in 1950.[49] The hydrogels investigated in these studies were able to change their volume in a reversible and reproducible manner by more than one order of magnitude, which corresponds to the largest volume change known for solid state materials. These studies, as well as the following work by T. Tanaka[47] at the beginning of the 1980s, inspired the development of other hydrogels with sensitivity to particular environmental parameters and the application of these materials in medicine, for pharmaceuticals, in sensors and for tissue engineering. Meanwhile, the potential of these hydrogels to control or regulate the transport of fluids was investigated and demonstrated through the first macrovalve for flow control at the milliliter and liter range.[50] However, the major disadvantage of hydrogel-based valve principles was their slow response.[47]

The first technologically relevant application of smart hydrogels to flow control was achieved with the pH-sensitive microvalve presented by Beebe *et al.* in 2000,[46] which relied on the swelling of a hydrogel in response to a change in pH, and used *in situ* photo patterning of the hydrogel in a microfluidic channel. The authors achieved a relatively fast valve response (less than 10 seconds), and demonstrated the use of hydrogel valves for autonomous microflow control. Since the swelling of hydrogel is diffusion controlled, the response time of these valves is mainly determined by the characteristic size of the actuator[42] and scales favorably at small length scales as discussed in Section 6.2. Specifically functionalized hydrogels have been developed for applications as microvalves in response to different parameters.[27,53–57] Current applications for biocompatible smart gels include artificial muscles,[36,58,60–62] and drug delivery.[37,63–67,69]

In a fundamentally different approach, thermally responsive polymer solutions have also been used as the working fluid in microfluidic devices. These polymer solutions solidify reversibly upon heating, so that selective heating of specific areas of a microfluidic system then causes localized gel formation to temporarily block a channel. Shirasaki and co-workers[72] used external infrared lasers for heating and localized gel formation and channel blockage for cell sorting; however, this approach depends significantly on transparency and absorbance of the materials involved. In a slightly different approach, electric heaters were integrated into a microflow system to achieve rapid and controlled heating of the fluid and subsequent[53] valving, taking advantage of the fast time scales for thermal diffusion in microdevices as discussed in Section 6.2.

6.4 THERMALLY RESPONSIVE FLUIDS FOR MICROFLOW CONTROL

6.4.1 *Temperature Responsive Materials*

Temperature responsive fluids (TRF) or temperature responsive hydrogels are among the most commonly studied and used materials for environmentally responsive microsystems.[26–35] Temperature responsive hydrogels react to the local temperature through phase transition or swelling and this thermo sensitivity can be used as an actuation method. Based on their reversible actuation mechanisms, smart hydrogels have been widely used in microvalves and microsensors as mentioned earlier,[27,75]. Recently, these materials have gained increasing attention because of their potential application to biomedical systems and devices for tissues engineering, cell handling and drug delivery.[28,29,37,65,66,76–87]

Thermally responsive fluids that undergo a phase change have a critical solution temperature (CST) at which the phase transition occurs, which can take place in form of a sol-gel transition upon heating or cooling. This transition occurs through interaction of polymer and water molecules in the case of TRFs exhibiting a critical phase separation temperature; otherwise, phase transition at a critical gelation temperature can be caused by intermolecular forces between polymer groups in the presence of water in form of hydrogen bonding and hydrophobic interactions due to the presence of hydrophobic groups such as methyl, ethyl and propyl.[29] Among the first group, poly(*N*-iso-propylacrylamide) (PNIPAAm) and also poly(*N*, *N*-diethylacrylamide) (PDEAAm) are the most extensively used materials.[88] The fundamental behavior of NIPAAm has been extensively studied and the possibility to form copolymers of NIPAAm by using other monomers has been demonstrated.[29,88,89] Gelation based on sol-gel transition of polymer solutions through hydrophobic/hydrophilic interaction on the other hand can be achieved with certain types of block copolymers made of poly-ethylene-oxide (PEO) and poly-propylene-oxide (PPO). Pluronics (or Poloxamers) and Tetronics consisting of PEO-PPO block copolymers belong to this group and are commercially available and commonly used to form TRFs.

6.4.2 Properties of Pluronic Solutions

6.4.2.1 Phase Transition of Pluronic Solutions

Pluronics are biocompatible symmetric triblock copolymers of poly(ethylene oxide)$_x$-poly(propylene oxide)$_y$-poly(ethylene oxide)$_x$; $P(EO)_x$–$P(PO)_y$–$P(EO)_x$. These amphiphilic copolymers have hydrophilic $EO_{(x)}$ and hydrophobic $PO_{(y)}$ units where x and y can take on specific values depending on the particular kind of Pluronic. Aqueous Pluronic solutions undergo reversible gel formation at a specific critical solution temperature that depends on the polymer concentration, its molecular weight and the polymer architecture.[29,32,90,91,93–95] The sol-gel transition of Pluronic solutions occurs through three dimensional packing of polymer micelles into a lattice, to achieve a hydrophilic/hydrophobic balance.[34]

Above the critical micelle concentration (CMC) the polymers aggregate into multi molecular structures with a hydrophobic central core (PO) and a hydrophilic shell (EO) facing the external medium.[96,97] At higher concentrations, above the critical gel concentration (CGC) the micelles can take on a cubic lattice structure.[34] Elevated temperatures lead to dehydration of the hydrophobic core and also increase the hydrophobicity of the hydrophilic blocks, which eventually leads to gel formation.[92,98] The aggregation and phase behavior of Pluronics have been extensively studied using several techniques including light scattering, NMR, SANS, SAXS, DSC, microscopy, and rheology.[34,99–104]

Pluronic F127, or Poloxamer 407, has previously been used for several applications in drug delivery. Its chemical formula is $(PEO)_{106}(PPO)_{70}(PEO)_{106}$, it is a nonionic surfactant, and in aqueous solution it can thermo-reversibly change from its liquid phase into a gel structure at a gelation temperature T_G, which depends on its concentration. Aqueous solutions of Pluronic F127 are non-Newtonian fluids with thermo-thickening and shear thinning properties.[26,99] The rheological properties of Pluronic F127 solutions including the effect of salts on gelation as well as its biocompatibility will be further discussed.

6.4.2.2 Rheological Characterization of Pluronic Solutions

Pluronic F127 solutions are thermo thickening materials. This means that upon heating their viscosity increases gradually over a wide range of temperatures until to their gel point, where the viscosity appears to increase dramatically. At higher concentration of Pluronic, gel forms at lower temperature. Stoeber *et al.* have shown this behavior using rheometry for different concentrations of Pluronic F127.[26] Continuous cone and plate vioscometry characterizes this phase transition under conditions that are relevant for the application to fluids streaming in microfluidic channels. This change in viscosity or phase change is completely reversible and after one thermal cycle, the shear viscosity returns to its

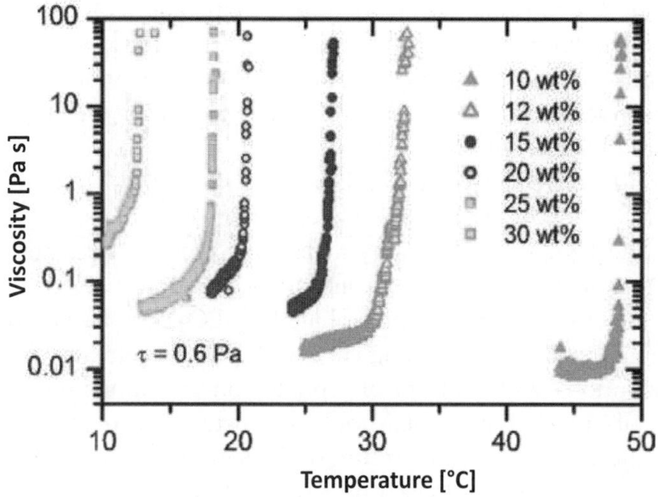

Figure 6.1. Viscosity as a function of temperature for different concentrations of Pluronic F127 in water from cone and plate viscometry at controlled shear stress (0.6 Pa s).[26] For color reference, see page 265.

original value. The phase transition between liquid and gel occurs over a narrow temperature range (0.5°C) as shown in Fig. 6.1.

Pluronic F127 solutions also show pseudo plastic flow behavior (i.e. shear thinning). Their viscosity occurs to decrease for increasing shear stress at a constant temperature as previously shown using rheometry.[26]

6.4.2.3 *Effect of Ions on the Gelation Temperature of Pluronic Solutions*

The effect of salt on micellization and also on the gelation temperature of Pluronic F127 has previously been shown.[105–108] In general, a lowering of the micellization and gelation temperature is observed upon addition of salt except for salts with large polarizable anions, such as SCN^-, and I^-, where an initial increase in the gelation temperature is often encountered. The effect is often discussed in terms of "salting in" and "salting out", of "structure-making" and "structure-breaking", additives. The addition of salt is expected to further dehydrate PPO blocks and induce some dehydration of the PEO blocks.[108] Almgren *et al.* have used a simple but effective approach to qualitatively predict the effects of additives: type one, those that increase the polarity difference between the solvent and the polymer will favor phase separation and decrease the micellization temperature. Salts such as NaCl, $MgSO_4$ and KF are depleted in the immediate vicinity of the polymer with its low polarizability, and increase the polarity of water. Phase separation is facilitated and so micellization occurs at a lower temperature. For salts of type 2, such as NaI or NaSCN on the other hand, the large, polarizable anion seems to partition slightly in favor of the polymer, increasing its polarity and thereby reducing the polarity difference. The polymer becomes more compatible with the solvent, and the micellization temperature is increased. It has been observed that the effect of salt on changing the gelation temperature is also consistent with the "salting-in" and the "salting-out" effect and the behavior of different types of salts on the micellization temperature is identical to the effect on the gelation temperature of Pluronic solutions.[108,109]

It has also been reported by Pandit *et al.* that the gel formation ability of Pluronic F127 can be lost at elevated concentrations of some salts.[109] The authors showed this effect for several salts by measuring the gel formation temperature and gel melting temperature for a solution of Pluronic F127. Bazargan *et al.* investigated the effect of sodium phosphate (Na_3PO_4) on the gel formation of a 15 wt% Pluronic F127 solution, using cone and plate viscometry.[110] The viscosity of Pluronic solutions with sodium phosphate concentrations ranging from 0 to 0.253 mol/L of solution was measured at constant shear stress (τ = 0.6 Pa) while slowly increasing the temperature of the sample at a rate of 2°C / minute. Figure 6.2 shows the measured viscosity as a function of temperature.

For each sample the viscosity increased first slowly until to a certain temperature from which on the viscosity occurred to increase by several orders of magnitude over only a few degrees Celsius in the same way as in the absence of salt. According to the measurements shown in Fig. 6.2 the gel formation temperature of 26.5°C for a 15 wt% solution of Pluronic F127 decreases to 19.8°C for a concentration of 0.253 mol/L of sodium phosphate.

At sodium phosphate concentrations above C_{sp} = 0.253 mol/L, no gel formation was observed using cone and plate viscometry at τ = 0.6 Pa s, which is consistent with the work by N. K. Pandit and J. Kisaka.[109] However, Fig. 6.3

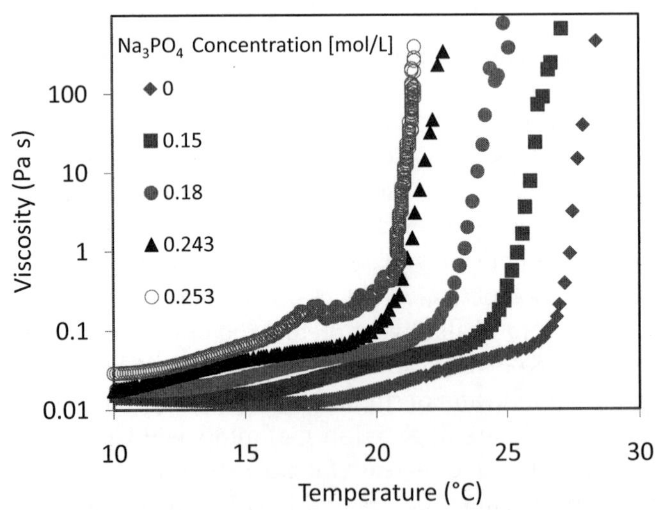

Figure 6.2. Viscosity as a function of temperature for different concentrations of sodium phosphate in a 15 wt% Pluronic F127 solution in water from cone and plate viscometry at controlled shear stress (0.6 Pa s).[110] For color reference, see page 265.

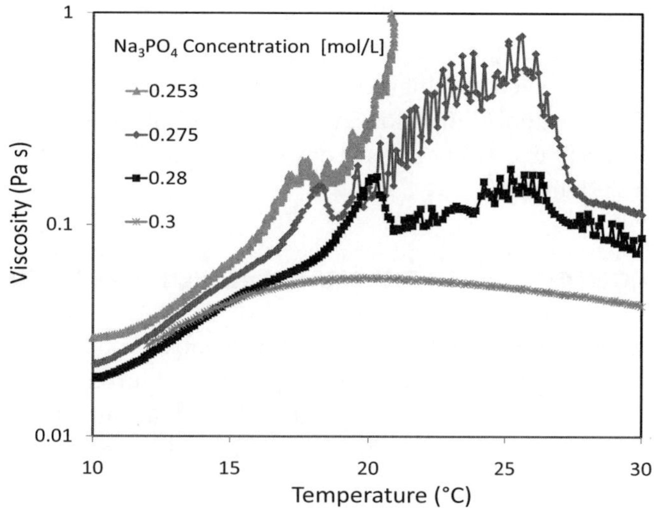

Figure 6.3. Viscosity as a function of temperature for high concentrations of sodium phosphate in a 15 wt% Pluronic F127 solution in water from cone and plate viscometry at controlled shear stress (0.6 Pa s).[110] For color reference, see page 266.

reveals a slight increase in viscosity between T = 18°C and T = 27°C at $C_{sp} = 0.253$ mol/L to no significant change in viscosity at $C_{sp} = 0.300$ mol/L. The apparent slight increase in viscosity between these two sodium phosphate concentrations including the high noise level of the measurement might indicate the formation of a very soft gel over a narrow temperature range. Therefore, above a certain concentration of salt in a 15 wt% Pluronic F127 solution, the gel formation ability is lost. This effect has been captured in the phase diagram in Fig. 6.4.

6.4.2.4 Biocompatibility of Pluronic Solutions

Pluronic F127 solutions have been extensively studied as a potential drug delivery system due to their excellent biocompatibility[59] and ease of injection.[111] These block copolymers are among the very few synthetic polymeric materials approved by the U.S. Food and Drug Administration for use as food additives and pharmaceutical ingredients.[70] The micelles produced from Pluronic block copolymers have been found to increase the bioavailability of drugs. In recent years, these systems have been evaluated as potential candidates in gene therapies and vaccination strategies.[70] However, Khattak *et al.* showed that Pluronic F127 solutions at high concentrations, 15 to 20 wt%, can be problematic.[68] The authors measured the effect of F127 on viability and proliferation of human liver carcinoma cells at different concentrations, i.e. for both liquid and gel formulations over 5 days. Even though at low concentrations (up to 5 wt%) cells proliferated and maintained high viability, at concentrations of 10 wt% and above the authors observed a significant decrease in cell viability, and no cell proliferation was evident after 5 days, while high concentrations (15 to 20 wt%), and in the gel phase, Pluronic F127 was reported to lead to complete cell death within 5 days[68]; however, addition of hydrocortisone led to a significant improvement in cell viability. In a different study, Wasan *et al.* showed that high concentration of Pluronic F127 led to notable toxicity after intraperitoneal injection in rats.[59] Since Pluronic gels release drugs within periods rarely exceeding several days, they can still be considered a good material for drug delivery.[52]

6.5 FLOW CONTROL USING THERMALLY RESPONSIVE FLUIDS

The following simple and effective microflow control concepts use thermally responsive fluids, in particular aqueous solutions of Pluronic, as the transport medium for the fluid or solid sample of interest. That means that the thermally responsive fluid is present as the carrier fluid in all channels of the microfluidic system.

6.5.1 *Active Valving*

6.5.1.1 *Principle of an Active Thermal Hydrogel Valve*

In a very simple valving approach, a flow conduit is blocked by transforming the thermally responsive fluid into a hydrogel within the microchannel. Heaters can be integrated into the channel network so that selective activation of a particular heater locally generates heat, which increases the fluid temperature beyond the gel formation temperature of the thermally responsive fluid. This then leads to localized gel formation as illustrated in Fig. 6.5. As the fluid is present everywhere in the microfluidic network, localized heating and subsequent gel formation can be

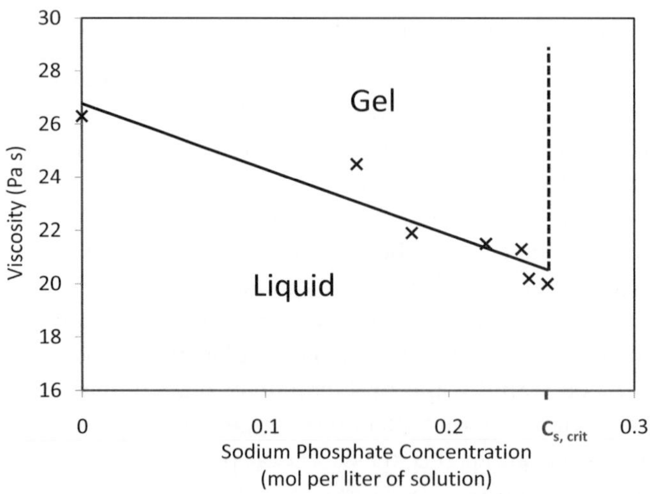

Figure 6.4. Phase diagram for a 15 wt% Pluronic F127 solution in water and sodium phosphate.[110]

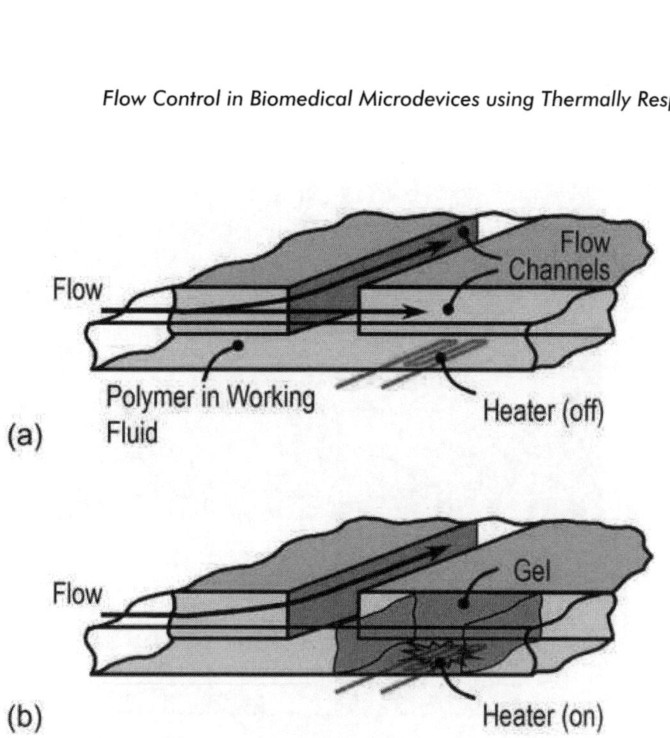

Figure 6.5. A schematic illustration of the principle of an active valve using a thermally responsive fluid in a microfluidic network. (a) Fluid from one channel is diverted into two channels at a channel bifurcation; (b) activating an integrated heater leads to localized gel formation in the corresponding microchannel, which subsequently blocks this channel to flow. For color reference, see page 266.

achieved in an arbitrary location and at any time in the microfluidic system.[72] As gel formation of Pluronic solutions is completely reversible, the hydrogel returns to its liquid phase upon cooling, and the gel block disappears, opening the channel again to flow.

6.5.1.2 Microdevice Design and Fabrication

The device shown in Fig. 6.6 was fabricated using standard MEMS technology. The fabrication process for this test device should be considered as one possible example process. When designing a specific process for a given application, process compatibility will other devices of a mictoTAS needs to be considered as well.

Figure 6.7 shows the schematic of the cross section of the device in Fig. 6.6. The silicon substrate with the heaters and the glass substrate with the flow channels are fabricated separately, and bonded together using anodic bonding. A silicon wafer is coated with a thin film (0.2 μm) of silicon nitride using low pressure chemical vapor deposition (LPCVD). This film is patterned by photolithography and plasma etching to open areas that will be used for routing of electrical leads and heaters. A subsequent silicon wet etch forms 4.5 μm deep recesses in the silicon substrate, following which the silicon nitride hard mask is removed by plasma etching. Another silicon nitride layer for electric insulation is deposited onto the substrate using LPCVD, followed by an LPCVD thin film of doped polysilicon with a sheet

(a)

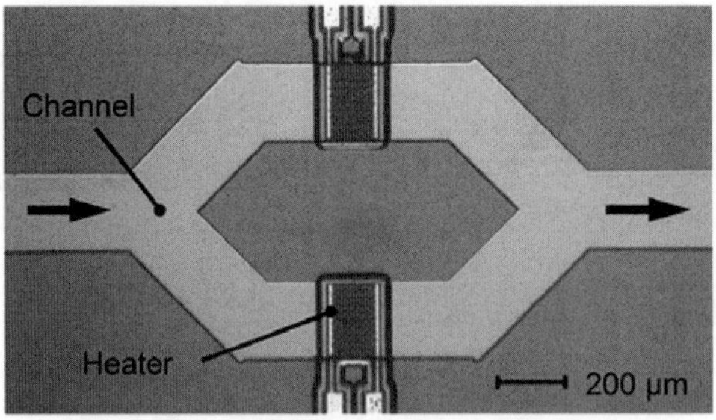

(b)

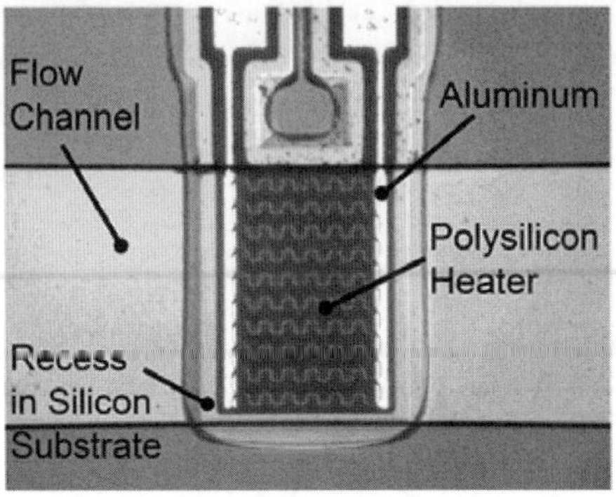

Figure 6.6. (a) Top view (through the transparent glass lid) of a microfluidic channel system with heaters integrated into recesses in the channel bottom (silicon substrate). All flow channels are 10 μm deep and 200 μm wide. (b) Detailed view of an integrated heater with aluminum leads and several parallel polysilicon heating elements.

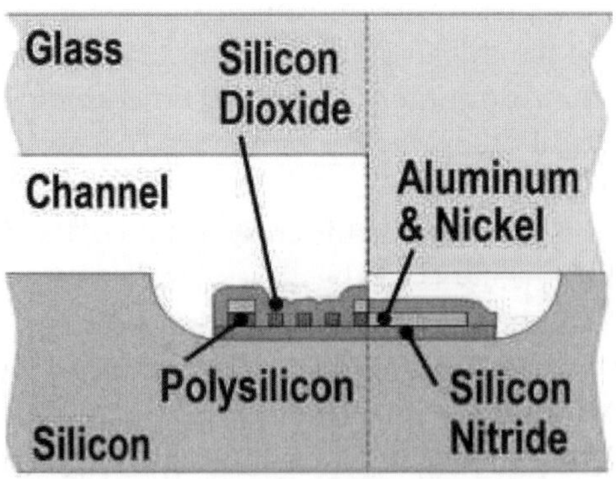

Figure 6.7. A schematic cross-section of the microfluidic device shown in Fig. 6.6. The left-hand side shows the cross-section across a heater in the channel, while the right-hand side shows a cross-section further away from the channel, through an aluminum lead. For color reference, see page 267.

resistance 210Ω. The polysilicon layer is then patterned using wet etching to form the resistive heaters, followed by a plasma etch of the silicon nitride to expose the silicon substrate in all areas that do not require electric insulation. A 0.6 μm thick film of aluminum and 40 nm of nickel are then sputter deposited for a sheet resistance of 0.18Ω and structured using wet etching, followed by sintering in forming gas. The electrical components are then encapsulated in a 0.96 μm thick layer of LPCVD silicon dioxide. The silicon dioxide is then removed from all areas except above the electrical components using wet chemistry; simultaneously, the bond pads are being exposed.

The Pyrex® glass is coated with a thin film of amorphous silicon as an adhesion layer for photoresist. After patterning the photoresist and the exposed amorphous silicon using plasma etching, the microfluidic channels are etched into the glass using wet chemistry. After removal of photoresist and amorphous silicon, both substrates are thoroughly cleaned, aligned and bonded using anodic bonding.

All flow channels measure 10 μm in depth, and the heaters have an electric resistance of 1.5 kΩ, while they span the entire width of the channel and measure about 160 μm in flow direction.

6.5.1.3 Valve Performance

A 15 wt% Pluronic F127 solution was driven through the device at a constant flow rate using a syringe pump. At an ambient temperature of 20°C a temperature increase of 7°C was required to reach the gel temperature of 27°C according to Fig. 6.1. One of the two heaters shown in Fig. 6.6 was activated by applying 100 ms long 400 mW pulses. Each pulse provided sufficient heat to raise the fluid temperature in the vicinity of the heater to form a local gel that subsequently blocked the flow channel.

For flow observation, the fluid was seeded with 0.7 μm large neutrally bouyant fluorescent polystyrene particles that were small enough not to disturb the flow field. The particles in the flow were imaged using a CCD camera and an inverted stage epifluorescence microscope to collect images of the tracer particles at a frame rate of 30 Hz. Figure 6.8 shows the fluorescent seed particles at the bifurcation of the microchannel in Fig. 6.6. The sequence of images was then used for particle image velocimetry (PIV) to determine the flow field in the microfluidic channel as a function of time.[73,74]. PIV uses the cross correlation of the pixel gray scale values of corresponding interrogation regions of image pairs. The cross correlation peak for each region then indicates the fluid displacement in this region and is be transformed into a value for the local velocity. The combined information of velocity vectors of the entire flow field then yields the velocity field.

Figure 6.9 shows the velocity field of the flow right before (a) and immediately after (b) activating the electric heater. For all experiments, valve closing occurred in less than 33 ms according to the data form PIV. Valve opening was expected to occur slower because it relies on passive heat dissipating into the silicon substrate. Despite this, the time to open the valve was also much shorter than 33 ms, the

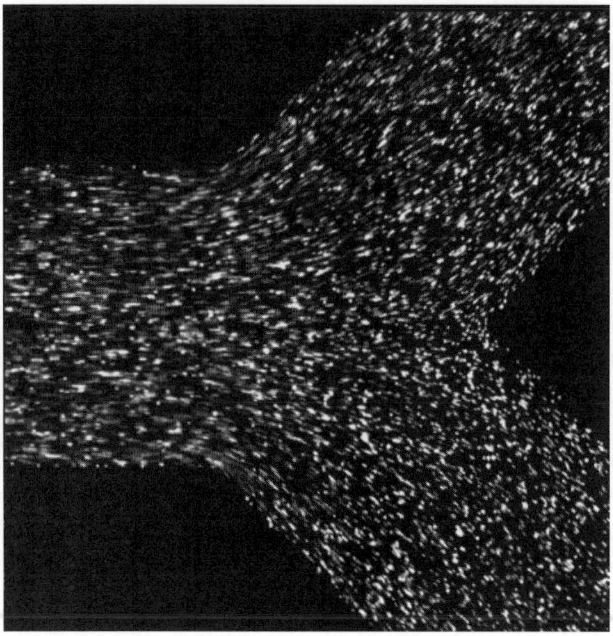

Figure 6.8. An image of the 0.7 μm large fluorescent seed particles in the microchannels shown in Fig. 6.6.

time between image frames. Valve closing and opening has been demonstrated for 0.25 - 4 Hz cycles of 100 ms to 1 s long 400 mW pulses.

The time scale for thermal diffusion across the height of the channel limits the dynamic response of this valve. Assuming the thermal diffusivity of water $\alpha = 1.41 \cdot 10^{-7}$ m^2/s as a good approximation for the thermal diffusivity of the Pluronic solution, the thermal time scale $t_d \sim H^2/\alpha$ for heat diffusion across a channel of height $H = 10$ μm is less than 1 ms as discussed in Section 6.2. Optimizing the pulse shape of the heating voltage applied to the heaters should yiels a valve responses below 1 ms.

6.5.2 Passive Valving

6.5.2.1 Concept for Passive Flow Control

In addition to active flow control in microfluidic systems using Pluronic solutions, their usefulness for passive flow control has been demonstrated as well.[26,53,72] The concept for active flow control discussed in the previous section relies on a significant temperature change of the Pluronic solution that leads to a rapid transition across the phase transition temperature, where rapid heating results in rapid gel formation, and cooling leads to liquefaction.

A concept for passive flow control using Pluronic solutions uses a small amount of heat generated in the fluid through viscous heating. Viscous heating describes heat generated through internal friction in a fluid that is in motion,

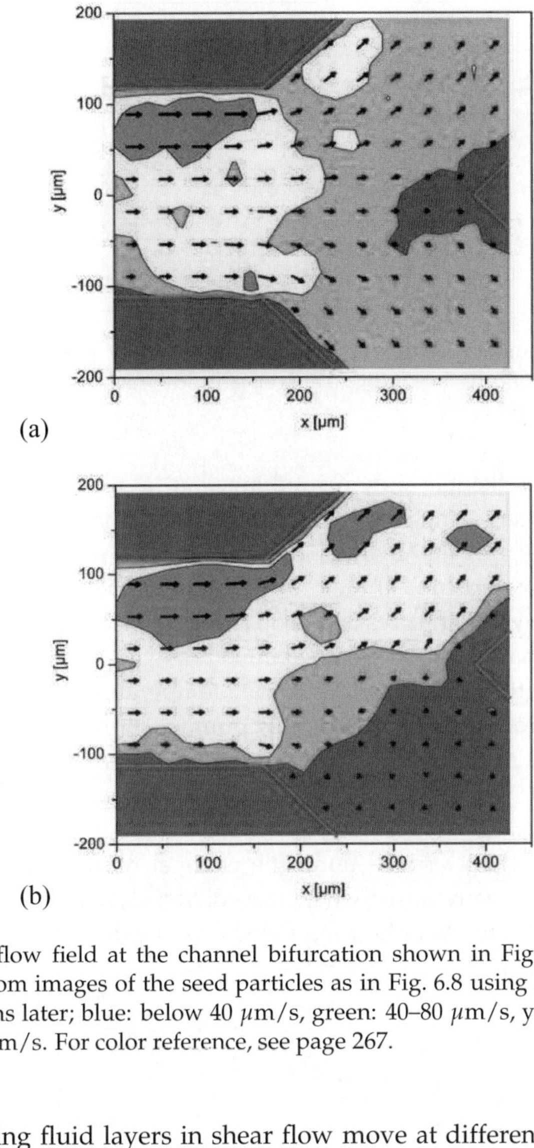

Figure 6.9. The flow field at the channel bifurcation shown in Fig. 6.6. The velocities were evaluated from images of the seed particles as in Fig. 6.8 using PIV; (a) before valve actuation, (b) 33 ms later; blue: below 40 μm/s, green: 40–80 μm/s, yellow: 80–120 μm/s, orange: 120–160 μm/s. For color reference, see page 267.

where neighboring fluid layers in shear flow move at different speeds. The rate of viscous heating Φ can be approximated by $\mu\,(\dot{\gamma})^2$ with the shear rate $\dot{\gamma}=\partial U/\partial y$ indicating the change in velocity perpendicular to the flow direction. This means that at high shear rates, heat will be produced the fluid temperature will increase. If the Pluronic solution is already near the phase boundary, only a small increase in temperature is required for a transition from the liquid phase to the soft gel phase.

6.5.2.2 Demonstration of Passive Valving

A straight microchannel with a rectangular cross section was fabricated through deep reactive ion etching (DRIE) of a trench into silicon followed by anodic bonding of a Pyrex lid to the silicon substrate. The device was connected to a

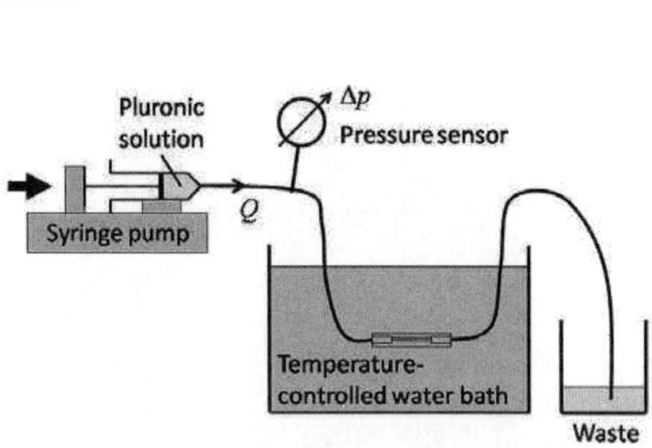

Figure 6.10. Experimental setup for the demonstration of passive flow control using thermally responsive Pluronic solutions at constant flow rate in microchannels. For color reference, see page 268.

syringe pump with flexible polymer tubing, and a gauge pressure sensor was attached between the syringe pump and the microchannel as shown in Fig. 6.10. The fluid drained from the channel through a piece of tubing to atmosphere. The gauge pressure recorded therefore corresponded to the pressure along the microchannel, neglecting the pressure drop along the wide flexible tubing on either end of the microchannel. The microchannel was submerged in a temperature-controlled water bath to achieve well-defined temperature conditions.

A 13 wt% Pluronic F127 solution was pumped through the microchannel at a flow rate $Q = 0.5\ \mu\text{L/min}$ using a syringe pump. The device temperature was maintained at 27°C with the temperature-controlled water bath, which is well below the gel formation temperature of 31°C of the Pluronic solution. After initiating the flow, the pressure along the channel reached a constant value as shown in Fig. 6.11. The pressure then increased, first slowly, and then more rapidly at a nearly constant rate. After holding ice against the device, the pressure rapidly decreased.

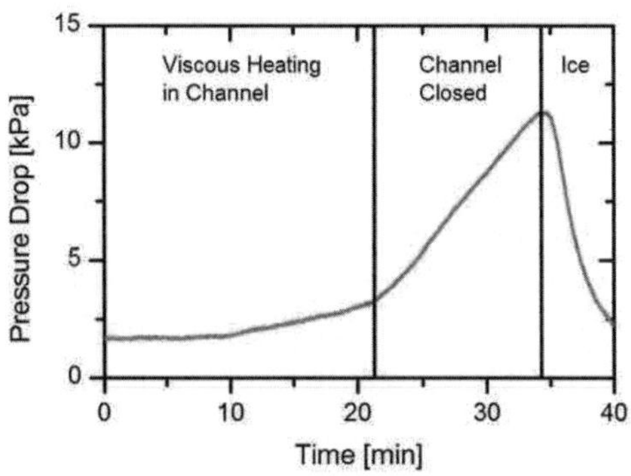

Figure 6.11. Pressure recorded along a 10 mm long microchannel with rectangular cross section (150 μm × 100 μm) for a 13 wt% Pluronic F127 solution at 27°C driven through the channel at a flow rate 0.5 μL/min. At around 34 minutes, ice was held against the device to lower its temperature.

6.5.2.3 Discussion of the Passive Flow Control Mechanism

The pressure readings shown in Fig. 6.11 and their significance for the performance of the device can be interpreted as follows: Viscous heating in the high-shear regions near the channel walls led to localized gel formation, which slowly reduced the channel cross section and led to a slow increase in pressure along the channel, from around 10 minutes to around 20 minutes in Fig. 6.11. At 22 minutes, the flow channel was entirely blocked, and the subsequent rapid increase in pressure at nearly constant rate corresponds to fluid accumulating in the flexible tubing upstream from the blocked channel. This interpretation is supported by the fact that cooling the device using ice led to a rapid decrease in pressure, as a result of the gelled fluid in the microchannel turning liquid and releasing the pressure through the channel.

The mechanism of flow-induced gel formation can also be explained through scaling arguments. For flow in a pipe of diameter D, the maximum strain rate $\dot{\gamma} \sim Q/D^3$ occurs at the wall and is proportional to the volumetric flow rate Q. For a constant flow rate, the rate of viscous dissipation therefore becomes $\Phi \sim \mu/D^6$. This means that if viscous heating leads to gel formation on the channel wall, this reduces the effective diameter of the channel D, and the viscosity μ of the Pluronic solution also increases because of the heating. Both effects lead to more viscous heating until the entire channel is blocked. The time to complete channel blockage is reduced for a device temperature that is closer to the gel formation temperature of the solution. This is shown by the measurements in Fig. 6.12. Here, the time to complete channel blockage decreases dramatically as the temperature of the water bath is increased while the flow rate is kept constant at 1 μL/min. In addition, the time to channel blockage is decreased by a factor around 2 compared with the experiment in Fig. 6.11, which was performed at half the flow rate, corresponding to half the amount of viscous heat dissipation.

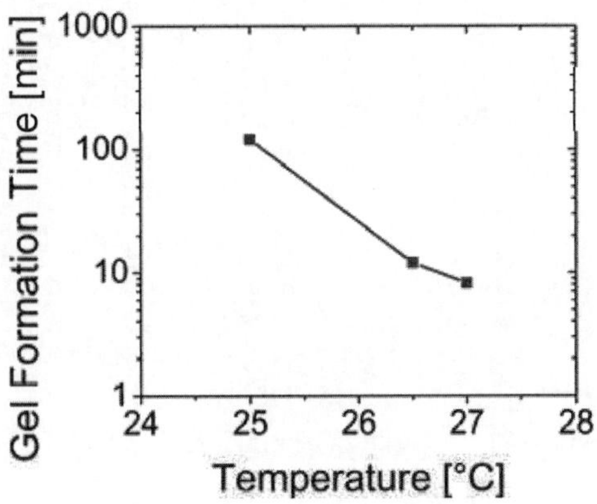

Figure 6.12. Time for complete channel blockage caused by shear-induced gel formation in a microchannel for a 13 wt% Pluronic F127 solution driven through the channel at a flow rate 1.0 μL/min at different temperatures.

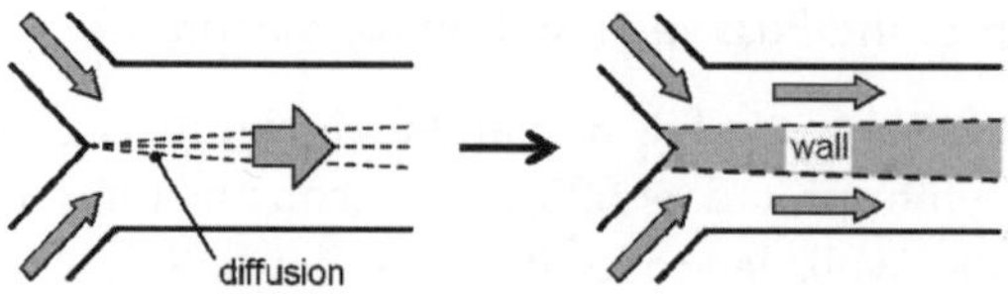

Figure 6.13. Concept of wall formation in a microfluidic channel at the diffusive interface between two fluids.

6.5.3 Cross-Channel Transport

6.5.3.1 Concept of Cross-Channel Transport using a Gel Wall

Figure 6.13 shows the concept of formation of a wall between two fluids in a microfluidic channel. Diffusion of compounds occurs across the interface between both fluids. Presence of one compound in the other stream leads to formation of a wall in the center of the channel. If further the material system and the flow condition are chosen appropriately, this wall can be removed on one side and regenerated on the other side at a constant rate leading to a steady displacement of the wall perpendicular to the flow velocity of the fluid. During wall generation, if objects such as compounds, particles or cells are incorporated into the wall, these objects are slowly moved from one fluid stream into the other stream.

6.5.3.2 Generation of a Gel Wall in a Microchannel

Figure 6.14 shows a microfluidic device with two inlet channels, one for a Pluronic F127 solution and one for a saline solution. The channels were fabricated in

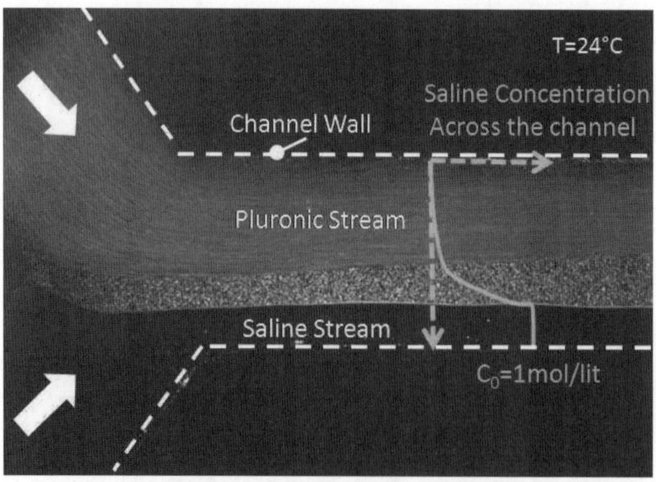

Figure 6.14. Top view of a microfluidic device. Pluronic solution and saline solution are introduced in a 10 μm high microchannel at $T_a = 24°$C. Gel formation occurs in the center of the channel. The Pluronic stream is seeded with fluorescent particles. For color reference, see page 268.

polydimethylsiloxane (PDMS) using standard soft lithography techniques; then the PDMS was bonded to a glass substrate.

The 15 wt% Pluronic solution has a gel temperature of 26.5°C, which is above the ambient temperature T_a = 24°C. A gel wall forms in the center of the channel where sodium phosphate from the saline stream diffuses into the Pluronic stream. This leads to a locally reduced gel formation temperature for the Pluronic solution according to the phase diagram in Fig. 6.4. Fig. 6.14 also shows the distribution of sodium phosphate ions across the channel. The ions diffuse from the saline stream across the gel wall into the Pluronic solution. The ion concentration in the wall near the Pluronic stream is the minimum concentration at which gel formation can occur at ambient temperature. The concentration inside the wall near the saline stream corresponds to the critical sodium phosphate concentration above which gel formation cannot occur. At steady state, sodium phosphate diffuses from the saline stream into the Pluronic stream at a constant flux. This increases the sodium phosphate concentration in the wall near the saline stream above the critical concentration and therefore the saline side of the wall liquefies. At the same time, the steady diffusion of ions into the Pluronic stream increases the ion concentration in this stream near the wall above the necessary concentration for gel formation at ambient temperature $T_a = 24$°C. As a consequence, the wall is permanently removed near the saline solution, while it is simultaneously rebuilt in the Pluronic stream. This leads to a steady motion of the wall across the microchannel, as documented through the PIV measurement in Fig. 6.15. The cross-channel velocity of the wall is nearly constant at $v = 5$ μm/s, and it is insensitive to the magnitude of the pressure driving the fluids and the concentration of the saline solution. This suggests that the motion of the wall is defined by the gel removal rate near the saline solution.

Figure 6.16 shows the wall thickness along the flow channel for different experimental conditions. In all cases, the wall thickness increases almost linearly

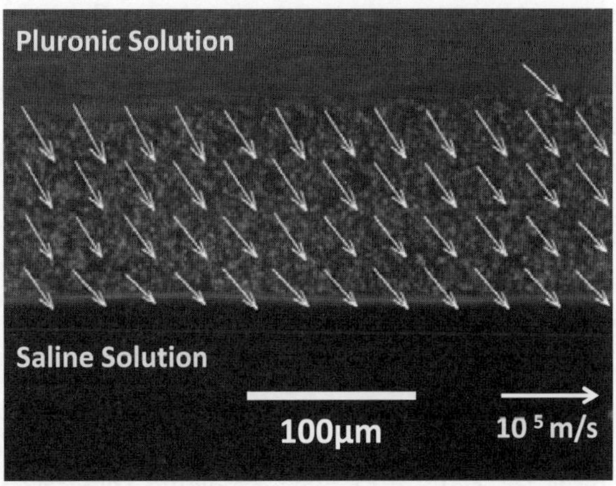

Figure 6.15. Velocity field of a section of the wall from PIV.

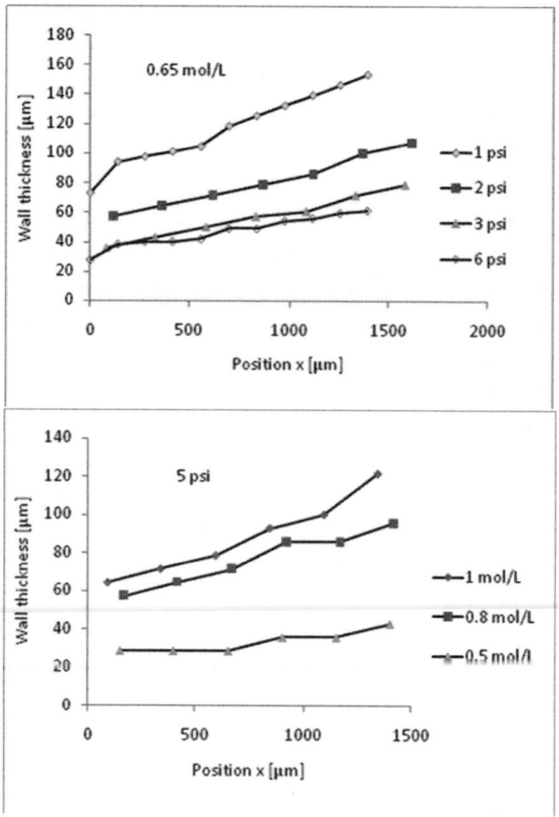

Figure 6.16. Thickness of the gel wall along the channel for different experimental conditions. For color reference, see page 269.

over the first 1.5 mm along the channel. The wall thickness is smaller for higher pressures driving the two fluids for the same sodium phosphate concentration in the saline stream. Higher pressures lead to faster flow and therefore to a faster convective transport of ions with the Pluronic stream after diffusing across the wall. As a consequence, gel formation in the Pluronic stream occurs further downstream compared to a slower flow. This leads to the formation of a thinner gel wall. Similarly, a higher sodium phosphate concentration leads to a thicker gel wall. Here, a larger ion concentration variation across the channel leads to a higher flux of ions into the Pluronic stream and therefore to a wider wall.

6.5.3.3 Discussion of the Wall Motion

While the wall moves at a constant velocity across the channel, the thickness of the wall can be set through the flow rate of the fluids and through the saline concentration. This allows selecting a well-defined transit time for objects transported across the wall. This is especially attractive for biological objects such as living cells that are transported from the Pluronic stream into the saline stream. During this transition, the cells experience an increasing concentration of sodium phosphate, until they arrive in an environment with the maximum concentration. Controlling the transit time for these cells therefore defines the temporal rate of increase in ion concentration to which the cells are being exposed. This concept therefore represents a tool to control the rate of change of ion concentration around living cells for investigating cell response to temporal ion gradients. In addition, this concept can be used to limit the rate of change in ion concentration that a cell is submitted to in order to prevent cell damage.

A similar influence of ions on the gel formation temperature of Pluronic solutions has been observed for other compounds than sodium phosphate. Correspondingly, other compounds can be employed to achieve a similar cross-channel transport as described here for sodium phosphate, so that this concept is not limited to the system Pluronic/sodium phosphate.

6.6 CONCLUSIONS

Three different flow control mechanisms based on thermally responsive polymer solutions have been discussed, all of which are based on the presence of the thermally responsive polymer Pluronic in the working fluid of the microfluidic system. All of these concepts use very simple fabrication techniques that allow easy integration into lab-on-a-chip devices. All concepts include no moving parts making them very robust to failure.

First, selective heating of specific regions in microfluidic devices was used to reversibly transform the polymer solution into a gel, which then blocked the microfluidic channel. This concept was demonstrated with a heating element integrated into the wall of a 10 μm deep channel. The fast valve response (1 ms possible) was hereby limited by thermal diffusion. Future extension of this concept to deeper channels and faster valve response could possibly be achieved with Joule heating within the fluid.

Passive flow control was achieved for ambient temperatures slightly below the gelation temperature of the Pluronic solution. Viscous heating generated within the flow led to successive gel formation inside a microfluidic channel until the channel was blocked to flow. Channel blockage occurred faster for ambient temperatures closer to the gel temperature. This effect could serve as a mechanism for automatic flow dosage in response to the ambient temperature.

Furthermore, a cross-channel transport concept has been developed in which a gel wall is formed in the center of a microfluidic channel at the diffusive interface between a Pluronic solution and a sodium phosphate solution. The wall is generated on one of its sides and it is dissolved on its second side, leading to a steady motion across the channel. During wall generation, particles or living cells can be included in the wall, which then move at the same velocity as the wall and are transported from one fluid stream into the second fluid stream. It is possible to maintain a concentration gradient of diffusive compounds across the wall and the transition time can be adjusted through the flow rates of the fluid streams and the concentration of the compound that leads to a lower gel formation temperature of the Pluronic solution (here: sodium phosphate).

The careful design of a microfluidic system for use of a simple thermally responsive fluid such as an aqueous Pluronic solution allows a wide range of flow control concepts for biomedical applications. However, all concepts presented here require the polymer Pluronic be present in the working fluid of the microfluidic system. Future work could therefore explore hydrogel-based flow control concepts that allow separating the hydrogel from the fluid to be controlled.

References

1. E. M. Purcell, Electricity and magnetism, *American Journal of Physics*, **34**, 830 (1966).
2. T. M. Squires and S. R. Quake, Microfluidics: Fluid physics at the nanoliter scale, *Reviews of Modern Physics*, **77**, 977–1026 (2005).
3. H. Bruus, *Theoretical Microfluidics*, Oxford University Press, USA (2008).
4. P. Gravesen, J. Branebjerg and O. S. Jensen, Microfluidics—a review, *J. Micromech. Microeng.*, **3**, 168 (1993).
5. G. Karniadakis, A. Besk'ok and N. R. Aluru, *Microflows and Nanoflows: Fundamentals and Simulation*, Springer Verlag (2005).
6. T. S. Wiedmann, H. Herrington, C. Deye and D. Kallick, Distribution and diffusion of sodium taurocholate and egg phosphatidylcholine aggregates in rat intestinal mucin, *Pharm. Res.*, **18**, 1489–1496 (2001).
7. P. Tabeling, *Introduction to Microfluidics*, Oxford University Press, USA (2005).
8. J. M. Ottino and S. Wiggins, Introduction: mixing in microfluidics, *Philosophical Transactions A*, **362**, 923 (2004).
9. R. F. Ismagilov, A. D. Stroock, P. J. A. Kenis, G. Whitesides and H. A. Stone, Experimental and theoretical scaling laws for transverse diffusive broadening in two–phase laminar flows in microchannels, *Appl. Phys. Lett.*, **76**, 2376 (2000).
10. N. T. Nguyen and Z. Wu, Micromixers—a review, *J. Micromech. Microengineering*, **15**, R1 (2005).
11. V. Hessel, H. L'owe and F. Sch'onfeld, Micromixers—a review on passive and active mixing principles, *Chemical Engineering Science*, **60**, 2479–2501 (2005).
12. S. Shoji and M. Esashi, Microflow devices and systems, *J. Micromech. Microengineering*, **4**, 157 (1994).
13. D. Bosch, B. Heimhofer, G. M'uck, H. Seidel, U. Thumser and W. Welser, A silicon microvalve with combined electromagnetic/electrostatic actuation, *Sensors and Actuators A: Physical*, **37**, 684–692 (1993).
14. I. Chakraborty, W. Tang, D. Bame and T. Tang, MEMS micro-valve for space applications, *Sensors and Actuators A: Physical*, **83**, 188–193 (2000).
15. C. A. Rich and K. D. Wise, A high-flow thermopneumatic microvalve with improved efficiency and integrated state sensing, *J. Microelectromech. Syst.*, **12**, 201–208 (2003).
16. A. A. Deshmukh, D. Liepmann and A. P. Pisano, Characterization of a micro–mixing, pumping, and valving system, in *Proceedings of the 11th International Conference on Solid State Sensors and Actuators (Transducers' 2001)*, 10–14 (2001).
17. M. Koch, A. Evans, A. Brunnschweiler, A. Evans and A. Brunnschweiler, *Microfluidic Technology and Applications*. Research Studies Press Baldock (2000).
18. M. J. Madou, *Fundamentals of Microfabrication: The Science of Miniaturization*. CRC (2002).

19 W. S. N. Trimmer, Microrobots and micromechanical systems, *Sensors and Actuators,* **19**, 267–287 (1989).

20 E. C. Peters, F. Svec and J. M. J. Fréchet, Thermally responsive rigid polymer monoliths, *Adv. Mater.,* **9**, 630–633 (1997).

21 C. Yu, S. Mutlu, P. Selvaganapathy, C. H. Mastrangelo, F. Svec and J. M. J. Fréchet, Flow control valves for analytical microfluidic chips without mechanical parts based on thermally responsive monolithic polymers, *Anal. Chem.,* **75**, 1958–1961 (2003).

22 R. Liu, Q. Yu, J. Bauer, J. Moore and D. Beebe, Hydrogel microvalves fabricated using in–situ polymerization, in *Technical Digest of the Solid–State Sensor and Actuator Workshop, Hilton Head Island, SC, USA,* 2000, 222–226.

23 E. T. Carlen and C. H. Mastrangelo, Electrothermally activated paraffin microactuators, *Microelectromechanical Systems, Journal of,* **11**, 165–174 (2002).

24 E. T. Carlen and C. H. Mastrangelo, Surface micromachined paraffin—actuated microvalve, *Microelectromechanical Systems, Journal of,* **11**, 408–420 (2002).

25 J. Melin, G. Gimenez, N. Roxhed, W. van der Wijngaart and G. Stemme, A passive 2-dimensional liquid sample micromixer, in *Proceedings of Micro Total Analysis Systems,* **3**, 5–9 (2004).

26 B. Stoeber, C. M. J. Hu, D. Liepmann and S. J. Muller, Passive flow control in microdevices using thermally responsive polymer solutions, *Phys. Fluids,* **18**, 053103 (2006).

27 K. W. Oh and C. H. Ahn, A review of microvalves, *J. Micromech. Microengineering,* **16**, R13 (2006).

28 C. B. Liu, C. Y. Gong, Y. F. Pan, Y. D. Zhang, J. W. Wang, M. J. Huang, Y. S. Wang, K. Wang, M. L. Gou and M. J. Tu, Synthesis and characterization of a thermosensitive hydrogel based on biodegradable amphiphilic PCL-Pluronic (L35)—PCL block copolymers, *Colloids Surf. Physicochem. Eng. Aspects,* **302**, 430–438 (2007).

29 E. S. Gil and S. M. Hudson, Stimuli—reponsive polymers and their bioconjugates, *Progress in Polymer Science,* **29**, 1173–1222 (2004).

30 R. Yoshida, K. Sakai, T. Okano and Y. Sakurai, Pulsatile drug delivery systems using hydrogels, *Adv. Drug Deliv. Rev.,* **11,** 85–108 (1993).

31 A. S. Hoffman, Applications of thermally reversible polymers and hydrogels in therapeutics and diagnostics, *J. Controlled Release,* **6**, 297–305 (1987).

32 L. E. Bromberg and E. S. Ron, Temperature-responsive gels and thermogelling polymer matrices for protein and peptide delivery, *Adv. Drug Deliv. Rev.,* **31**, 197–221 (1998).

33 M. Malmsten and B. Lindman, Self-assembly in aqueous block copolymer solutions, *Macromolecules,* **25**, 5440–5445 (1992).

34 K. Zhang and A. Khan, Phase behavior of poly (ethylene oxide)-poly (propylene oxide)-poly (ethylene oxide) triblock copolymers in water, *Macromolecules,* **28**, 3807–3812 (1995).

35 B. H. Lee, Y. M. Lee, Y. S. Sohn and S. C. Song, A thermosensitive poly (organophosphazene) gel, *Macromolecules,* **35**, 3876–3879 (2002).

36 R. Kishi, H. Ichijo and O. Hirasa, Thermo-responsive devices using poly (vinyl methyl ether) hydrogels, *J. Intell. Mater. Syst. Struct.,* **4**, 533 (1993).

37 A. S. Huffman, A. Afrassiabi and L. C. Dong, Thermally reversible hydrogels: II. Delivery and selective removal of substances from aqueous solutions, *J. Controlled Release,* **4**, 213–222 (1986).

38 A. Suzuki and T. Tanaka, Phase transition in polymer gels induced by visible light, (1990).

39 F. Iemma, U. G. Spizzirri, F. Puoci, R. Muzzalupo, S. Trombino, R. Cassano, S. Leta and N. Picci, pH-Sensitive hydrogels based on bovine serum albumin for oral drug delivery, *Int. J. Pharm.*, **312**, 151–157 (2006).

40 B. Johnson, D. Niedermaier, W. Crone, J. Moorthy and D. Beebe, Mechanical properties of a pH sensitive hydrogel, in *Society for Experimental Mechanics, Annual Conference Proceedings*, (2002).

41 M. Lei, A. Salim, R. A. Siegel and B. Ziaie, A hydrogel—actuated microvalve for smart flow control, in *26th Annual International Conference of the IEEE Engineering in Medicine and Biology Society (IEMBS'2004)*, pp.2041–2044 (2004).

42 A. Richter, G. Paschew, S. Klatt, J. Lienig, K. F. Arndt and H. J. P. Adler, Review on hydrogel–based pH sensors and microsensors, *Sensors*, **8**, 561–581 (2008).

43 J. Y. Park, H. J. Oh, D. J. Kim, J. Y. Baek and S. H. Lee, A polymeric microfluidic valve employing a pH–responsive hydrogel microsphere as an actuating source,' *J. Micromech Microengineering*, **16**, 656 (2006).

44 R. H. Liu, Q. Yu and D. J. Beebe, Fabrication and characterization of hydrogel-based microvalves, *Microelectromechanical Systems, Journal of*, **11**, 45–53 (2002).

45 A. Baldi, Y. Gu, P. E. Loftness, R. A. Siegel and B. Ziaie, A hydrogel-actuated environmentally sensitive microvalve for active flow control, *Microelectromechanical Systems, Journal of*, **12**, 613–621 (2003).

46 D. J. Beebe, J. S. Moore, J. M. Bauer, Q. Yu, R. H. Liu, C. Devadoss and B. H. Jo, Functional hydrogel structures for autonomous flow control inside microfluidic channels, *Nature*, **404**, 588–590 (2000).

47 T. Tanaka, Collapse of gels and the critical endpoint, *Phys. Rev. Lett.*, **40**, 820–823 (1978).

48 W. Kuhn, O. K'unzle and A. Katchalsky, Verhalten polyvalenter Fadenmolekelionen in L'osung, *Helv. Chim. Acta*, **31**, 1994–2037 (1948).

49 W. Kuhn, B. Hargitay, A. Katchalsky and H. Eisenberg, Reversible dilation and contraction by changing the state of ionization of high–polymer acid networks, (1950).

50 K. F. Arndt, D. Kuckling and A. Richter, Application of sensitive hydrogels in flow control, *Polym. Adv. Technol.*, **11**, 496–505 (2000).

51 R. Bashir, J. Hilt, O. Elibol, A. Gupta and N. Peppas, Micromechanical cantilever as an ultrasensitive pH microsensor, *Appl. Phys. Lett.*, **81**, 3091 (2002).

52 K. Al–Tahami and J. Singh, Smart polymer based delivery systems for peptides and proteins, *Recent Patents on Drug Delivery and Formulation*, **1**, 65–71 (2007).

53 B. Stoeber, D. Liepmann and S. Muller, Microflow control using thermally responsive triblock copolymers, in *Proceedings of the Seventh International Symposium on Micro Total Analysis Systems*, 183–186 (2003).

54 A. Richter, D. Kuckling, S. Howitz, T. Gehring and K. F. Arndt, Electronically controllable microvalves based on smart hydrogels: magnitudes and potential applications, *J. Microelectromech. Syst.*, **12** (2003).

55 A. Richter, S. Howitz, D. Kuckling, K. Kretschmer and K. F. Arndt, Automatically and electronically controllable hydrogel based valves and microvalves–design and operating performance, in *Macromolecular Symposia*, 447–456 (2004).

56 J. Wang, Z. Chen, M. Mauk, K. S. Hong, M. Li, S. Yang and H. H. Bau, Self–actuated, thermo–responsive hydrogel valves for lab on a chip, *Biomed. Microdevices*, **7**, 313–322 (2005).

57 C. Yu, S. Mutlu, P. Selvaganapathy, C. H. Mastrangelo, F. Svec and J. M. J. Fréchet, Flow control valves for analytical microfluidic chips without mechanical parts based on thermally responsive monolithic polymers, *Anal. Chem.*, **75**, 1958–1961 (2003).

[58] M. Suzuki, Amphoteric polyvinyl alcohol hydrogel and electrohydrodynamic control method for artificial muscles, *Polymer Gels: Fundamentals and Biomedical Applications*, pp. 221 (1991).

[59] K. M. Wasan, R. Subramanian, M. Kwong, I. J. Goldberg, T. Wright and T. P. Johnston, Poloxamer 407–mediated alterations in the activities of enzymes regulating lipid metabolism in rats, *J. Pharm. Pharm. Sci.*, **6**, 189–197 (2003).

[60] K. Kajiwara and S. B. Ross–Murphy, Synthetic gels on the move, *Nature*, **355**, 208–209 (1992).

[61] Y. Osada, H. Okuzaki and H. Hori, A polymer gel with electrically driven motility, (1992).

[62] Y. Ueoka, J. Gong and Y. Osada, Chemomechanical polymer gel with fish–like motion, *J. Intell. Mater. Syst. Struct.*, **8**, 465–471 (1997).

[63] K. Park and H. Park, Smart hydrogels, *Concise Polymeric Materials Encyclopedia*, 1476–1478 (1999).

[64] K. Park, W. S. W. Shalaby and H. Park, *Biodegradable Hydrogels for Drug Delivery*. CRC (1993).

[65] H. Ichijo, O. Hirasa, R. Kishi, M. Oowada, K. Sahara, E. Kokufuta and S. Kohno, Thermo–responsive gels, *Radiat. Phys. Chem.*, **46**, 185–190 (1995).

[66] B. Jeong, S. W. Kim and Y. H. Bae, Thermosensitive sol–gel reversible hydrogels, *Adv. Drug Deliv. Rev.*, **54**, 37–51 (2002).

[67] H. Okuzaki and Y. Osada, A chemomechanical polymer gel with electrically driven motility, *J. Intell. Mater. Syst. Struct.*, **4**, 50 (1993).

[68] S. F. Khattak, S. R. Bhatia and S. C. Roberts, Pluronic F127 as a cell encapsulation material: utilization of membrane–stabilizing agents, *Tissue Eng.*, 11, 974–983 (2005).

[69] Y. Osada and M. Hasebe, Electrically activated mechanochemical devices using poly-electrolyte gels, *Chem. Lett.*, **14**, 1285–1288 (1985).

[70] X. Xiong, K. Tam and L. Gan, Synthesis and aggregation behavior of pluronic F127/Poly (lactic acid) block copolymers in aqueous solutions, *Macromolecules*, **36**, 9979–9985 (2003).

[71] Y. Shirasaki, H. Makazu, K. Tashiro, S. Ikeda, T. Sekigushi, S. Shoji, S. Tsukita and T. Funatsu, A novel biomolecule sorter using thermosensitive hydrogel in micro flow system, in *Proceedings of the Micro Total Analysis Systems*, 925–927 (2002).

[72] B. Stoeber, Z. Yang, D. Liepmann and S. J. Muller, Flow control in microdevices using thermally responsive triblock copolymers, *Microelectromechanical Systems, Journal of*, **14**, 207–213 (2005).

[73] C. Willert and M. Gharib, Digital particle image velocimetry, *Exp. Fluids*, **10**, 181–193 (1991).

[74] J. Westerweel, Fundamentals of digital particle image velocimetry, *Measurement Science and Technology*, **8**, 1379 (1997).

[75] J. W. Gardner, V. K. Varadan and O. O. Awadelkarim, *Microsensors, MEMS, and Smart Devices*. John Wiley & Sons (2001).

[76] T. Kissel, Y. Li and F. Unger, ABA-triblock copolymers from biodegradable polyester A-blocks and hydrophilic poly (ethylene oxide) B-blocks as a candidate for in situ forming hydrogel delivery systems for proteins, *Adv. Drug Deliv. Rev.*, **54**, 99–134 (2002).

[77] B. Jeong, Y. H. Bae, D. S. Lee and S. W. Kim, Biodegradable block copolymers as injectable drug–delivery systems, *Nature*, **388**, 860–862, (1997).

[78] E. Ruel–Gariépy and J. C. Leroux, In situ-forming hydrogels-review of temperature-sensitive systems, *European Journal of Pharmaceutics and Biopharmaceutics*, **58**, 409–426, (2004).

79 B. Jeong, D. S. Lee, J. I. Shon, Y. H. Bae and S. W. Kim, Thermoreversible gelation of poly (ethylene oxide) biodegradable polyester block copolymers, *Journal of Polymer Science Part A: Polymer Chemistry,* **37**, 751–760 (1999).

80 M. Ebara, M. Yamato, M. Hirose, T. Aoyagi, A. Kikuchi, K. Sakai and T. Okano, Copolymerization of 2-carboxyisopropylacrylamide with N-isopropylacrylamide accelerates cell detachment from grafted surfaces by reducing temperature, *Biomacromolecules,* **4**, 344–349 (2003).

81 K. Nakajima, S. Honda, Y. Nakamura, F. López-Redondo, S. Kohsaka, M. Yamato, A. Kikuchi and T. Okano, Intact microglia are cultured and non-invasively harvested without pathological activation using a novel cultured cell recovery method, *Biomaterials,* **22**, 1213–1223 (2001).

82 M. Yamato, C. Konno, A. Kushida and M. Hirose, Release of adsorbed fibronectin from temperature-responsive culture surfaces requires cellular activity, *Biomaterials,* **21**, 981–986 (2000).

83 M. A. Nandkumar, M. Yamato, A. Kushida, C. Konno, M. Hirose, A. Kikuchi and T. Okano, Two-dimensional cell sheet manipulation of heterotypically co-cultured lung cells utilizing temperature-responsive culture dishes results in long-term maintenance of differentiated epithelial cell functions, *Biomaterials,* **23**, 1121–1130 (2002).

84 K. Uchida, K. Sakai, E. Ito, O. Hyeong Kwon, A. Kikuchi, M. Yamato and T. Okano, Temperature-dependent modulation of blood platelet movement and morphology on poly (N–isopropylacrylamide)-grafted surfaces, *Biomaterials,* **21**, 923–929 (2000).

85 A. Chilkoti, M. R. Dreher, D. E. Meyer and D. Raucher, Targeted drug delivery by thermally responsive polymers, *Adv. Drug Deliv. Rev.,* **54**, 613–630 (2002).

86 D. Meyer, B. Shin, G. Kong, M. Dewhirst and A. Chilkoti, Drug targeting using thermally responsive polymers and local hyperthermia, *J. Controlled Release,* **74**, 213–224 (2001).

87 B. Jeong, K. M. Lee, A. Gutowska and Y. H. An, % Thermogelling biodegradable copolymer aqueous solutions for injectable protein delivery and tissue engineering, *Biomacromolecules,* **3**, 865–868 (2002).

88 H. Schild, Poly (N-isopropylacrylamide): experiment, theory and application, *Progress in Polymer Science,* **17**, 163–249 (1992).

89 T. M. Swager and K. Kuroda, Thermally responsive polymers and their applications, (2003).

90 Y. Qiu and K. Park, Environment-sensitive hydrogels for drug delivery, *Adv. Drug Deliv. Rev.,* **53**, 321–339 (2001).

91 A. Gutowska, J. Seok Bark, I. Chan Kwon, Y. Han Bae, Y. Cha and S. Wan Kim, Squeezing hydrogels for controlled oral drug delivery, *J. Controlled Release,* **48**, 141–148 (1997).

92 M. Bohorquez, C. Koch, T. Trygstad and N. Pandit, A study of the temperature–dependent micellization of Pluronic F127, *J. Colloid Interface Sci.,* **216**, 34–40, (1999).

93 W. Brown, K. Schillen and S. Hvidt, Triblock copolymers in aqueous solution studied by static and dynamic light scattering and oscillatory shear measurements: influence of relative block sizes, *J. Phys. Chem.,* **96**, 6038–6044 (1992).

94 B. Jeong, Y. Choi, Y. Bae, G. Zentner and S. Kim, New biodegradable polymers for injectable drug delivery systems, *J. Controlled Release,* **62**, 109–114 (1999).

95 B. Jeong, Y. H. Bae and S. W. Kim, Drug release from biodegradable injectable thermosensitive hydrogel of PEG–PLGA–PEG triblock copolymers, *J. Controlled Release,* **63**, 155–163 (2000).

96 M. Guzálin, M. Aberturas, F. Garcia and J. Molpeceres, Gelatine gels and polyoxyethylene-polyoxypropylene gels: Comparative study of their properties, *Drug Dev. Ind. Pharm.,* **20**, 2041–2048 (1994).

97 B. K. Lau, Q. Wang, W. Sun and L. Li, Micellization to gelation of a triblock copolymer in water: Thermoreversibility and scaling, *Journal of Polymer Science Part B: Polymer Physics,* **42**, 2014–2025 (2004).

98 J. Escobar-Chávez, M. López-Cervantes, A. Na'ık, Y. Kalia, D. Quintanar-Guerrero and A. Ganem-Quintanar, Applications of thermo–reversible pluronic F–127 gels in pharmaceutical formulations, *J. Pharm. Pharmaceut. Sci.,* **9**, 339–358 (2006).

99 T. J. Pappas and L. A. Holland, Fluid steering in a microfluidic chip by means of thermally responsive phospholipids, *Sensors Actuators B: Chem.,* **128**, 427–434, (2008).

100 C. Wu, T. Liu, B. Chu, D. K. Schneider and V. Graziano, Characterization of the PEO-PPO-PEO triblock copolymer and its application as a separation medium in capillary electrophoresis, *Macromolecules,* **30**, 4574–4583 (1997).

101 S. Senkow, S. K. Mehta, G. Douhéret, A. H. Roux and G. Roux-Desgranges, Aqueous solutions of some amphiphilic poly (ethylene oxide)–poly (propylene oxide)-poly (ethylene oxide) triblock copolymers. A thermodynamic study over a wide concentration range at temperatures between 288.15 and 328.15 K, *Physical Chemistry Chemical Physics,* **4**, 4472–4480 (2002).

102 Y. Su, J. Wang and H. Liu, Formation of a hydrophobic microenvironment in aqueous PEO– PPO– PEO block copolymer solutions investigated by Fourier Transform Infrared Spectroscopy, *J Phys Chem B,* **106**, 11823–11828 (2002).

103 A. Caragheorgheopol, H. Caldararu, I. Dragutan, H. Joela and W. Brown, Micellization and micellar structure of a poly (ethylene oxide)/poly (propylene oxide)/poly (ethylene oxide) triblock copolymer in water solution, as studied by the spin probe technique, *Langmuir,* **13**, 6912–6921, (1997).

104 E. Ricci, M. Bentley, M. Farah, R. Bretas and J. Marchetti, Rheological characterization of Poloxamer 407 lidocaine hydrochloride gels, *European Journal of Pharmaceutical Sciences,* **17**, 161–167 (2002).

105 N. Pandit, T. Trygstad, S. Croy, M. Bohorquez and C. Koch, Effect of salts on the micellization, clouding, and solubilization behavior of pluronic F127 Solutions, *J. Colloid Interface Sci.,* **222**, 213–220 (2000).

106 Y. S. Rhee, Y. H. Shin, C. W. Park, S. C. Chi and E. S. Park, Effect of flavors on the viscosity and gelling point of aqueous poloxamer solution, *Arch. Pharm. Res.,* **29**, 1171–1178 (2006).

107 B. C. Anderson, S. M. Cox, A. V. Ambardekar and S. K. Mallapragada, The effect of salts on the micellization temperature of aqueous poly (ethylene oxide)–b–poly (propylene oxide)–b–poly (ethylene oxide) solutions and the dissolution rate and water diffusion coefficient in their corresponding gels, *J. Pharm. Sci.,* **91**, 180–188 (2002).

108 M. Almgren, W. Brown and S. Hvidt, Self–aggregation and phase behavior of poly (ethylene oxide)–poly (propylene oxide)–poly (ethylene oxide) block copolymers in aqueous solution, *Colloid & Polymer Science,* **273**, 2–15 (1995).

109 E. B. Jorgensen, S. Hvidt, W. Brown and K. Schillen, Effects of salts on the micellization and gelation of a triblock copolymer studied by rheology and light scattering, *Macromolecules,* **30**, 2355–2364 (1997).

110 V. Bazargan and B. Stoeber, Formation of temporary separating walls in microfluidic devices, in *6th International Conference on Nanochannels, Microchannels, and Minichannels (ICNMM2008),* (2008).

111 A. V. Kabanov, E. V. Batrakova and V. Y. Alakhov, Pluronic®block copolymers as novel polymer therapeutics for drug and gene delivery, *J. Controlled Release,* **82**, 189–212, (2002).

Chapter Seven

Application of MEMS in Drug Delivery: The Dynamic Between Biocompatibility and Biofunctionality

Gabriela Voskerician

Department of Biomedical Engineering, Case Western Reserve University, Cleveland, Ohio-44122, USA
E-mail: gxv4@case.edu

The 21st century approach to patient pharmacological treatment is a projection of our consumer oriented society. The patient population demands more efficient and effective therapies, with less hospitalization time and fewer physician visits, and the opportunity to lead a normal life, reflected by limited therapy related constraints. Achieving such goals significantly surpasses the capabilities of traditional oral or injectable treatments. Such need opens the avenue to drug delivery solutions that present versatile, dynamic, and reliable control over prescribed conventional treatments. The present work intends to address the critical aspects of biocompatibility and biofunctionality as they relate to the development of novel drug delivery solutions, here exclusively exemplified by micro-electromechanical systems for clinical therapy (BioMEMS).

7.1 INTRODUCTION

7.1.1 *Current Therapies in Drug Delivery*

Factors such as need for long-term treatment, narrow therapeutic window, complex dosing schedule, combination therapy, individualized or emergency dosing regimen, and labile active agents are presently outside of the capabilities of conventional drug administration.[1] Drug delivery systems such as tablets or injections typically result in a drug delivery profile marked by an initial sharp increase in concentration above the therapeutic range, followed by a relatively rapid decrease into the non-therapeutic range.[2] Thus, the drug concentration remains within the therapeutic range for a very short period of time, often too short, necessitating subsequent administration. In the case of potent drugs, a concentration signifi-

Biomaterials for MEMS, Edited by M. Chiao and J.-C. Chiao

www.panstanford.com
978-981-4241-46-5

cantly above the therapeutic level could induce toxicity leading to serous clinical complications. In other circumstances, such as the case of administered antibiotics, a minimum concentration must be maintained for an extended period. If the antibiotic concentration falls below the acceptable therapeutic level, the intended therapeutic effect is not reached.

Many of the novel drugs made available to the patient population, if delivered orally or through an injection regiment must be formulated by taking into account the limitations of such treatment routes.[3] The traditional oral administration leads to poor site availability of the active drug component due to:

(a) exposure to the harsh intestinal environment and potential for proteolytic enzyme degradation,
(b) loss of drug into the intestinal lumen reducing the drug availability at the level of the intestinal epithelium,
(c) further reduction in active drug availability as it travels through the circulatory system, followed by subsequent diffusion through various tissue before it reaches the intended location.

In the case of injectible administration, the exposure to the harsh intestinal environment and drug loss in the lumen are certainly avoided, however, reduction of drug availability as a result of its path through the circulatory system and subsequent tissues until reaching the intended location still represents a significant limitation. Further, patient compliance for injectible regiments is generally low, an overwhelming preference being given to oral administration which does not require a regular hospital/physician's office visit or pain associated with intravenous administration.

7.1.2 The BioMEMS Solution

The current drug therapy limitations are being addressed through the development of medical devices, such as MEMS, which can protect labile active ingredients, precisely control drug release kinetics, deliver multiple doses, and eliminate frequent injections.[1] Traditionally, BioMEMS represent the family of medical devices manufactured using microfabrication techniques and materials similar to those used by the semiconductor industry. Presently, the BioMEMS definition has been broadened to include medical devices manufactured using adapted microfabrication techniques and alternative materials to those employed by the semiconductor industry. These techniques and materials usually fall outside of the traditional micromachining of silicon which has defined the semiconductor industry focus.

The field of controlled release has emerged to particularly address the shortcomings of oral and injectible routes of administration. It initially focused on developing systems which would lead to a nearly constant release of drug over an extended period of time, independent of local biological factors such as pH.[4] Much of this work involved the manipulation of biodegradable polymers systems to achieve the desired constant drug release profile. Several successful approaches have been commercialized in the past decade, notably Gliadel®, poly(anhydride) wafers that release Carmustine (BiCNU®) at a constant rate asthe polymer degrades, developed for the treatment of malignant brain tumors. An additional example is Norplant® which contains an anti-contraceptive drug that is being released at a constant rate via diffusion, as the polymer rods containing the drug depot degrade over time.

The technological development presently permits the access to delivery devices which ensure a constant drug release. Thus, expectations of addressing the more complex pulsatile and dynamic mode drug delivery set the stage for the next technological development in drug delivery. These needs are rooted into clinically relevant treatment regiments. Once more, the technological development took advantage of known polymeric systems and their properties to fulfill such need. Advances related to discretely modulating a polymer's mechanical properties in response to external stimuli such as change in electric or magnetic fields, exposure to ultrasound, light or enzymes, and changes in pH or temperature have lead to some success.[5,6] Unfortunately, several limitations to complete control over the release regiment must be noted:

(a) stimuli application usually leads to a "burst" in drug release, yet, removal of the source does not stop the release completely,
(b) in some cases, technology has not been capable of miniaturizing the stimulus source or such source might be too expensive for the development of a commercially viable device,
(c) the need for a completely implantable rather than trans-cutaneous drug delivery system which has the capability of delivering in a controlled and dynamic manner a desired and complex drug regiment over extended periods of time eludes the present technological reality.

Langer's research group at Massachusetts Institute of Technology (MIT) has assumed an active role in advancing a viable solution to such complex requirements. Their technological strides have materialized into using microfabrication technology to initiate the development of an implantable multi-array BioMEMS which allows for remotely activated release of complex drug regiments.

7.1.3 The Host-device Continuum

A fundamental requirement for any *in vivo* implantable drug delivery system is its reliable performance over time. A plethora of biomaterials and mechanical devices have been successfully implanted in the body for decades.[7–9] The development of biomaterials for in vivo use is proof of the ability of the body to tolerate long-term implants of non-living materials, under certain circumstances. But, in the case of BioMEMS, tolerance of the device, otherwise known as biocompatibility, represents only one aspect of the required biological performance. Normal biological reactions such as protein adsorption, cellular adhesion, and fibrous encapsulation constitute a welcome event in the case of biomaterial tissue replacements or reinforcement. However, their potential effect on the biofunctionality of the device, particularly, activation and release, must be carefully considered, as such modulating events could lead to deviations from the intended therapeutic effect.[10–12] Hence, the successful development of BioMEMS drug delivery solutions must address two fundamental aspects of the host-device continuum, the effect of the BioMEMS on the body (biocompatibility) and the impact of the body on the function of the BioMEMS device (biofunctionality).[13]

7.2 BIOMEMS IN DRUG DELIVERY: THE STATE OF THE FIELD

7.2.1 Acceptance of BioMEMS

BioMEMS represent a powerful platform for the controlled delivery of drugs. Through this avenue, potent therapeutic agents whose temporal administration is vital to their efficacy and whose effect is naturally amplified by the human body could be released in situ. The inherent material and design flexibility of such platform along with its relative ease of programmability leads to enhanced control over drug release profiles compared to traditional routes and means of drug administration. Ideally, the sensing components allow drug(s) release based on cues received from relevant physiological changes in the environment surrounding the device, such as pH, temperature, analyte concentration, etc.(Table 7.1).[1,14] However, since their initial introduction as a potential solution to versatile drug release regiments, commercially, there has been a low adoption of these devices (Table 7.1). Limitations such as inadequate understanding of clinical needs in the process of design development; technical complexity associated with material selection and fabrication process, as well as mode of action; reliable drug release profiles; and complex FDA approval process have been recognized as significant hurdles to commercialization of BioMEMS.[15]

In the process of BioMEMS medical device development for drug delivery, two distinct categories of devices have emerged: transdermal and totally implantable. Understanding and acknowledging their distinct anatomical placement in reference to the body is vital in subsequent evaluation and optimization of their biocompatibility and biofunctionality.

7.2.2 The state of the BioMEMS Field

7.2.2.1 Transdermal BioMEMS for Drug Delivery

The concept of using skin as an administration route for modern therapy management has drawn considerable attention in recent years. The option of transdermal drug delivery is attractive because it eliminates the gastrointestinal tract degradation and further liver metabolization, both effects associated with oral drug delivery. Such approach also eliminates the pain and inconvenience of the intravenous injections. A number of methods have been developed to address

Table 7.1 Adoption of BioMEMS Technology in Medical Device Commercialization.

Advantages	Challenges
• smaller devices using microfabrication techniques similar to those used to create silicon computer chips • lower cost due to batch fabrication • the ability to incorporate sensing, signal conditioning, and actuating functions in close proximity or on the same substrate • superior functionality (resolution, sensing, etc.)	• inadequate understanding of clinical needs • technical complexity associated with material selection, fabrication, incorporation of fluidics, and internal mode of action • unproved *in vivo* short and long term reliability as a result of inadequate system design requirements (has the response of the material to the body and that of the body to the material thoroughly addressed during the device design phase?) • difficulties in testing, calibration, and packaging • high entry and development costs • complex FDA approval process

Table 7.2 Transdermal BioMEMS (select examples).

Device	Principle and Materials	Designation	Ref.
Transdermal microprojection delivery system	• drug coated on solid microneedle array • titanium • laser patterned	Pase II clinical trials; delivery of PTH for osteoporosis	http://zosanopharma.com
BioMEMS microneedle array	• drug coated on solid microneedle array • silicon • etching using SF_6/O_2	Research stage	17
BioMEMS microneedle array	• drug released through hollow microneedles using a syringe • silicon • etching using SF_6/O_2	Research stage	13
BioMEMS microneedle array	• drug released through hollow microneedles using a micropump • silicon • polysilicon molding process	Research stage	19

transdermal drug delivery such as chemical penetration enhancers, iontophoresis, and sonophoresis .[15] However, the efficiency of transdermal drug delivery by these methods has been limited by the low permeability of the human skin. It has been recognized that stratum corneum, in particular, is responsible for the low drug transport efficiency through the skin due to its inherent impermeability. An approach to by-passing the trandermal drug delivery limitations imposed by the stratum corneum is the BioMEMS microneedle technology. This alternative uses microneedles to penetrate the skin past the stratum corneum (approximately 10–15 μm), thus, reaching the level of the epidermis, a significantly more desirable drug delivery location. Traditionally, a microneedle array system for drug delivery comprises of an assembly of micron-scale needles (solid or hollow) etched from silicon and able to penetrate the stratum corneum to produce micron-scale channels that project into the underlying tissue layers. [16] Alternative materials and methods to silicon etching have also been developed to control the cost associated with the clean-room environment required by silicon based technologies. Zosano Pharma has developed a microneedle array patch that allows the delivery of parathyroid hormone (PTH) at a predictable rate over a period of several hours (Table 7.2) Phase II clinical trials are in progress to investigate the efficacy of such solution in the treatment of osteoporosis. A number of academic research groups have also been involved in the development of microneedle arrays for drug delivery (Table 7.2). Xie *et al* evaluated in vitro the release of bovine serum albumin (BSA) off chitosan loaded films coated on the surface of a microneedle array.[17] Teo*et al* highlighted the difficulties in microneedle architecture in their study focused on an in vivo inslulin release microneedle array system.[18] Finally, Zahn*et al* investigated the in vitro release of a fluid through hollow microneedles providing a proof of principle of microneedle array delivery.[19] The drug release evaluations performed in all cases were on the order of several hours.

7.2.2.2 Totally Implantable Drug Delivery Microsystems

Mechano-active systems(moving parts): Pumps (P1-pg.568)

Transdermal patches operate based on the release of a fixed amount of drug at fixed intervals of time, regardless of the body physiological requirements. Thus, the need of combining miniaturization and drug therapy delivered at the optimal time and concentration to a specific site has led to the development of the BioMEMS micropumps (Table 7.3).

Medtronic had pioneered the Minimed®implantable micropump providing a pulsatile, radio-controlled injection rate into the intraperitoneal cavity. While technical information on the system is not readily available, Amer *et al* had qualified the micropump as a BioMEMS system. To ensure real-time monitoring, the sensing element and the cannula must be exchanged every three days.[20]

Alza Corporation had developed Viadur™on behalf of Crescendo Pharmaceuticals for the long term delivery of leuprolide acetate, a testosterone suppressor for the treatment of prostate cancer. A miniaturized osmotic pump

Table 7.3 Mechano-active Systems:Pumps (select examples).

Device	Principle and Materials	Designation	Ref.
Minimed®	• delivery through a catheter • materials not identified	Commercial	20
Viadur™	• delivery through an osmotic pump • titanium casing	Commercial	http://www.pslgroup.com
MIP and DebioSTAR™	• pumping mech anism which compresses the drug reservoir for controlled release • silicon and glass • piezoelectric ceramic disc • titanium connectors	Under development	http://www.debiotech.com

is fully implanted subcutaneously ensuring an up to 12 month drug release (http://www.pslgroup.com).

Finally, Debiotech's technology has materialized into several product development paths, of particular interest being MIP and DebioSTAR ™,both being envisaged as implantable systems, and, presently under development. The technology appears to be able to claims to deliver a drug over several weeks to several months. The technology allows for additional surface chemistry to be added to the membrane in order to delay the delivery over a prolonged period of time (http://www.debiotech.com).

Mechano-passive systems (no moving parts), tissue contactings ChipsRx had proposed an implantable single reservoir device employing a novel method of controlling bioavailability and pharmacokinetics. The release mechanism involved the use of electrical stimulation of a conducting polymer which would in turn control a swellable hydrogel. Unfortunately, the development of this device has been suspended (Table 7.4).

pSIVIDIA has used MEMS bulk and surface machining in addressing the controlled release features of their proprietary technology. In this case, drug holding reservoirs with membrane barriers of precise nanoporosity were created to control the drug release profiles (Table 7.4).

The BioSilicon ™proprietary technology of pSIVIDIA has materialized into Medidur™which is presently undergoing Phase III clinical trials to treat diabetic macular edema. Medidur ™is intended to deliver the fluocinolone acetonide, a corticosteroid, to the retina for up to three years. A double-blinded randomized multi center clinical trial including 956 patients from US, Canada, Europe,

Table 7.4 Mechano-passive systems (select examples).

Device	Principle and Materials	Designation	Ref.
ChipsRx	• electrical stimulation • conducting polymer/hydrogel	Development stopped	***
pSIVIDIA	• drug release through controlled nano-porosity	Clinical trials	***
MicroCHIPS	• thermal energy dissolution • titanium casing • silicon-based micro-array reservoirs	Under development	P7

and India is ongoing, with a targeted new drug application (NDA) submission in 2010. The fabrication involves silicon nanostructuring to create nano-pores within the material. The BioSilicon ™allows for the creation of pores that are tens of atoms in width. The pores can be loaded with drugs including small molecules, peptides, proteins and vaccines. The platform technology appears to allow for precise control of nanoprorosity during the nanostructuring process (http://www.pSIVIDIA.com).

In contrast, MicroCHIPS has used MEMS technology to create an array of independent reservoirs pioneering the concept of selective thermal energy dissolution of multi-array drug holding reservoirs. The membranes sealing each reservoir are composed of platinum and titanium layers which could be removed by local resistive heating from an applied current. The technology allows for a customized therapeutic regimen programmed into the device and modified as necessary.[2] The microchip, along with the wireless communication hardware, power supply, and electrical components are embedded and hermetically sealed within the device. The present overall size is that of a cardiac defibrillator, however, it is suggested that incorporation of custom electronic components would lead to a significant reduction in the device size (Table 7.4).

Mechano-passive systems (no moving parts), blood contacting Micromachining technology rooted within the semiconductor industry had been employed since early 1980s in micromachining bare metal stents for prevention of cardiovascular occlusions. Initially intended for temporary prevention of vessel closure, at the turn of the 21st century stents have become part of mainstream treatment in occlusive cardiovascular disease. Published clinical outcomes recognized that in the presence of bare metal stents (BMS), restenosis occurs as a result of neointimal hyperplasia (growth of scar tissue within the stent) due to the proliferation and migration of vascular smooth muscle cells. This phenomenon was clinically evident within the first 6–9 months after stent placement, and occured in response to strut-associated injury and inflammation.[21] In addition to restenosis, the actual stent placement procedure caused exaggerated endothelial injury and inflammation, rendering both the stent and vessel highly thrombogenic.[22,23] A fibrinogen layer covered the stent surface further inducing platelet activation and thrombosis.[23,24,25]

In 2001, drug-eluting stents (DES) were introduced as a strategy to minimize restenosis and requirement for reintervention. The currently available polymer-coated stents contain antiproliferative agents which elute locally at placement location to prevent neointimal hyperplasia. At two-year follow-up using both angiography and ultrasound, the clinical safety of DES was further established with minimal late lumen loss observed.[26] A recent pooled analysis demonstrated a 74% reduction in the risk of target lesion revascularization for both sirolimus-eluting stents (SES) and paclitaxel-eluting stents (PES) compared to BMS.[27] At present, 90% of all stents placed in the United States and Europe are DES. Despite the enthusiasm that resulted with the advent of DES, incomplete endothelialization and stentthrombosis continue to plague these devices. Despite therapeutic regimens post-placement, late stent thrombosis (LST), defined as occurring > 30 days post-stent insertion, remains a significant complication in patients with DES. Late stent thrombosis carries a 45% mortality rate.[28,29]

Conor Medsystems (CMS) had proposed a solution to enhance the performance of stent technology by developing a stent platform which is capable of flexible and controllable pharmacokinetics profiles. Micromachining technology has been used to create strategically placed drug depots within the depth of the stent wall. The deep reservoirs of CMS stent were designed to provide greater dose capacity than thin surface coatings (e.g., DES), allowing these stents to potentially deliver more drug for an extended period of time, if required. Presently, the technology is fairly new, and independent clinical feedback is further required. Recently, in a 1,700 patients clinical trial, Krucoff et al reported that there were no significant benefits to using CMS stents compared to the traditional DES.[30]

7.3 THE DYNAMIC BETWEEN BIOCOMPATIBILITY AND BIOFUNCTIONALITY

7.3.1 Overview

The onset of the inflammatory and wound healing response resulting from the placement of BioMEMS in situ affects their biofunctionality. Protein adsorption, cellular adhesion, and fibrous encapsulation are normal inflammatory and wound healing phenomena which occur following implantation of any material or device.

Consequently, the role of the *in vivo* environment as it interacts with a BioMEMS device must be understood and acknowledged. Critical medical device development limitations imposed by the intended *in vivo* placement must be recognized and incorporated early into the idea and prototype design and development. The fundamental hurdle encountered by BioMEMS in reaching a commercial position today is directly linked to the initial development of these devices, at times taking place in an alternative reality where biology is inconsequential, because the role of the biological environment seems to be neglected.

7.3.2 Tissue Biocompatibility and Effect on Biofunctionality

Inflammation, wound healing, foreign body response, and fibrosis are recognized as principal phases of the tissue or cellular host response to injury, here, as a result of BioMEMS implantation (Fig. 7.1).[31,32] Inflammation represents the reaction of the vascularized living tissue to the local injury created at the time of implantation. The subsequent onset of wound healing leads to the formation of the granulation tissue, characterized by angiogenesis and presence of fibroblasts (Fig. 7.2).[13] Finally, the foreign body reaction, represented by macrophages and foreign body giant cells (FBGC), follows leading to fibrosis and fibrous encapsulation. The fibrous tissue encapsulates the BioMEMS, isolating the implant from the local tissue environment.[12,13] The size, shape, length of time of intended use, chemical and physical properties of the BioMEMS drug delivery systems may be responsible for variations in the intensity and duration of the inflammatory and/or wound-healing process.[33–36]

The inflammation and wound healing response can be divided into several fundamental stages: acute inflammation, chronic inflammation, formation of granulation tissue, foreign body reaction, and fibrous capsule development (Fig. 7.1).[13]

The acute phase of the inflammatory response occurs immediately after tissue injury, and is of relatively short duration, lasting from minutes to days, depending on the extent of the injury. It is characterized by plasma protein adsorption and the migration of leukocytes from the microcirculation, including poly-morphonuclear

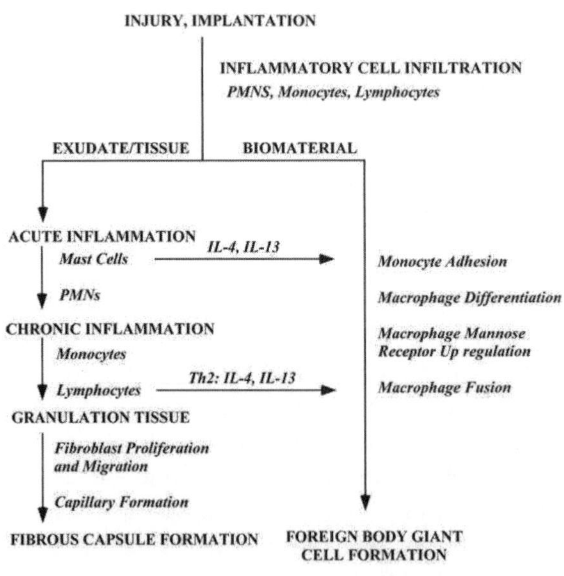

Figure 7.1. The host response to injury. The injury and the presence of the implanted device jointly induce an inflammatory and wound healing response, characterized by the following phases: acute inflammation, chronic inflammation, granulation tissue and fibrous capsule formation. During this process, macrophages and foreign body giant cells undergo "frustrated phagocytosis" in an attempt to breakdown the device.

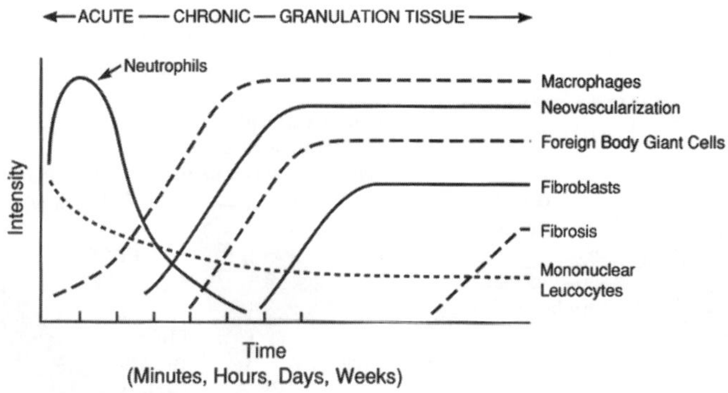

Figure 7.2. The temporal modulation of the cellularity present in the various stages of the inflammatory and wound healing response. The intensity and time variables are dependent upon the extent of the injury created by implantation and the size, shape, topography, and chemical/physical properties of the device.

leukocytes (PMN), monocytes, and lymphocytes (Fig. 7.2). These inflammatory cells actively migrate from the vasculature in response to chemotactic factors present at the implant site. Protein-rich fluid (exudate) accompanies this cellular movement. Further, increased vascular permeability facilitates this movement resulting in accumulation of cells and exudate, and it is the result of several mechanisms including endothelial contraction, cytoskeletal reorganization, leukocyte-mediated endothelial cell injury and leakage from regenerated capillaries.[13] Along with the device itself, the extent of tissue injury proportionally contributes to the overall extent of the inflammatory response. Placement is also critical, as more destruction/injury of tissue in general leads to a more active acute inflammatory phase.

The predominant cell type within the exudate during the acute phase is the poly-morphonuclear leukocyte (PMN), also called a neutrophil. Its major role is to attack and digest (phagocytose) bacteria, tissue debris and the foreign material, so that wound healing can proceed.[37–39]. Although BioMEMS are not generally phagocytosed by PMNs or macrophages due to the disparity in size, certain events in phagocytosis are known to occur. While the implant size may prevent its total ingestion by macrophages, they will attach to the device and undergo what is termed "frustrated phagocytosis".[13] This process does not involve engulfment of the BioMEMS, but does cause the extracellular release of leukocyte products (lysosomal enzymes, proteases and free radicals) in an attempt to degrade the surface contact material (Fig. 7.3).[13] In general, the number of PMNs throughout the implantation time is indicative of negative compatibility or even toxic effect induced by the device leading to unsatisfactory overall biocompatibility.[13,39] A stable large PMN population over extended periods of time suggests a continued cellular migration from the vascular system since the lifetime of the PMN is relatively short (48 hours) leading to an unresolved inflammatory phase, a sign of poor biocompatibility of the BioMEMS.

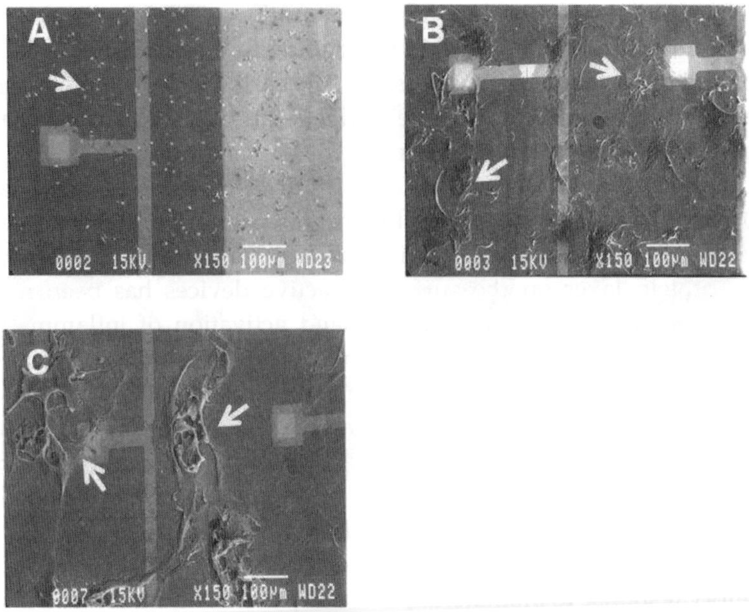

Figure 7.4. BioMEMS for drug delivery, multi-array reservoir system. The inflammatory and wound healing response induced due to the presence of an injury and device implantation is characterized by the migration of blood monocytes from the ruptured vessels to the site of implantation where they become tissue macrophages (A). Through cellular fusion (B), tissue macrophages that had adhered onto the surface of the device migrate towards each other forming foreign body giant cells (C). These cells have been known to have similar tissue characteristics as the individual adherent macrophages.

Regardless of the mode of placement, transdermal, intravascular or tissue based, the blood-device interaction is responsible for the formation of micro-thrombi and activation of complement factors as well as phagocytic cells.[49,50] The presence of the micro-thrombi, along with that of the phagocytic cells is responsible for the generation of a local environment that distorts the normal (uninjured) biochemistry.[13] Consequently, if the BioMEMS action is designed to respond to local cues in determining the volume of drug to be released, distortions in the level and presence of targeted analytes which might lead to an erroneous device response must be taken into account.

Granulation tissue is the hallmark of healing inflammation, and actually begins within 24 hours of the implantation injury. At this point, fibroblasts and vascular endothelial cells proliferate at the implant site, and begin to form a histologically pink granular tissue which is called granulation tissue. The granulation tissue displays neovascularization and a large number of proliferating fibroblasts. Macrophages are also present within granulation tissue and are considered to be central to the coordination of reorganization and reparative events due to their release of additional bioactive agents such as growth factors and cytokines.[13,51] During this stage, angiogenesis occurs by budding of pre-existing vessels, where new endothelial cells organize themselves into capillaries. Further, the fibroblasts become very active in synthesizing proteoglycans and collagen.[51]

Granulation tissue and the formation of the FBGCs characterize the next stage, which is known as the foreign body reaction. This is considered a normal wound healing response to "inert" biocompatible devices, and the absence of mononuclear cells, including lymphocytes and plasma cells, indicates the resolution of the chronic phase.[31,53] The observed foreign body reaction can be altered by the implant geometry and surface topography. The number of macrophages, FBGCs, fibroblasts and new capillaries will vary depending on surface characteristics such as smoothness, or conversely, roughness. It is probable that this reaction consisting of macrophages and FBGCs will persist at the tissue-material interface for extended periods of time (months to years).[13] The long term presence of phagocytic cells in the local environment of active devices is another factor which may be responsible for the diminished or absent long term *in vivo* performance of BioMEMS in drug delivery.

Once the local tissue inflammation response has determined that the device cannot be ingested or expelled, it is effectively walled off from the local tissue environment by complete fibrous encapsulation (Fig. 7.5).[13] The thickness of the fibrous capsule around the material has been used as a measure of the biocompatibility[13].The nature of this capsule is determined mainly by macrophage mediation in response to particulates or other agents originating from the material.

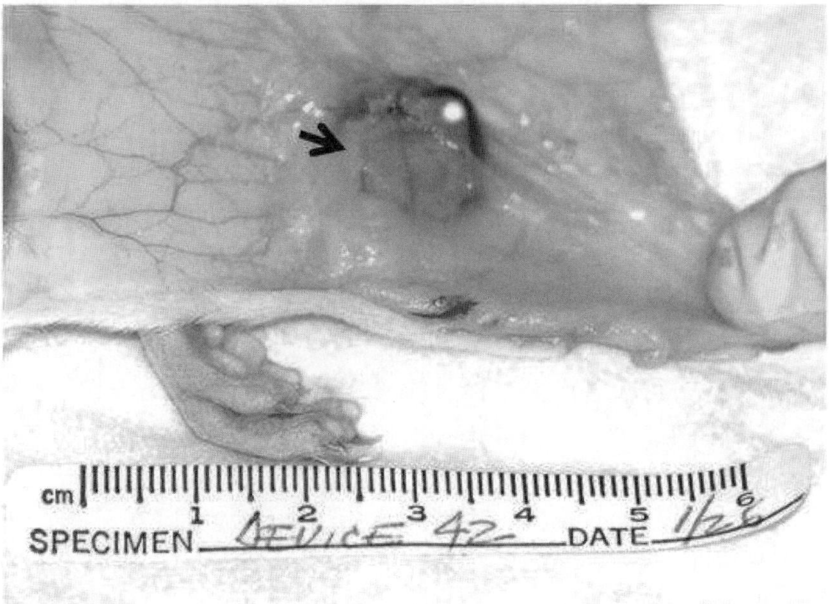

Figure 7.5. BioMEMS multi-array prototype implanted subcutaneously in a rodent model. The local environment had encapsulated the device in an attempt to isolate it from the body (28 days post-implantation). For color reference, see page 269.

Relative motion at the implant site is another aspect influencing the macrophages to produce fibrogenic agents that result in thicker capsules.[31,54] Type III collagen is predominant, especially in early stages of wound healing and its synthesis by fibroblasts is increased as the network of collagen fibers are oriented parallel to the implant surface, forming a capsule of concentric layers around the implant.[55] However, with time, the amount of type III collagen decreases as it is replaced by type I collagen, which is the primary collagen composition surrounding the implanted devices, another measure of the extent of biocompatibility.[55,56]

The nature of the tissue environment is incompletely understood, and when distorted through local exudation of fluid and addition of highly metabolizing leukocytes, tissue correlation to blood chemistry changes may be difficult to evaluate. Over extended periods of implantation, cellular adhesion, local hypoxia, and decreased pH may adversely affect the BioMEMS mode of action, either by modulating the initiation of drug release or affecting the therapeutic effect of the released drug(s)(Fig. 7.6). Granulation tissue and neovascularization may mitigate against such effects in maximizing solute/analyte supply, alternatively, fibrous capsule formation around the device may in fact partition the BioMEMS from the surrounding biological environment fully impairing its action.

It has been recognized that the mature fibrous capsule developed around any *in vivo* implanted BioMEMS drug delivery device retards the transport of even low-molecular-weight drug molecules. Sharkawy*et al.* pointed out that the fibrous capsule represents a dual transport barrier, a diffusion barrier imposed of the densely packed collagen fibers of the capsular tissue and a perfusion barrier imposed by capsule tissue avascularity.[57] First, the dense capsular tissue is responsible for significantly reducing the fluid volume available for transport and increasing the tortuosity and steric hindrance, leading to lower effective diffusion through the capsular tissue. For BioMEMS devices whose action is dependent on the presence of a certain analyte concentration to initiate drug release, the fibrous capsule induces a significant delay, at best, of the analyte to be detected/measured,

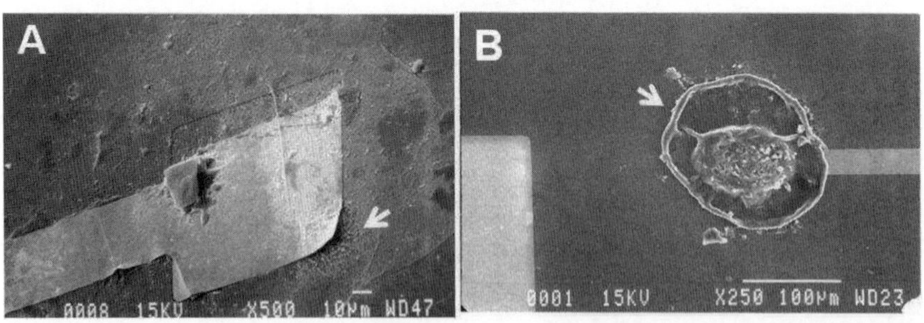

Figure 7.6. BioMEMS prototypes explanted at 21 days post-subcutaneous implantation in a rodent model: (A) the dielectric used to hermetically seal the reservoir membrane had been degraded leading to breach of the reservoir seal, thus, premature drug release; (B) a FBGC has developed atop the membrane sealing one of the drug reservoirs, thus, impairing the drug release.

as the diffusion through the fibrous capsule towards the active sensing surface is either reduced or absent.[11,57,58] Second, limited capsular vascularity results in poor drug perfusion, leading to significant increases in the delivery time, in the best of circumstances, and possible accumulation of drug within the local environment with limited/absent distribution.[59]

When tissue represents the implantation site, the surface exposed BioMEMS material directly interacts with the local biological environment being subject to sustained attack and degradation by tissue, such as hydrolytic enzymes and monocyte-induced free radical damage (Fig. 7.6).[13] Aside from affecting BioMEMS mode of action, there may be local toxicity effects of surface degradation products that have to be considered.[13]

7.3.3 Hemocompatibility and Effect on Biofunctionality

Placement of BioMEMS drug delivery systems in the blood stream leads to adsorption of proteins on the device surface, particularly fibrinogen and von Willebrand's factor, which can lead to adhesion and activation of platelets, the first step in the initiation of thrombus on the surface of any implanted material/device.[60]

The coagulation cascade is initiated by platelets deposition, recruitment, and degranulation at the surface of the biosensor, which is, then, amplified by leukocyte and erythrocyte deposition. In combination with Factor XII, the initiated coagulation creates a dense biofouling layer that could significantly diminish the intended device action. Further, surface coagulation occurs even in the presence of an anticoagulant. In addition, intravascular placement of devices leads to distorted haemodynamic flow, with local vortex formation. Similar to the intravascular effect of an anchored thrombus, the intravascular presence of a BioMEMS results in elevated shear forces and collisions by platelets that aggregate onto the surface. Over time, platelet aggregation leads to thrombus formation and possible thromboembolism, if the device is maintained in the intravascular environment for extended periods of time. In spite of all these apparently staggering drawbacks, it is the field of implantable stents for the treatment of cardiovascular disease which leads the commercialization of MEMS based therapies.

7.4 BIOMEMS DESIGN PARAMETERS AFFECTING BIOCOMPATIBILITY AND BIOFUNCTIONALITY

7.4.1 Material selection

Traditionally, the substrate used in BioMEMS fabrication has been silicon, however, presently, the material selection has been significantly expanded to include glass, metals (titanium), and polymers. The alternative material selection in the fabrication of BioMEMS has been prompted by the poor biocompatibility displayed by silicon compared to materials such as titanium, polymethylmethacrylate

(PMMA) and polydimethylsiloxane (PDMS), in addition to the silicon's limited versatility of mechanical properties.

The question then arises — what is the acceptable level of biocompatibility which should be displayed by the component materials of a BioMEMS? The simple answer to such question rests with the critical interactive aspect of each device with the local biological environment, the contacting surface. It is the device contacting surface that is responsible for transducing its structural make-up to direct or influence the response of proteins, cells, and the organism as a whole. The local biological environment reads the surface structure and responds to it. Certainly, the task becomes more difficult in the case of devices which release bi-products locally, leach substances that could affect the local or systemic environment or expose, as a result of their mechanism of action, a "new" contact surface to an active inflammatory and wound healing site. Specific to BioMEMS developed for drug delivery applications, following drug release from reservoirs, "new" surfaces are exposed to the local environment; indeed, for example, silicon walls of an empty reservoir, now exposed, represent a "new" surface to the biological environment, thus, inducing an associated inflammatory response(Fig. 7.7).

The comparative analysis on the effect of "new" surface exposure must be evaluated within the context of the miniaturized device and not normalized to an irrelevant macroscopic control device of similar surface contact material, yet, of significant size disparity. This approach becomes even more important in circumstances where arrays of BioMEMS reservoirs are opened all at once, exposing to the local biological environment a large "new" contact surface (Fig. 7.8).[56,61] The newly exposed area will induce a typical staged inflammatory response initiated by protein adsorption and cellular migration and activation, thus, enhancing the ongoing inflammatory response subsequent of the initial injury and placement of the device in the local biological environment.

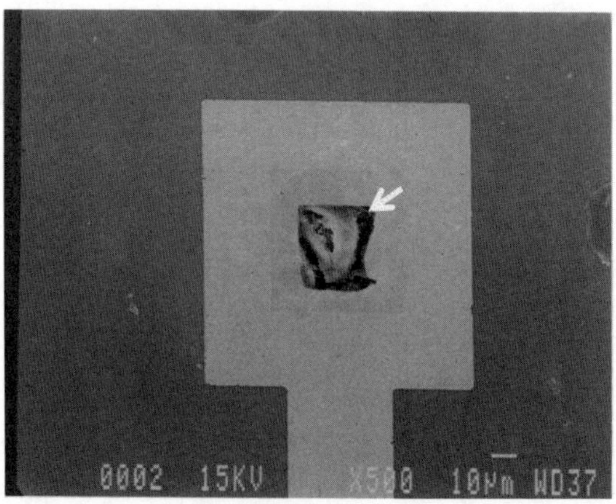

Figure 7.7. BioMEMS multi-array prototype. The membrane of a drug sealed reservoir had undergone dissolution. Following the release of the drug from the reservoir, the inner surface of the reservoir is now exposed to the local biological environment (top view).

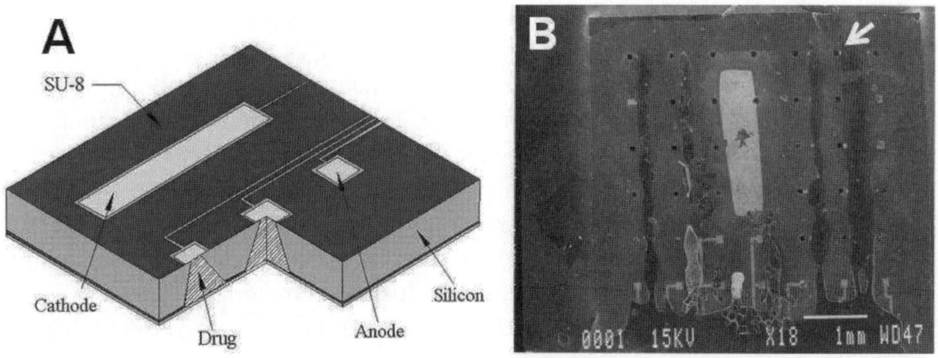

Figure 7.8. BioMEMS micro-array prototype. The inner-reservoir surface area is larger than it appears from a top view (A). The activation of the entire multi-array reservoir system leads to dissolution of the reservoir sealing membrane, release of the drug, and the onset of a "new" inflammatory response initiated by the exposed surface of the inner reservoir (B). For color reference, see page 270.

Consequently, in the design and development of BioMEMS particular attention must be given to the component materials which during the life of the device will become surface contact materials, as they could adversely affect the extent of the inflammatory and wound healing response, and, by association, modulate the therapeutic outcome of the BioMEMS outside the intended window of operation. The state of the BioMEMS field described in Section 2 of this work clearly identifies the use of titanium as a surface contacting material in devices that have reached or are very close to reaching commercialization. Historically, the inert nature of this material has rendered it useful in muskulo-skeletal applications which require good biocompatibility and enhanced load bearing support. Other materials, such as PMMA and PDMS are presently under investigation to determine their potential for combining their proven biocompatibility with MEMS based micromachining technologies.

Finally, there is no exact biocompatibility sliding scale related to the BioMEMS component materials, but, some pre-development guidance could be provided based on the knowledge acquired through previous and present investigations:

(a) Determine which material components would present surface contact exposure to the local biological environment,
(b) Evaluate the biocompatibility of BioMEMS surface contacting materials in reference to recognized biocompatible materials standards,
(c) Evaluate the in vivo inflammatory and wound healing response of the critical surface contact materials in a simple animal model, such as the subcutaneous tissue of rats.

7.4.2 Fabrication Methods

BioMEMS are fabricated by adapting techniques developed for the semiconductor industry. The platform technologies span from basic surface etching to fabrication of integrated complex programmable and structural elements. Table 7.1 outlines the advantages of MEMS based fabrication technologies. Silicon and glass have been the two major materials used in the micromachining of MEMS, mostly due to their traditional use in the semiconductor industry as well as their suitability for a variety of transducing applications.[14] Yet, the biological interface requires a level of biocompatibility which surpasses that of silicon and glass. Further, the inherent advantages of the polymer platform with respect to their material state (hard, glassy or soft, and rubbery), established sterilization methods, and flexibility of processing led to associated inclusion of materials such as silicone rubber, polycarbonate, polyimide, and isobornyl acrylate as part of the BioMEMS candidate materials.

Consequently, fabrication methods have expanded from the "hard" MEMS technologies (bulk, surface, and high aspect ratio micromachining) to "soft" manufacturing methods such as soft lithography and other polymer processing techniques.[14] The fabrication process could affect the biocompatibility of a medical device. The use of corrosive organic solvents and other processing components which could leave behind even a trace amount of highly toxic residue will affect the biocompatibility of the overall BioMEMS, if these surfaces are exposed to the local biological environment. The traditional etching process involves the use of either of the three highly biologically toxic etchants, potassium hydroxide (KOH), ethylene diamine pyrochatechol (EDP) or tetramethyl ammonium hydroxide (TMAH). These etchants attack silicon along preferred crystallographic directions. Masking materials such as silicon nitride and photoresists such as SU-8 have also been used.[14] Voskerician *et al* reported on the in vivo performance of silicon and silicon based materials in a rodent model. It was determined that silicon and SU-8 induced an elevated inflammatory and wound healing response compared to silicon nitride.[53] In contrast, "soft" micromachining of PDMS or PMMA, also known as micromolding does not require the use of highly corrosive compounds in the fabrication process. These polymers are biocompatible, cost effective, and lead to BioMEMS components with excellent sealing properties (PDMS) and elastic properties, which are absent in the case of silicon based MEMS.[14]

Thus, the choice of fabrication methods should not simply be related to the choice of material, but, also to the expectation of acceptable overall biocompatibility of the surface contacting the biological environment. As in the case of material selection, some guidance on fabrication selection vis-à-vis biocompatibility has been included:

(a) Based on material selection and surface contacting exposure, select fabrication methods which would tend to not compromise the overall biocompatibility of the BioMEMS,
(b) If a "hard" fabrication method must be employed, ensure that the final cleaning of the device removes all toxic residues or a secondary biocompatible layer is bonded at the contact surface to prevent associated toxicity, without impairing the overall activity of the BioMEMS,

(c) Evaluate in vivo, in the subcutaneous tissue of a rodent model, the top three fabrication methods intended for the surface contacting material and select the method which combines acceptable biocompatibility, reliability, and cost effectiveness.

7.5 BioMEMS IN DRUG DELIVERY: THE REALITY

Biocompatibility and biofunctionality represent the cornerstones onto which medical device development rests. The biological environment is not the culprit, it represents the defense mechanism which protects the body from harm and injury. It is time to acknowledge that the present BioMEMS technology is not developed enough to symbiotically integrate a device, from the perspective of biocompatibility and biofunctionality, within the body. Yet, the question remains — today, what can be achieved to enhance the clinical success of BioMEMS drug delivery systems?

The notion of acceptable material biocompatibility and biofunctionality changes depending on whether a device is employed as a temporary versus a permanent fixture within the biological environment. Transdermal BioMEMS represented by drug delivering microneedle array systems (Section 2.2.1) are attractive for a number of reasons. First, these devices require no surgical procedure, thus, meeting an increased patient compliance. The simple pressing of the microneedle array into the epidermis ensures intended location of delivery. Second, following the suggested treatment, which does not exceed several hours at most, the device is removed. In such circumstance less aggressive guidelines of material biocompatibility are considered compared to those in place for the totally implantable BioMEMS. The short duration that the trandermal BioMEMS are in contact with the biological environment leads to an abbreviated acute inflammatory and wound healing response, not nearly as developed and prolonged as in the case of total implantation. Voskerician et al evaluated the in vivo biocompatibility of silicon based materials and determined that when fully implanted, silicon did induce an elevated inflammatory response, but, adverse toxicity was excluded.[53] Thus, in the case of short-term use, such as the concept employed by most of the microneedle array BioMEMS, MEMS based materials and technologies are still widely acceptable as a result of the short exposure duration, even though the exposure is usually repeated (not a one-time use). Third, recognizing and understanding the acute inflammatory changes brought into the biological environment by the presence of the transdermal microneedles could be incorporated into the drug delivery treatment to ensure that therapeutic levels are not compromised and the desired biofunctionality maintained.

Disadvantages to the implementation of BioMEMS microneedle array technology involve potential fracture of the microneedles during skin penetration, thus, exposing the biological environment to a permanent foreign body presence of micro- or nano-scale which will initiate an associated inflammatory and wound healing response. Further, potential occlusion of channeled microneedle arrays during placement could lead to impaired drug release profiles, thus affecting the biofunctionality of the device.

The totally implantable devices intended for long term use in situ also present some advantages which render them attractive for specific clinical applications. While patient compliance is curbed by the prospect of implantation/explantation of non-biodegradable BioMEMS, the surgical placement of these devices at the desired anatomical location and the remote actuation aspect (transcutaneously wireless, thus no skin penetration) could enhance patient compliance. However, the biocompatibility and biofunctionality of these devices must be carefully considered, subject discussed at length in previous sections. In turn, the biofunctionality will reflect the material and fabrication selection. It has been suggested that the local release of anti-inflammatory drugs, outside or part of the intended therapeutic profile of the BioMEMS could extend the biofunctionality of the device by delaying the formation of a mature fibrous capsule by curbing the inflammatory response. Such approach would provide a solution to the difficult aspect of impaired drug transport as a result of fibrous encapsulation.

Other aspects of MEMS design must also be considered in the development of these devices, specifically, those related to proper biofunctionality. The presence/absence of moving parts within a BioMEMS system has been a controversial subject. Theoretically, the absence of moving parts (see system proposed by Langer *et al*) would ensure superior reliability compared to systems such as implantable pumps, where the performance of the device relies on concerted functionality of a number of mechanical parts. Further, elaborate delivery of drugs involving complex mechano-active systems must be housed which, certainly, increases the size of the overall BioMEMS and decreases the storage space available to various drugs. Presently, most of the MEMS based proposed solution used for clinical therapeutic applications do not advance highly complex mechano-active solutions, as already captured in Section 2.2. Yet, mechano-active systems must not be outright discredited. The complexity of the mechano-active system must be considered within the context of the solution. If such approach is justifiable and achievable, thus, providing superior biofunctionality (impeccable control over delivery profiles and over timing), then, it should be at least considered. Also associated with the development of these devices, the mode of drug release employed could also impact the biological environment. Some emerging BioMEMS technologies have proposed an electrochemical method of dissolution of the metal based membrane sealing the drug reservoir.[56,61]. The question which, then, arises is two-fold. First, do the products resulting from the dissolution process lead to an elevated local inflammatory response, supplemental to that initiated by the injury and presence of the device at the location of interest? There does not seem to be an unequivocal agreement on this matter. Certainly, it depends on the type, concentration, and extent of time that of such dissolution products are being released. Second, there is a concern that local application of a current in initiating membrane dissolution of drug release could lead to an elevated focal temperature in the area of the implant which could denature proteins, affect cellular activity, impact the therapeutic benefit of some of the drugs released, and possibly render necrotic some of the immediate tissue, much like in the case of a deep tissue burn. In the case of the technology emerging from Langer's laboratory, diligent research has shown that the concern was not founded.[61]

Finally, the development of any technology for commercialization within the healthcare domain must meet the standards set forth by the Food and Drug Administration (FDA). While it is not the intention of this work to discuss such matters, it is critical during the process of product design and development to refer to the appropriate FDA guidelines and address potential designation issues (biologic, drug or device in the context of combination products) in a timely fashion. Such approach could certainly lead to possible deviations from the original path to development and commercialization, yet, save time and resources, overall.

Indeed, the reality is that in BioMEMS development, as in the development of any medical device, a balancing act must be performed where acceptable biocompatibility and biofunctionality is more desirable than superior biocompatibility and compromised biofunctionality or vice-versa. Any medical device development involves a series of key stages that take a technology from concept to commercialization. Briefly, these stages include:

(a) Design, materialize the concept into drawings, identify preliminary materials, and functionality modes,
(b) Development, build a series of prototypes to encompass the different fabrication methods, materials, and modes of action,
(c) Testing, test the developed prototypes in a clinically relevant setting,
(d) FDA submission, submit the appropriate documentation to support the medical device designation,
(e) Early stage production, production of the medical device while seeking FDA approval,
(f) Transfer to production, once FDA approval is achieved, the manufacturing process is scaled up and full production commences,
(g) Intellectual property, while the original concept must be protected prior to initiating the aforementioned stages, supplemental intellectual property usually emerges during the product development process which should also be protected.

While the process appears logical and simple to follow, the development of any medical device is not straight forward. It is a perpetual winding road with numerous round-abouts and forks, which demands attention to detail, elegant solutions, careful planning, and clear understanding of the FDA final product designation. In particular, in the case of BioMEMS, the role of biocompatibility and biofunctionality is paramount. Thus, actively addressing specific questions and concerns early in the process of development of these devices could be effective in avoiding extensive delays and staggering costs or an unexpected FDA labeling

Table 7.5 The Road Map: Impact of Biocompatibility and Biofunctionality on the Path Towards Product Development of BioMEMS for Drug Delivery.

Stage	Task	Primary Aspect Addressed
Design	• Identify principle mode of action	• Biofunctionality
	• Identify component material	• Biocompatibility
	• Establish the role of each material within the device	• Biocompatibility
	• Identify solutions to device packaging to ensure appropriate in vivo performance and desired shelf-life	• Biofunctionality and biocompatibility
	• Identify fabrication techniques	• Biocompatibility
	• Identify list of therapeutic drugs to be tested, in the context of the desired clinical application	• Biofunctionality and biocompatibility
Development	• Select fabrication techniques which are reliable, cost effective, and scalable	• Biocompatibility
	• Identify materials in direct contact with the biological environment	• Biocompatibility
	• Select material choices by reviewing their evaluated biocompatibility within the scientific literature	• Biocompatibility
	• Refine the mode of action	• Biofunctionality
	• Select relevant therapeutic drugs	• Biofunctionality and biocompatibility
	• Create prototypes	• Biofunctionality and biocompatibility
Testing	• Test the mechanical/other mode(s) of action without and with drug loading, initially bench-top, yet a clinically relevant setting (saline/plasma solution, presence of cells atop the device, etc.)	• Biofunctionality and biocompatibility
	• Test the *in vitro* and in vivo biocompatibility of individual components intended for direct contact with the biological environment	• Biocompatibility
	• Test the *in vitro* and in vivo biocompatibility of the built prototype	• Biocompatibility
	• Initiate aging test to evaluate overall device packaging reliability, stability of loading methods, and therapeutic stability of the drug(s)	• Biofunctionality

in product designation. The successful completion of design, development, and testing stages mentioned above set the tone for the remainder of the process. Table 7.5 summarizes some important aspects of biocompatibility and biofunctionality within the pre-FDA submission process which should be addressed.

7.6 CONCLUDING REMARKS

MEMS technological advances rooted within the semiconductor industry have opened the road to the development of BioMEMS for drug delivery applications which target critical clinical needs such as diabetes, vascular disease, and cancer. Indeed, the desire of developing a treatment solution which would restore the life of patients suffering of diabetes or vascular disease, as well as those battling cancer, has motivated engineers, biologists, and physicians to put forth a concerted effort through advancing purely industrial technologies to clinical viability.

The development of many materials intended for clinical applications has found its origin in the industrial field, materials created without a biological application in mind. As these materials enter the realm of intended clinical application, selected for their flexibility of processing, mechanical properties or mechano-active/electro-chemical aspect, the questions of impact on the host response (biocompatibility) and of in situ functionality (biofunctionality) become central.

The emergence of the bionanotechnology, at which level fundamental biology is defined, advances the micro-scale to a nano-level. The nano-world relates to that of the extracellular matrix and its components, which defines the realm of tissue engineering and advances the concept of tissue regeneration. The natural progression from BioMEMS to BioNEMS (Bio-Nano-ElectroMechanical Systems) is taking place today. The bionanotechnology brings together multiple approaches, materials, and systems all directed towards the understanding, control, and construction of molecular structures exhibiting new useful properties and functions. It is the hope of the interdisciplinary scientific community that bionanotechnology will integrate these properties and functions within the local biological environment leading the effort towards advanced regenerative therapies.

References

1 Staples M., Daniel K., Cima M. J., and Langer R., Application of micro- and nano-electromechanical devices to drug delivery. *Pharm. Res.* (2006).

2 Sbiaa Z., MEMS fabricated chip for an implantable drug delivery device. *Conf. Proc. IEEE Eng. Med. Biol. Soc.* **1**, 5621–4.

3 Tao S. L., and Desai T. A., Micromachined devices: The impact of controlled geometry from cell-targeting to bioavailability. *J. Control Release*, **109**, 127–38 (2005).

4 Santini J. T., Jr, Richards A. C., Scheidt R. A., Cima M. J., and Langer R. S., Microchip technology in drug delivery. *Ann. Med.* **32**, 377–9 (2000).

5 Langer R., and Kost J., Real time response polymeric delivery systems. *Ann. NY Acad. Sci.* **618**, 330–4 (1991).

6 Edelman E. R., Kost J., Bobeck H., and Langer R., Regulation of drug release from polymer matrices by oscillating magnetic fields. *J. Biomed. Mater. Res.* **19**, 67–83 (1985).

7 Kokubo T., Kim H. M., Kawashita M., and Nakamura T., Bioactive metals: Preparation and properties. *J. Mater. Sci. Mater. Med.* **15**, 99–107 (2004).

8 Flanagan T. C., and Pandit A., Living artificial heart valve alternatives: A review. *Eur. Cell Mater.* 6, 28–45; discussion 45 (2003).

9 Bond R., Rerkasem K., AbuRahma A. F., Naylor A. R., and Rothwell P. M., Patch angioplasty versus primary closure for carotid endarterectomy. *Cochrane Database Syst. Rev.*, CD000160 (2004).

10 Wisniewski N., Moussy F., and Reichert W. M., Characterization of implantable biosensor membrane biofouling. *Fresenius J. Anal. Chem.* **366**, 611–21 (2000).

11 Wisniewski N., Klitzman B., Miller B., and Reichert W. M., Decreased analyte transport through implanted membranes: Differentiation of biofouling from tissue effects. *J. Biomed. Mater. Res.* **57**, 513–21 (2001).

12 Moussy F., and Reichert W. M., Biomaterials community examines biosensor biocompatibility. *Diabetes Technol. Ther.* 2, 473–7 (2000).

13 Voskerician G., Biological design criteria for implantable electrochemical devices, Case Western Reserve University (2004).

14 Ziaie B., Baldi A., Lei M., Gu Y., and Siegel R. A., Hard and soft micromachining for BioMEMS: Review of techniques and examples of applications in microfluidics and drug delivery. *Adv. Drug Deliv. Rev.* **56**, 145–72 (2004).

15 Polla D. L., Erdman A. G., Robbins W. P., Markus D. T., Diaz-Diaz J., Rizq R., Nam Y., Brickner H. T., Wang A., and Krulevitch P., Microdevices in medicine. *Annu. Rev. Biomed. Eng.* **2**, 551–76 (2000).

16 Coulman S., Allender C., and Birchall J., Microneedles and other physical methods for overcoming the stratum corneum barrier for cutaneous gene therapy. *Crit Rev. Ther. Drug Carrier Syst.* **23**, 205–58 (2006).

17 Xie Y., Xu B., and Gao Y., Controlled transdermal delivery of model drug compounds by MEMS microneedle array. *Nanomedicine* **1**, 184–90 (2005).

18 Teo M. A., Shearwood C., Ng K. C., Lu J., and Moochhala S., In vitro and in vivo characterization of MEMS microneedles. *Biomed. Microdevices* **7**, 47–52 (2005).

19 Zahn J. D., Deshmukh A., Pisano A. P., and Liepmann D., Continuous on-chip micropumping for microneedle enhanced drug delivery. *Biomed. Microdevices* **6**, 183–90 (2004).

20 Amer S., and Badawy W., An integrated platform for bio-analysis and drug delivery. *Curr. Pharm. Biotechnol.* **6**, 57–64 (2005).

21 Moliterno D. J., Healing Achilles–sirolimus versus paclitaxel. *N. Engl. J. Med.* **353**, 724–7 (2005).

22 Gawaz M., Neumann F. J., Ott I., May A., and Schomig A., Platelet activation and coronary stent implantation. Effect of antithrombotic therapy. *Circulation* **94**, 279–85 (1996).

23 Caramori P. R., Lima V. C., Seidelin P. H., Newton G. E., Parker J. D., and Adelman A. G., Long-term endothelial dysfunction after coronary artery stenting. *J. Am. Coll. Cardiol.* **34**, 1675–9 (1999).

24 Eagle K. A., Berger P. B., Calkins H., Chaitman B. R., Ewy G. A., Fleischmann K. E., Fleisher L. A., Froehlich J. B., Gusberg R. J., Leppo J. A., Ryan T., Schlant R. C., Winters W. L., Jr, Gibbons R. J., Antman E. M., Alpert J. S., Faxon D. P., Fuster V., Gregoratos G., Jacobs A. K., Hiratzka L. F., Russell R. O., and Smith S. C., Jr., ACC/AHA guideline

update for perioperative cardiovascular evaluation for noncardiac surgery—executive summary a report of the American College of Cardiology/American Heart Association Task Force on Practice Guidelines (Committee to update the 1996 Guidelines on Perioperative Cardiovascular Evaluation for Noncardiac Surgery). *Circulation* **105**, 1257–67 (2002).

25 Wilson S. H., Fasseas P., Orford J. L., Lennon R. J., Horlocker T., Charnoff N. E., Melby S., and Berger P. B., Clinical outcome of patients undergoing non-cardiac surgery in the two months following coronary stenting. *J. Am. Coll. Cardiol.* **42**, 234–40 (2003).

26 Sousa J. E., Costa M. A., Sousa A. G., Abizaid A. C., Seixas A. C., Abizaid A. S., Feres F., Mattos L. A., Falotico R., Jaeger J., Popma J. J., and Serruys P. W., Two-year angiographic and intravascular ultrasound follow-up after implantation of sirolimus-eluting stents in human coronary arteries. *Circulation* **107**, 381–3 (2003).

27 Hill R. A., Dundar Y., Bakhai A., Dickson R., and Walley T., Drug-eluting stents: An early systematic review to inform policy. *Eur. Heart J.* **25**, 2–19 (2004).

28 McFadden E. P., Stabile E., Regar E., Cheneau E., Ong A. T., Kinnaird T., Suddath W. O., Weissman N. J., Torguson R., Kent K. M., Pichard A. D., Satler L. F., Waksman R., and Serruys P. W., Late thrombosis in drug-eluting coronary stents after discontinuation of antiplatelet therapy. *Lancet* **364**, 1519–21 (2004).

29 Joner M., Finn A. V., Farb A., Mont E. K., Kolodgie F. D., Ladich E., Kutys R., Skorija K., Gold H. K., and Virmani R., Pathology of drug-eluting stents in humans: Delayed healing and late thrombotic risk. *J. Am. Coll. Cardiol.* **48**, 193–202 (2006).

30 Krucoff M. W., Kereiakes D. J., Petersen J. L., Mehran R., Hasselblad V., Lansky A. J., Fitzgerald P. J., Garg J., Turco M. A., Simonton C. A., 3rd, Verheye S., Dubois C. L., Gammon R., Batchelor W. B., O'Shaughnessy C. D., Hermiller J. B., Jr, Schofer J., Buchbinder M., and Wijns W., A novel bioresorbable polymer paclitaxel-eluting stent for the treatment of single and multivessel coronary disease: primary results of the COSTAR (Cobalt Chromium Stent with Antiproliferative for Restenosis) II study. *J. Am. Coll. Cardiol.* **51**, 1543–52 (2008).

31 Kirkpatrick C. J., Krump-Konvalinkova V., Unger R. E., Bittinger F., Otto M., and Peters K., Tissue response and biomaterial integration: the efficacy of in vitro methods. *Biomol. Eng.* **19**, 211–7 (2002).

32 Ratner B. D., Replacing and renewing: Synthetic materials, biomimetics, and tissue engineering in implant dentistry. *J. Dent. Educ.* **65**, 1340–7 (2001).

33 Quinn C. P., Pathak C. P., Heller A., and Hubbell J. A., Photo-crosslinked copolymers of 2-hydroxyethyl methacrylate, poly(ethylene glycol) tetra-acrylate and ethylene dimethacrylate for improving biocompatibility of biosensors. *Biomaterials* **16**, 389–96 (1995).

34 Gerritsen M., Kros A., Sprakel V., Lutterman J. A., Nolte R. J., and Jansen J. A., Biocompatibility evaluation of sol-gel coatings for subcutaneously implantable glucose sensors. *Biomaterials* **21**, 71–8 (2000).

35 Chen X., Matsumoto N., Hu Y., and Wilson G. S., Electrochemically mediated electrodeposition/electropolymerization to yield a glucose microbiosensor with improved characteristics. *Anal. Chem.* **74**, 368–72 (2002).

36 Ward W. K., Slobodzian E. P., Tiekotter K. L., and Wood M. D., The effect of microgeometry, implant thickness and polyurethane chemistry on the foreign body response to subcutaneous implants. *Biomaterials* **23**, 4185–92 (2002).

37 Wagner J. G. and Roth R. A. Neutrophil migration mechanisms, with an emphasis on the pulmonary vasculature. *Pharmacol. Rev.* **52**, 349–74 (2000).

38 Williams M. A. and Solomkin J. S. Integrin-mediated signaling in human neutrophil functioning. *J. Leukoc. Biol.* **65**, 725–36 (1999).

39 Rinder C., and Fitch J., Amplification of the inflammatory response: Adhesion molecules associated with platelet/white cell responses. *J. Cardiovasc. Pharmacol.* 27 Suppl 1, S6–12 (1996).

40 Gray J. J., The interaction of proteins with solid surfaces. *Curr. Opin. Struct. Biol.* **14**, 110–5 (2004).

41 Hoffman A. S., Non-fouling surface technologies. *J. Biomater. Sci. Polym. Ed.* **10**, 1011–4 (1999).

42 Werner C., and Jacobasch H. J., Surface characterization of polymers for medical devices. *Int. J. Artif. Organs.* **22**, 160–76 (1999).

43 Hallab N. J., Bundy K. J., O'Connor K., Clark R., and Moses R. L., Cell adhesion to biomaterials: Correlations between surface charge, surface roughness, adsorbed protein, and cell morphology. *J. Long Term Eff. Med. Implants* **5**, 209–31 (1995).

44 Courtney J. M., Lamba N. M., Sundaram S., and Forbes C. D. Biomaterials for blood-contacting applications. *Biomaterials* **15**, 737–44 (1994).

45 Ratner B. D., and Bryant S. J., Biomaterials: Where we have been and where we are going. *Annu. Rev. Biomed. Eng.* **6**, 41–75 (2004).

46 Tang L., and Eaton J. W. Fibrin(ogen) mediates acute inflammatory responses to biomaterials. *J. Exp. Med.* **178**, 2147–56 (1993).

47 Hu W. J., Eaton J. W., Ugarova T. P., and Tang L., Molecular basis of biomaterial-mediated foreign body reactions. *Blood* **98**, 1231–8 (2001).

48 Collier T. O., and Anderson J. M., Protein and surface effects on monocyte and macrophage adhesion, maturation, and survival. *J. Biomed. Mater. Res.* **60**, 487–96 (2002).

49 Kao W. J., Evaluation of protein-modulated macrophage behavior on biomaterials: Designing biomimetic materials for cellular engineering. *Biomaterials* **20**, 2213–21 (1999).

50 Hamilton J. A., Nondisposable materials, chronic inflammation, and adjuvant action. *J. Leukoc. Biol.* **73**, 702–12 (2003).

51 Janatova J., Activation and control of complement, inflammation, and infection associated with the use of biomedical polymers. *Asaio J.* **46**, S53–62 (2000).

52 Heming T. A., Dave S. K., Tuazon D. M., Chopra A. K., Peterson J. W., and Bidani A., Effects of extracellular pH on tumour necrosis factor-alpha production by resident alveolar macrophages. *Clin. Sci. (Lond)* **101**, 267–74 (2001).

53 Voskerician G., Shive M. S., Shawgo R. S., von Recum H, Anderson J. M., Cima M. J., and Langer R., Biocompatibility and biofouling of MEMS drug delivery devices. *Biomaterials* **24**, 1959–67 (2003).

54 Goodman S. B., The effects of micromotion and particulate materials on tissue differentiation. Bone chamber studies in rabbits. *Acta Orthop. Scand. Suppl.* **258**, 1–43 (1994).

55 Chavrier C. A., and Couble M. L., Ultrastructural immunohistochemical study of interstitial collagenous components of the healthy human keratinized mucosa surrounding implants. *Int. J. Oral Maxillofac. Implants* **14**, 108–12 (1999).

56 Voskerician G., Shawgo R. S., Hiltner P. A., Anderson J. M., Cima M. J., and Langer R., In vivo inflammatory and wound healing effects of gold electrode voltammetry for MEMS micro-reservoir drug delivery device. *IEEE Trans. Biomed. Eng.* **51**, 627–35 (2004).

57 Sharkawy A. A., Klitzman B., Truskey G. A., and Reichert W. M., Engineering the tissue which encapsulates subcutaneous implants. III. Effective tissue response times. *J. Biomed. Mater. Res.* **40**, 598–605 (1998).

58 Sharkawy A. A., Klitzman B., Truskey G. A., and Reichert W. M., Engineering the tissue which encapsulates subcutaneous implants. II. Plasma-tissue exchange properties. *J. Biomed. Mater. Res.* **40**, 586–97 (1998).

59 Sharkawy A. A., Klitzman B, Truskey G. A., and Reichert W. M., Engineering the tissue which encapsulates subcutaneous implants. I. Diffusion properties. *J. Biomed. Mater. Res.* **37**, 401–12 (1997).

60 Frost M. C., and Meyerhoff M. E., Implantable chemical sensors for real-time clinical monitoring: progress and challenges. *Curr. Opin. Chem. Biol.* **6**, 633–41 (2002).

61 Shawgo R. S., Voskerician G., Duc H. L., Li Y., Lynn A., MacEwan M., Langer R., Anderson J. M., and Cima M. J., Repeated in vivo electrochemical activation and the biological effects of microelectromechanical systems drug delivery device. *J. Biomed. Mater. Res. A* **71**, 559–68 (2004).

Chapter Eight

Polymer-Based Biocompatible Surface Coatings

Kai Yu, Guangzheng Gao and Jayachandran N. Kizhakkedathu*

Centre for Blood Research, Department of Pathology and Laboratory Medicine, University of British Columbia, Vancouver, BC, V6T 1Z3, Canada
E-mail: jay@pathology.ubc.ca

8.1 INTRODUCTION

The performance of a material intended for biomedical applications depends on its interfacial properties and reactions that occur when come in contact with biological fluids. Non-specific protein adsorption at the biomaterial interface is the first and critical event that initializes a cascade of host responses, including platelet activation, blood coagulation, and complement activation.[1,2] Many approaches have been used to prevent such non-specific interactions.[3–6]

Coating or immobilization of surfaces with biomacromolecules such as albumin[7–12] and anticoagulants like heparin[13–18] have been widely studied towards this purpose. Another approach to overcome this problem is to coat the surface with synthetic hydrophilic polymers[4,19] and this method has been used as a anti-fouling treatment for a number of applications including biosensors[20] and drug delivery systems.[21] It was demonstrated that such coatings frequently extend the life span of biomedical devices[22–23] and the circulation half life of drug delivery systems.[24–25]

Several factors that affect the protein-repelling properties of polymer thin films on surface include the similarity of interfacial free energies of the polymer with that of water, interaction of proteins with polymers through hydrophobic or charge interactions and environmental factors such as temperature and pH.[26–28] In the case of neutral hydrophilic polymer grafted surfaces, the steric barrier due to high conformational entropy of anchored chains is one of the contributing factor towards protein repulsion.[29–35] Other factors include, the structure of the polymer

Biomaterials for MEMS, Edited by M. Chiao and J.-C. Chiao

www.panstanford.com
978-981-4241-46-5

on the surface (linear vs. branched), chemistry of the polymers and molecular weight of the grafted chains.

This chapter is organized based on the application of different polymers and approaches used for the development of anti-fouling coatings.

8.2 NON-FOULING SURFACES BASED ON POLY(ETHYLENE GLYCOL)

Poly(ethylene glycol) (PEG) or polyethylene oxide (PEO) has been widely used for the development of non-fouling surfaces because of its capacity to resist the attachment of cells and proteins.[36] It is also nontoxic and nonimmunogenic.[36] PEG is composed of -CH_2-CH_2-O- repeating units with a hydroxyl group at each end of the polymer chain. The mechanisms of the antifouling ability of PEG are not fully understood. It is believed that the antifouling ability of PEG is due to the combined effects of steric repulsion, large excluded volume, configurational entropy, hydrated chain mobility, molecular weight and surface coverage.[4,19,37–39]

Several surface modification approaches have been reported for the development of PEG surfaces. We can classify these methods in terms of the attachment strategies; physical adsorption vs. chemisorption, self assembly vs. covalent attachment etc. Each method has its own advantages or disadvantages depending on the ease of the synthesis or its performance. The physical adsorption of PEG-containing copolymers,[40–48] oligo(ethylene glycol)-terminated self-assembled monolayers (SAMs),[49–65] grafted polymers based on PEG,[66–81] and surface initiated polymerization of a PEG-containing monomer[82–90] are the most commonly used approaches for the development of PEG surfaces. In the following section we will discuss different strategies adopted for the development of PEG grafted surfaces.

8.2.1 *Physical Adsorption of PEG-containing Copolymers*

8.2.1.1 Hydrophobic Adsorption

This method allows the attachment of PEG chains to a variety of surfaces solely based on the hydrophobicity of the material. One example is the use of triblock copolymer, polyethylene glycol-*b*-polypropylene oxide-*b*-polyethylene glycol (PEG-*b*-PPO-*b*-PEG)[40] that assemble spontaneously on hydrophobic surfaces through hydrophobic-hydrophobic interactions. This class of polymer is also known as poloxamers or pluronics and has been extensively studied[41–45] Liu and Bhatia reported an application of a commercially available, PEG-terminated triblock polymer (Pluronic^{TM}F108) to create nonadhesive domains on a variety of biomaterials that prevented cell adhesion for up to four weeks in culture medium.[45] They also demonstrated the ability to micropattern the growth of murine fibroblasts in 5% serum and retained cell-free domains for up to four weeks on polystyrene. The pluronics was also shown to adsorb to a variety of surfaces including tissue culture plates, methylated glass, silicone and polylactic-*co*-glycolic acid.[45]

8.2.1.2 Electrostatic Adsorption

In this method, electrostatic adsorption of charged polymers on oppositely charged surface was used for creating a non-fouling surface layers on metallic surfaces.[46–48] Hubbell and Spencer developed a class of copolymers based on poly(L-lysine)-*g*-poly(ethylene glycol) (PLL-*g*-PEG), which was found to spontaneously adsorb from aqueous solution onto several metal oxide surfaces, providing a high degree of resistance to protein adsorption.[46–47] The adsorbed copolymer layer formed a comb-like structure on the surface, with positively charged primary amine groups of the PLL bound to the negatively charged metal oxide surface due to electrostatic interaction, while the hydrophilic and uncharged PEG side chains were exposed to the solution phase. The resulting adsorbed layers were highly effective in reducing the protein adsorption from both individual protein solutions and blood serum. The amount of protein adsorbed from human serum on the modified surfaces was consistently below 1–2 ng cm^{-2}. Fibrinogen adsorption was reduced by 96-98% in comparison to the unmodified oxide surfaces.

8.2.2 Chemisorption of PEG Containing Thiol or Sulfide Groups

The spontaneous assembly of thiol or sulfide containing molecules on noble metal surfaces is a good method for surface modification.[49] This method has been extensively used in the modification of gold or gold coated surfaces. One great advantage of this method was the ease of preparation of surfaces; alkanethiols readily assemble when gold surface was exposed to alkanethiols either in solution or vapor phase.[49–50] Self assembled monolayers (SAMs) of alkanethiols, terminated with short oligomers of the ethylene glycol groups, ($HS(CH_2)_{11}$-$(OCH_2CH_2)_nOH$), were developed to resist the adsorption of several model proteins.[51–59]

Prime and Whitesides prepared monolayers from alkanethiols terminated with short oligomers of the ethylene glycol group ($HS(CH_2)_{11}(OCH_2$-$CH_2)_nOH$; n= 3-7) and studied their protein exclusion behavior[51] They found that these monolayers appeared to provide a general solution for controlling non-specific adsorption of proteins with a range of molecular weights and isoelectric points (pI), and under a wide range of solution conditions[52–53]

For a better understanding of the factors involved in the protein resistance of oligoether SAMs, Herrwerth and Grunze synthesized a series of oligoether-terminated alkanethiols with different oligoether backbones, different chain lengths and different alkyl terminals.[54–56] Their studies suggested that the penetration of water molecules to the interior of the SAMs was a necessary prerequisite for protein resistance. Three factors, including the hydrophilicity of the internal units, packing density and the hydrophilicity of the terminal group should be considered for making a non-fouling oligoether SAMs. It was demonstrated that an interior hydrophilic structure was necessary as evident from the fact that only the oligo(ethylene glycol) and oligo(trimethylene glycol) SAMs exhibited protein resistance. Oligo(propylene glycol) SAMs did not show protein repelling properties (Table 1). Another important factor was the packing density of SAMs.

Table 8.1 Amount of adsorbed fibrinogen on SAMs on gold and silver, measured by the ellipsometric thickness of the protein layer and normalized to the amount of fibrinogen adsorbed on a monolayer of hexadecanethiol (C16SH) on gold (=100%).[54]

	Protein Absorption[%]			Protein Absorption[%]	
	Au	Ag		Au	Ag
$EG_1OME(1)$	22	57	$EG_6OMe(10)$	0	35
$EG_2OH(2)$	0	0	$EG_6OEt(11)$	51	89
$EG_2OME(3)$	0	62	$EG_6OPr(12)$	69	100
$EG_3OH(4)$	0	0	$PRO_2OMe(13)$	52	87
$EG_3OME(5)$	0	37	$PRO_3OMe(14)$	49	80
$EG_3OEt(6)$	60	88	$PRO_4OMe(15)$	42	60
$EG_3Opr(7)$	79	93	$TRI_3OH(16)$	0	0
$EG_3OBu(8)$	77	98	$TRI_3OMe(17)$	0	52
$EG_6OH(9)$	0	0			

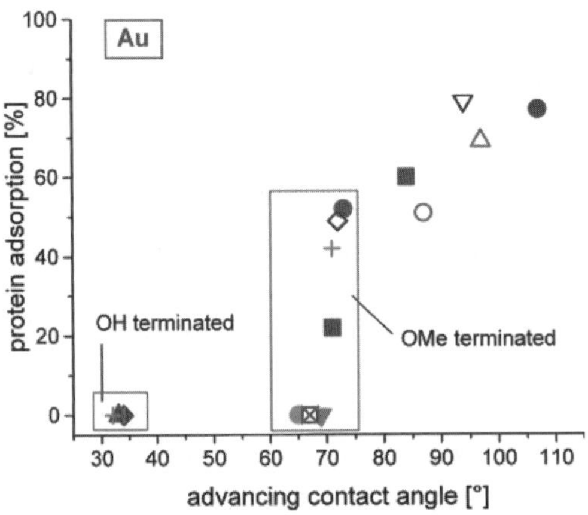

Figure 8.1. Amount of protein adsorption on a given oligoether SAM on gold normalized to the amount of protein adsorbed on a monolayer of hexa-decanethiol on gold (100%) versus advancing aqueous contact angle of the SAM. Symbols: red▲, EG_2OH; orange, EG_3OH; green+, EG_6OH; blue♦, TRI_3OH; blue ■, EG_1OMe; green▼, EG_2OMe; light blue •, EG_3OMe; red ×, EG_6OMe; blue□ , TRI_3OMe; red ×, PRO_2OMe; blue◊, PRO_3OMe; orange +, PRO_4OMe; purpl■, EG_3OEt; greeno, EG_6OEt; blue▽, EG_3OPr; green△, EG_6OPr; blue •, EG_3OBu. Images adapted from Ref. 54. For color reference, see page 270.

It was shown that SAMs with a relaxed lateral packing or with some defects or disorder was required for high protein resistance. The methoxy-terminated SAMs which form a highly ordered all-trans conformation lost non-fouling capacity due to high packing densities. The terminal groups of the SAMs were shown to play an equally important role (Fig. 8.1). For excellent protein resistance, the contact angles should not exceed 70°.[54]

Vanderah proposed a mechanism of protein rejection by oligo(ethylene glycol) modified surfaces in terms of changes in free energy of the system (ΔG_{system}, system = protein + surface) due to oligomer conformational constriction over an area greater than the contact area.[57] The adsorption of fibrinogen and bovine serum albumin onto polycrystalline Au coated with $HS(CH_2)_3O(CH_2CH_2O)_5CH_3$ was determined by surface plasmon resonance from bare Au (0% coverage) to the complete (~100% coverage) SAMs. It was found that both proteins exhibited similar adsorption curves with common onset (~60% coverage) and minimal protein adsorption. The reflection-absorption infrared spectroscopy showed that the SAMs used do not have widespread order indicating loosely packed layer but the bound molecules were uniformly distributed and fully screened the underlying substrate. Minimal protein adsorption corresponded to the maximum ΔG_{system} for a given compression. Their systematic study provides insights into the molecular level understanding of protein adsorption.

Li and Jiang investigated the effect of packing density and chains length on the anti-fouling properties of SAMs. They prepared the oligo(ethylene glycol) (OEG), $HS(CH_2)_{11}(EG)_nOH$ with different length ($n = 2, 4$, and 6), self-assembled monolayers on Au(111) surfaces.[58] Exposure of various OEG-SAMs to proteins established a correlation between OEG-SAMs packing structure on gold and protein resistance. It was shown that proteins (e.g. fibrinogen and lysozyme) were more adsorbed on the densely packed SAMs of $HS(CH_2)_{11}(EG)_2OH$ and $HS(CH_2)_{11}(EG)_4OH$. Besides the packing density, the number of ethylene glycol units was another factor influencing the protein resistance. $HS(CH_2)_{11}(EG)_6OH$ -SAMs generally resisted protein adsorption. It was suggested that both the hydration and the conformational flexibility of the ethylene glycol chains on OEG-SAMs were responsible for its non-fouling properties, consistent with the result of Grunze.

Liedberg *et al.* investigated the effect of conformation of SAMs on itsnti-fouling properties.[59] They prepared a variety of conformational gradients of alkanethiols with the structure HS-$(CH_2)_m$-X ($m = 15$; X = COOH, CH_2NH_2, or CH_3) and oligo(ethylene glycol)-terminated alkanethiols with the structures HS-$(CH_2)_{15}$-CO-NH-$(EG)_n$ ($n = 2$, 4, or 6), HS-$(CH_2)_{15}$-CO-NH-$(EG)_2$-$(CH_2)_2$-NH-CO-$(CH_2)_4$-biotin, and HS-$(CH_2)_{15}$-CO-NH-$(EG)_6$-CH_2-COOH on polycrystalline gold films. It was found that the amount of bound fibrinogen increased continuously on SAMs made from $(EG)_6$ to a significant amount on the $(EG)_2$. OEG tails with helical conformations ($(EG)_6$) appeared to be protein resistant. While OEG tails with disordered (in the mixing regime) and all-trans conformations ($(EG)_2$) seemed to have more non-specific protein adsorbption.

Although alkanethiolates have proved to be very useful surface-modifying agents for making anti-fouling surface, yet their overall impact was limited by their stability (oxidative and thermal). SAMs are fragile owing to their molecular scale thickness, defects in the SAM[60–62] and the propensity of the chemisorbed thiolate to oxidize.[63–64]

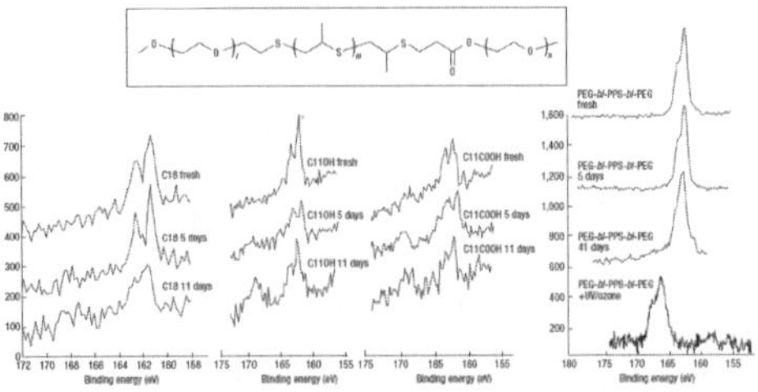

Figure 8.2. High-resolution S2p X-ray photoelectron spectra (in counts per second). These spectra are for $CH_3(CH_2)_17SH$ (1-octadecanethiol), $HOCH_2(CH_2)_{10}SH$ (mercapto-1-undecanol),and $HOOC(CH_2)_{10}SH$ (11-mercaptoundecanoic acid) exposed to ambient environmental conditions (fresh, 5 days, 11 days) and PEG-bl-PPS-bl-PEG (fresh, 5 days, 41 days and exposed to ultraviolet and ozone).Spectra comprise doublet structures due to the presence of the $S_{2p3/2}$ and $S_{2p1/2}$ peaks. In less than 2 weeks, thiols show significant oxidation, whereas the PEG-bl-PPS-bl-PEG only shows nominal oxidation near 6 weeks. The sample treated with ultraviolet and ozone demonstrates shifted oxidized S_{2p} spectra as positive control. Inset indicates chemical structure for PEG-bl-PPS-bl-PEG. Images adapted from Ref. 65.

Bearinger *et al.* explored an alternative structure to alkanethiolates, namely block copolymers comprising sulphides in a chemisorbing block (PEG_{17} $b-PPS_{25} - b-PEG_9$) where PPS = polypropylene sulfide, in an expectation of better resistance to oxidative instability.[65] These materials offer a more robust approach to passivate gold surfaces against protein adsorption and cell adhesion. The copolymer $PEG_{17} - b-PPS_{25} - b-PEG_9$ chemisorbed to form a 2.2 nm dense monolayer of 226 ± 26 ng cm^{-2}. The copolymeric monolayer was much more stable to oxidation than commonly used alkanethiolates (Fig. 8.2). Protein adsorption was greatly reduced (>95%), even when exposed to whole blood serum (>55 mg ml^{-1} protein concentration). The cell adhesion was also reduced over long culture durations (>97%) (Fig. 8.3). These results such that polysulphides containing copolymers are attractive alternatives to alkanethiolates with greater stability and allows greater flexibility in the design.

8.2.3 Covalent Grafting of Poly (ethyelene glycol)

Compared to oligo(ethylene glycol) SAMs, grafting of preformed poly(ethylene glycol) (PEG) provided more robust films and versatile architecture, and chemistry for the development of non-fouling surfaces. Several approaches have been reported for grafting PEG chains on solid surfaces such as metal, silicon, plastics to improve the biocompatibility relying on different coupling chemistry or techniques.

Unsworth and Brash used chain-end thiolated PEO assemblies to investigate the effect of PEO chain density on protein adsorption.[72] Their study showed that resistance to fibrinogen adsorption for 750 and 2000 MW (molecular weight) PEO layers increased with graft density of PEO chains and the least protein adsorption was observed at ~0.5 chains/nm^2. However, the protein adsorption increased

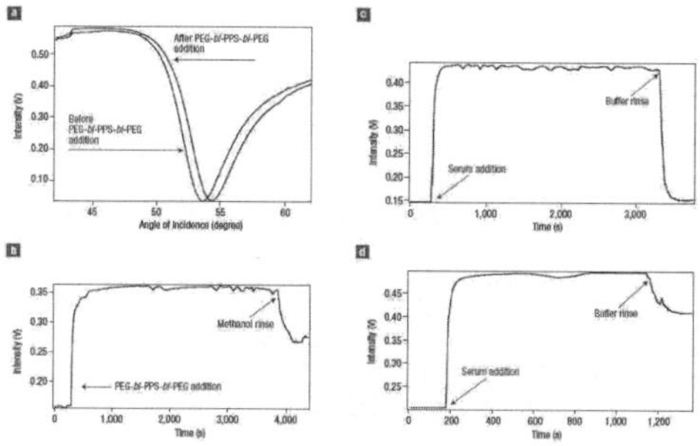

Figure 8.3. Surface plasmon resonance measurements. a, Intensity of reflected light against angle of incidence before and after deposition of PEG-*bl*-PPS-*bl*-PEG in methanol on a gold surface. Polymer adsorption induces a shift of the resonance angle,which is proportional to adsorbed mass. **b**, Shift of resonance angle observed in methanol as a function of time by monitoring the increase of reflectivity at about 1° gbelow the resonance angle.**c**,Intensity of reflected light against time for serum addition, followed by subsequent buffer wash, on PEG-*bl*-PPS-*bl*-PEG-coated gold surface.**d**, Intensity of reflected light against time for serum addition, followed by subsequent buffer wash, on gold surface. The chain density and architecture are two important factors which influence the anti-fouling properties. Images adapted from Ref. 65.

when the graft density was increased beyond 0.5 chains/nm^2. For 5000 MW PEO, the optimal chain density was found to be 0.5 chains/nm^2. Brash *et al.* suggested that at high chain density, the chemisorbed PEO was dehydrated giving a surface that was no longer protein resistant.

Sofia *et al.* tested surfaces covalently grafted with poly(ethylene oxide) polymers with different architecture, in linear and star form, on silicon membrane to determine their ability to prevent protein adsorption.[73] Modified surfaces were exposed to solutions of each of three proteins: cytochrome-*c*, albumin, and fibronectin. It was found that protein adsorption decreased almost exponentially with graft density of chains and it reached almost negligible adsorption at the highest grafting densities. In contrast to surfaces grafted with linear PEG, star PEG-grafted surfaces can prevent adsorption of larger proteins even though open spaces existed between bound molecules. The adsorption was prevented as long as the open spaces were smaller than the effective size of the protein. They also proposed a physical model of the surface to explain the protein adsorption resistance in terms of the spacing and degree of overlap of grafted PEO chains which lead to steric repulsion between surface and protein. In the case of linear PEG surfaces, a complete coverage of the surfaces with grafted polymer (chains were roughly half-overlapping (i.e., L $\approx R_G$linear; Rg = radius of gyration, L-distance between the grafting point)) was important for the prevention of protein adsorption. Since the star PEO-grafted surfaces have much more polymer segment concentration than linear chains of equivalent molecular weight or equivalent size (Fig. 8.4), they prevented the adsorption of proteins to surface more efficiently.

Zhu *et al.* reported a strategy for the formation of uniform and ultra-flat PEG monolayers on the atomically flat Si(111) surface with high grafting density

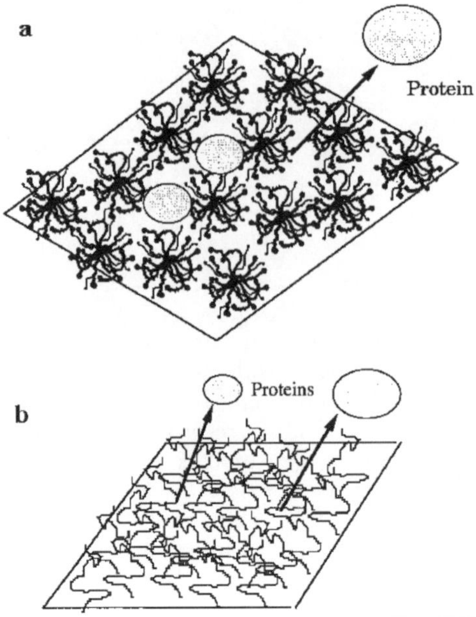

Figure 8.4. Depiction of protein adsorption on (a) PEO star and (b) linear PEG surfaces. Images adapted from Ref. 73.

(Fig. 8.5).[74] This method involved the direct interaction of the hydroxyl group on PEG with a chlorine-terminated silicon surface. It was found that with increasing molecular weight the PEG-grafting density was decreased. The measurement of film thickness revealed that surface-grafted PEG molecules were in a brush-like configuration at low molecular weight (MW = 200, 300) and a coil-like configuration at higher molecular weight (MW = 1000, 2000). These high-density PEG films showed excellent protein resistance.

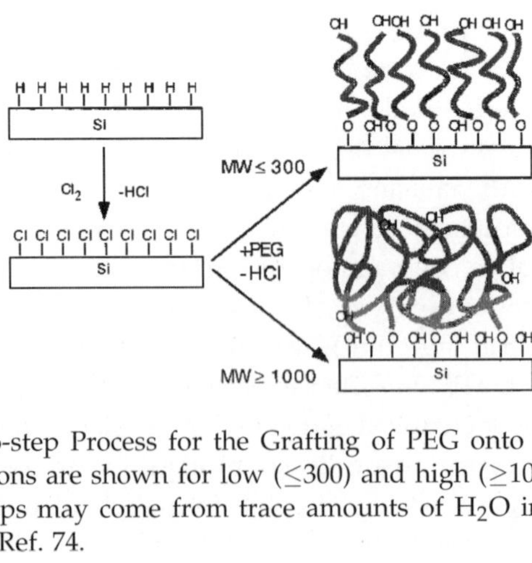

Figure 8.5. The Two-step Process for the Grafting of PEG onto Si(111). Two different molecular conformations are shown for low (≤300) and high (≥1000) molecular weights. The surface OH groups may come from trace amounts of H_2O in the reaction mixture. Images adapted from Ref. 74.

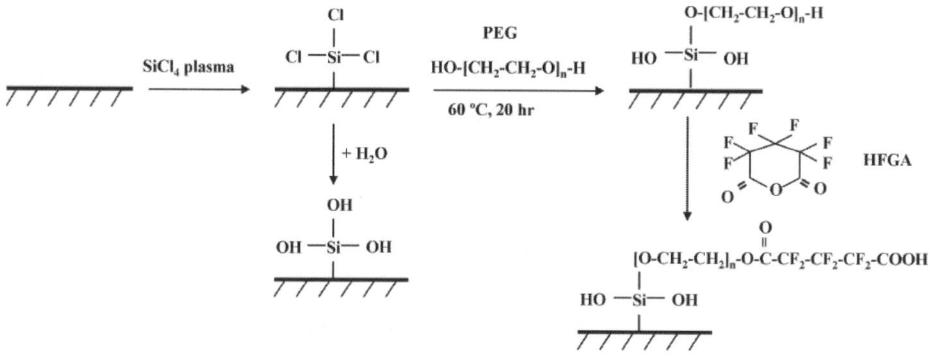

Figure 8.6. Reaction scheme for the surface modification of PA and PET by $SiCl_4$ plasma functionalization and PEG grafting. Also included are the reaction to convert Si-Cl to Si-OH and the derivatization of surface hydroxyl groups by HFGA. Images adapted from Ref. 75.

Denes *et al.* developed a simple cold plasma technique to functionalize the surfaces of polyamide (PA) and polyester (PET) for the grafting of polyethylene glycol chains.[75] The PA and PET surfaces were initially treated with silicon tetrachloride ($SiCl_4$) plasma, and then PEG was grafted (PA-g-PEG, PET-g-PEG) (Fig. 8.6). The surface coverage of PEG was controlled by changing the length of PEG chains or reaction time. Both PA-g-PEG and PET-g-PEG surfaces developed under optimal plasma conditions showed about 96% reduction in cell adhesion compared to the corresponding unmodified substrates.

In another report, poly(ethylene glycol) was covalently attached to the SU-8, epoxy-based negative photoresist used in microelectronic industry, to increase bio-functionality and improve the anti-fouling properties.[76] Different molecular weights and concentrations of PEG were used to produce polymer films of various grafting densities. X-ray photoelectron spectroscopy (XPS) analysis showed that an increase in concentration of PEG and immobilization time increases the grafting density of PEG. Protein adsorption and cell attachment were correlated with the PEG concentration or molecular weight.

Messersmith *et al.* explored the adhesive characteristics of 3,4-dihydroxyphenylalanine (DOPA), an key component of mussel adhesive proteins, to graft PEG onto surfaces to prepare protein resistant surfaces (Fig. 8.7).[77,78] Gold and titanium surfaces were modified by the adsorption of mPEG-DOPA and mPEG-MAPD (mPEG conjugated to the N-terminus of Ala-Lys-Pro-Ser-Tyr-Hyp-Hyp-Thr-DOPALys peptide) from solution. It was found that cell adhesion to mPEG-DOPA and mPEG-MAPD modified gold surfaces decreased by as much as 98% compared to the control surfaces. The measurement of human serum protein adsorption on mPEG-DOPA surface was related in terms of ethylene glycol (EG) surface density and mass of serum protein adsorbed. They demonstrated a threshold at $\sim$15–20 EG/nm^2 on the surface, above which the surface was substantially non-fouling (Fig. 8.8). The mPEG-DOPA and mPEG-MAPD modified surfaces also showed good resistance to cell attachment up to 14 days.

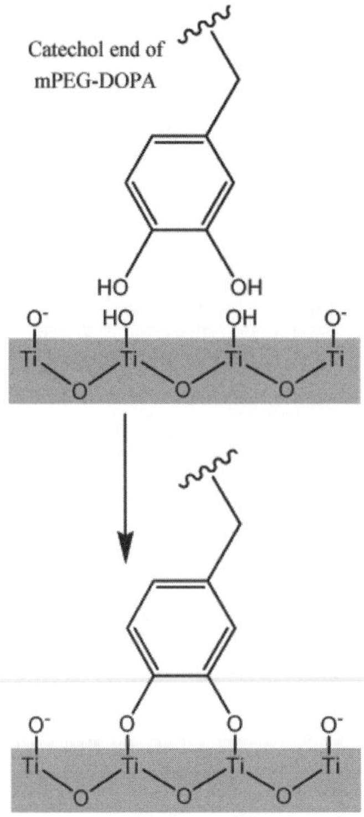

Figure 8.7. Proposed mechanism of mPEG-DOPA binding to TiO_2 surfaces, resulting in depletion of surface Ti-OH groups. Images adapted from Ref. 78.

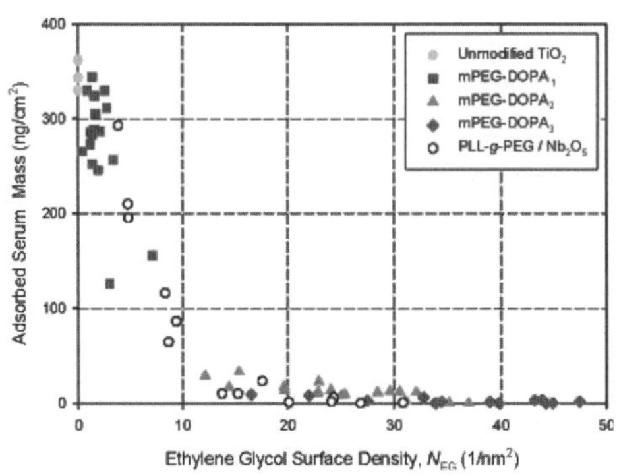

Figure 8.8. Ethylene glycol surface density versus adsorbed serum thickness. Images adapted from Ref. 78. For color reference, see page 271. For color reference, see page 271.

The other approaches used for grafting PEGs onto the surface involved the formation of aminosilanes, such as (3-aminopropyl) trimethoxysilane (APS) coatings on glass or silicon wafer, followed by the grafting of end functionalized PEG (e.g. epoxide or N-hydroxysuccinimide ester (NHS)) to the amine groups on the surface.[79–81]

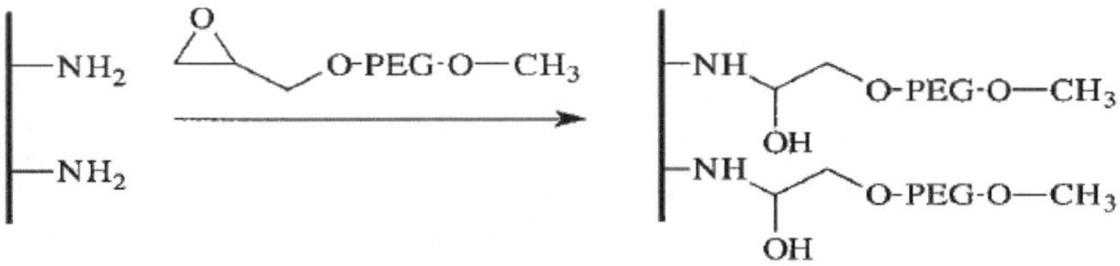

Figure 8.9. Reaction of aminopropylsilane (APS) modified surface with monomethoxy-epoxide-PEG. Images adapted from Ref. 80.

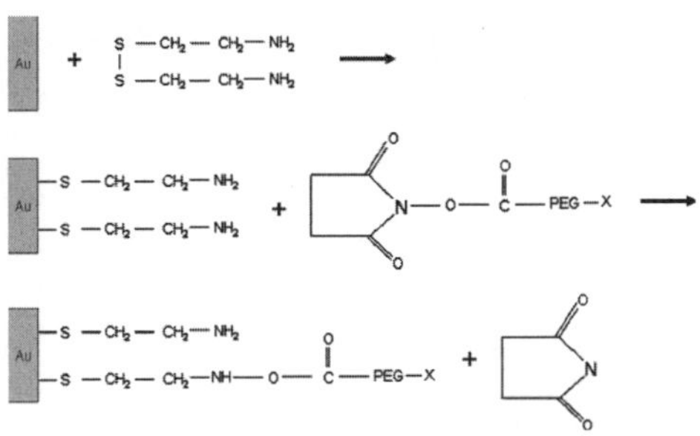

Figure 8.10. Formation of the PEG-Terminated Gold Surface. Images adapted from Ref. 81.

Malmsten *et al.* prepared PEG tethered surface through the reaction of aminopropylsilane (APS) modified surface with epoxide-PEG (Fig. 8.9). The grafting density was controlled by the reaction temperature. It was observed that with increasing PEG chain density the protein adsorption was decreased. At high interfacial chain density (~0.1 chain/nm^2) efficient protein rejection was observed for a number of proteins.

PEG-modified surfaces were also prepared by the coupling of PEG-NHS ester to the amine groups on the surface. Hillier *et al* prepared PEG tethered surfaces via this coupling chemistry (Fig. 8.10).[81] Cell adhesion studies illustrated the effectiveness of PEG-tethered surfaces in minimizing the cell adhesion compared to the control surfaces.

8.2.4 Surface Initiated Polymerization of PEG-containing Monomers

Surface-initiated polymerization (SIP) offers another complementary route for the in-situ synthesis of high surface density of PEGylated polymer surfaces that had shown exceptional protein resistance and long-term stability in a biological milieu.[82–90]

In this case a polyethylene glycol containing monomer was polymerized from the surface attached initiators. Since the growth of polymer was by the addition of small monomers to the growing polymer chain end from the surface, the graft density achieved was not limited[91] unlike the cases described in the previous section. In the case of surface modification using preformed polymers, there exist an inversion relationship between the graft density achieved and molecular weight of the chains due to the sterric crowding of the grafted chains.[92]

Chilkoti *et al.* reported the polymerization of oligo(ethylene glycol) methyl methacrylate (OEGMA) from an initiator functionalized alkanethiol SAM on gold using atom-transfer radical polymerization (ATRP) (Fig. 8.11).[82] It was found that the poly(OEGMA) brushes grafted on the gold surface reduced the adsorption of proteins from 100% serum to below the 1 ng cm^{-2}. They also examined the threshold value for the surface density of initiator required to create protein-resistant coatings.[83] Fibronectin (Fn) adsorption on the surface-initiated poly(OEGMA) coatings was below the detection limit of SPR (Fig. 8.12). The results suggested that SIP of OEGMA is a promising method to prepare protein-resistant coatings on different materials.

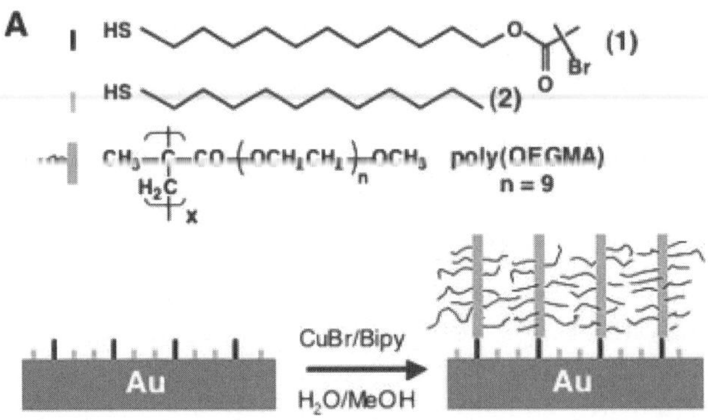

Figure 8.11. Schematic showing the SI-ATRP of OEGMA from a mixed SAM of an initiator-functionalized thiol **1** and a diluent thiol **2** on gold. Images adapted from Ref. 83

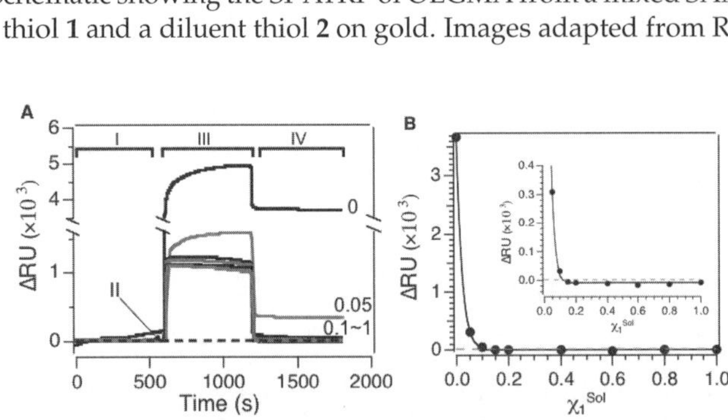

Figure 8.12. SPR chips were coated with poly(OEGMA) brushes grown from a series of mixed SAMs of **1** and UDT for a polymerization time of 40 min. A) SPR curves: after priming with PBS buffer for 10 min (region I), 0.5 mg mL^{-1} fibronectin was injected over the surface (at 10 min: region II) for 10 min (region III), followed by a 10 min rinse with PBS (region IV). B) The change in RU (DRU) is plotted against χ_1^{surf}. The inset is a magnified view of the response, and the curve is simply a guide to the eye. Images adapted from Ref. 83.

Messersmith *et al.* presented a comparative study on the antifouling properties of poly(ethylene glycol) (PEG)-based polymer coatings prepared by the attachment of preformed polymers and those prepared by surface-initiated polymerization.[84] OEGMA monomers containing side chains of 4, 9, and 23 EG units were used in surface-initiated ATRP to form polymer brushes. Greatly improved short-term antifouling properties for all three POEGMA modified samples were demonstrated by substantially reduced cell attachment for a period of three weeks as compared to bare Ti. The dependence of antifouling performance on EG side chain length was systemically tested and compared by 3T3 fibroblast cell adhesion assays. It was found that excellent cell fouling resistance continued with little dependence on EG side chain length up to three weeks but a complete cell coverage was reached at 7, 10 and 11 weeks respectively for POEGMA with side chain length 4, 9, and 23 EG units.

Kang *et al.* prepared heparin-coupled POEGMA-Si(111) hybrids via surface-initiated ATRP of oligo(ethylene glycol) monomethacrylate and subsequent immobilization of heparin(14 μg/cm^2) onto the POEGMA brushes.[85] The protein adsorption and platelet adhesion were significantly suppressed for heparinized POEGMA-Si hybrid surfaces. The resulting silicon surface exhibited significantly improved antithrombotic effects with a plasma recalcification time about 150 min. The persistence of high bioactivity of the immobilized heparin was attributed to the biocompatibility of the PEGMA units as well as their role as spacers in providing higher degree of conformational freedom for immobilized heparin in a hydrophilic environment.

Andruzzi and Ober prepared OEG$_n$-containing styrene-based homopolymer and block copolymer brushes with controlled architecture and high surface coverage using surface initiated nitroxide-mediated polymerization on a SiO_X surface (Fig. 8.13).[86] The preference for nitroxide mediated polymerization over ATRP was dictated by the fact that the former process does not utilize metal-based catalysts. Metal impurities are not always easy to eliminate from the polymerization product and are undesired in many biological applications. The biocompatibility of these polymer brushes was studied and compared with deposited assemblies of surface-bound OEGn-terminated silanes with selected chain lengths. Polymer brushes possessed superior abilities to inhibit protein and cell adhesion when compared to surface assemblies with the same OEG length. This was attributed to the greater thickness and surface coverage of polymer brushes compared to monolayers.

8.3 NON-FOULING SURFACES BASED ON ZWITTERIONIC GROUPS

Another important class of non-fouling surface treatments is based on phosphorylcholine (PC) functionalized materials.[93–108] This modification was based on the fact that zwitterionic phospholipids, which are major components of the outer cell membranes, have excellent non-thrombogenic property and biocomptability. The hydration of PC-based materials was thought to be the primary mechanism

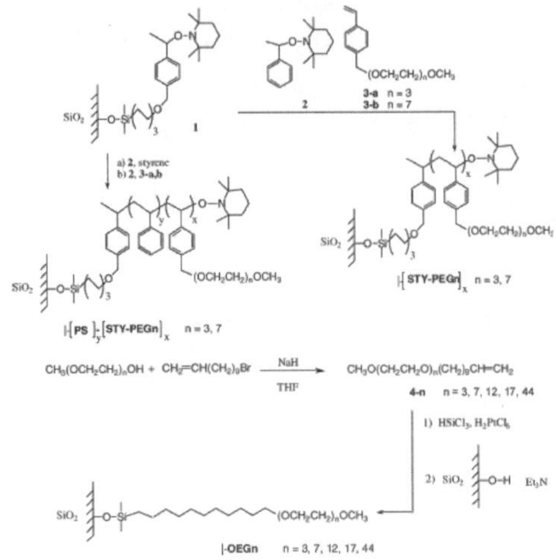

Figure 8.13. Synthesis of Tethered OEG$_n$-Containing Polymer Brushes and |-OEG$_n$ Assemblies on SiO$_X$.Images adapted from Ref. 84.

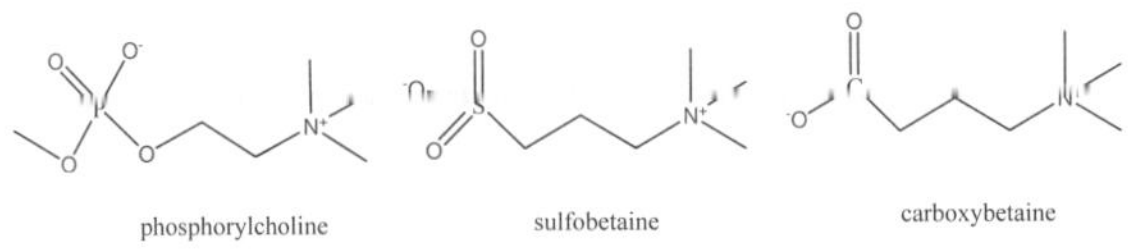

Figure 8.14. Chemical structure of phosphorylcholine (a), sulfobetaine (b) and cabocybetaine (c).

for their resistance to protein adhesion.[100,101] Zwitterionic structures similar to phosphorylcholine such as phosphobetaine, sulfobetaine, and carboxybetaine (Fig. 8.14) were also ideal for developing non-fouling surfaces when the surface density and chain length of zwitterionic groups were well controlled.[102–108]

We will discuss two important modification strategies (a) self assembled monolayers containing zwitterionic groups[93,94] and (b) surface initiated polymerization of a zwitterionic group containing monomers[97,102–104,106–108] in the following section. The physical adsorption of zwitterionic group based polymers was also used as a method to make non-fouling surfaces[105] but will not be discussed here.

8.3.1 Self-assembled Monolayers Containing Zwitterionic Groups

Holmlin and Whitesides fabricated SAMs terminated with charged groups on the gold surface to evaluate their ability to resist the nonspecific adsorption of proteins from aqueous buffer solutions.[93] Single-component SAMs with all positive groups (trimethylammonium group) and all negative groups (sulfonate group) adsorbed nearly full monolayers of fibrinogen and lysozyme. SAMs formed from a 1:1 mixture of thiols terminated with a negatively charged group and a positively

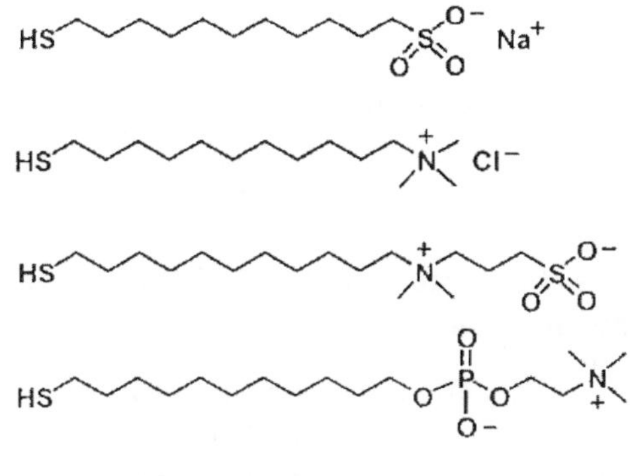

Figure 8.15. Single-component SAMs with single charges (SAMs of 1 or 2), zwitterionic mixed SAMs (SAM of 1 and 2 in a 1:1 ratio), and single-component zwitterionic SAMs (SAMs of 3 or 4). Images adapted from Ref. 93.

charged group adsorbed <1 % of a monolayer of protein (Fig. 8.15). Single-component zwitterionic SAMs composed of thiols terminated with *N, N*-dimethyl-amino-propane-1-sulfonic acid (-$N^+(CH_3)_2CH_2CH_2CH_2SO_3^-$) groups were substantially more effective in resisting the adsorption of fibrinogen and lysozyme from buffer solutions at physiological ionic strength and pH. These surfaces had better non-fouling properties than the single-component zwitterionic SAMs composed of thiols terminating in phosphoric acid 2-trimethylamino-ethyl ester (-$OP(O)_2^-OCH_2CH_2N^+(CH_3)_3$) (PC). The ability of zwitterionic surfaces to resist protein adsorption supports the notion that one of the essential characteristics in the design of a bio-inert surface was the electrical neutrality of the surface.

Chen and Jiang synthesized PC thiols using a modified synthetic route and formed the zwitterionic phosphorylcholine (PC) self-assembled monolayers on Au(111) (Fig. 8.22).[94] It was demonstrated that zwitterionic PC SAMs were highly resistant to protein adsorption. Balanced charge and minimized dipole were the two key factors for their non-fouling behavior. The experimental results showed that PC SAMs had very low protein adsorption when the nitrogen/phosphorus

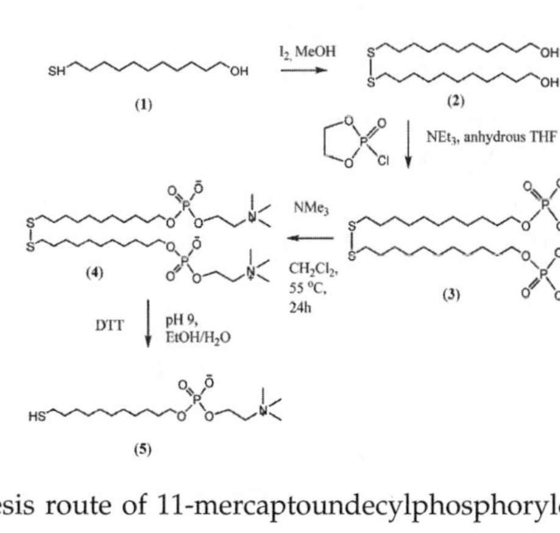

Figure 8.16. Synthesis route of 11-mercaptoundecylphosphorylcholine. Images adapted from Ref. 94.

(N/P) ratio was close to 1 and the charges were balanced. The molecular simulation results showed that PC head groups had packing densities similar to membrane lipids and preferred to have an anti-parallel orientation for the minimization of net dipole.

8.3.2 *Surface Initiated Polymerization of Zwitterionic Group Containing Monomers*

In this case surface initiated polymerization of zwitterion containing monomers were used for the surface functionalization. Feng and Zhu prepared biomimetic poly(2-methacryloyloxyethyl phosphorylcholine) (poly(MPC)) grafted surfaces with graft density 0.06–0.39 chains/nm^2 and chain length 5–200 monomer units on a silicon wafer surface.[102] They investigated the relationship between graft density and chain length on the non-fouling properties of grafted Poly(MPC) brushes. The fibrinogen adsorption from tris-buffered saline at pH 7.4 decreased significantly with increasing graft density and/or chain length of poly(MPC) and reached a level of <10 ng/cm^2 at graft density ≥ 0.29 chains/nm^2 and chain length ≥ 100 units, compared to ca. 570 ng/cm^2 for the unmodified surfaces (Fig. 8.17). The graft density appears to be more important than the chain length in determining the non-fouling properties of the surface.

Chio *et al.* described the formation of highly efficient non-biofouling polymeric thin films of poly((3-(methacryloylamino)propyl)-dimethyl(3-sulfopropyl)ammonium hydroxide), (poly(MPDSAH)).[103] The amount of proteins nonspecifically adsorbed onto the poly(MPDSAH) films was evaluated by surface plasmon resonance spectroscopy. The adsorption of proteins was <0.6 ng/cm^2 on the surfaces for all of the model proteins studied. In another report, Zhang and Jiang et al described the non-fouling properties of glass slides grafted with zwitterionic polymers (poly(sulfobetaine methacrylate) (polySBMA) and poly(carboxybetaine methacrylate) (polyCBMA) to resist the protein adsorption and the adhesion of mammalian cells (Fig. 8.18).[104] The grafted surfaces were synthesized from respective monomers using surface-initiated ATRP method. The surfaces with polySBMA or polyCBMA layers were shown to reduce fibrinogen adsorption to a level comparable with that of adsorption on PEG-like films (Fig. 8.19). The polySBMA or polyCBMA grafted glass surface were also highly resistant to the adhesion of bovine aortic endothelial cells.

In another study Chang and Chen *et al.* demonstrated the performance and stability of an antifouling surface with grafted zwitterionic sulfobetaine methacrylate.[106] The SBMA was grafted from a bromide-functionalized gold surface via surface-initiated ATRP to form well-defined polymer brushes. PolySBMA brushes showed excellent non-biofouling property and stability during the repeated protein adsorption tests and under a wide range of ionic strengths from 0.1 to 1.0 M, pH values from 7.4 to 11, and temperature cycling from 22 to 37°C in PBS buffer. The polySBMA grafted surface effectively reduced the plasma protein adsorption from platelet-poor plasma and was superior to that of surfaces

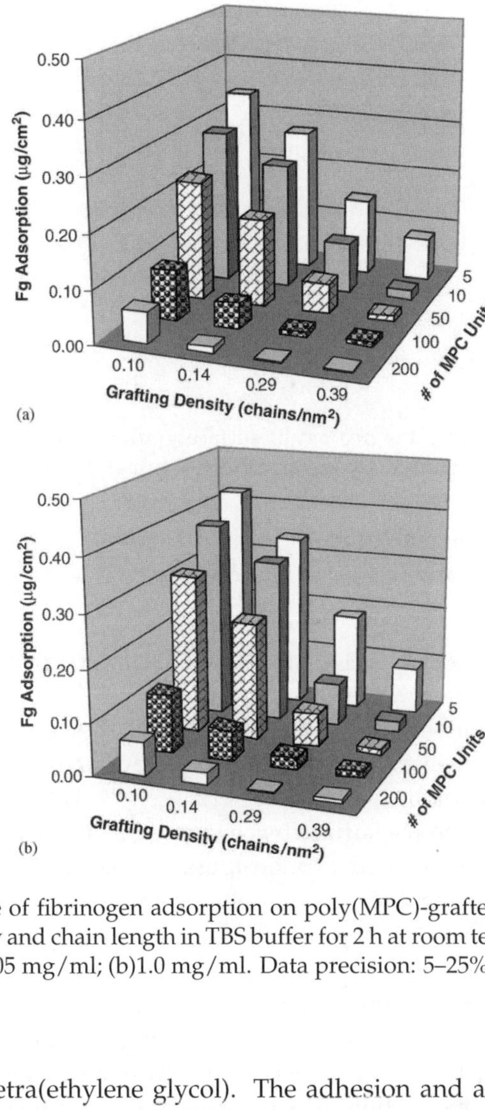

Figure 8.17. Profile of fibrinogen adsorption on poly(MPC)-grafted silicon surfaces with varying graft density and chain length in TBS buffer for 2 h at room temperature. Fibrinogen concentration: (a) 0.05 mg/ml; (b)1.0 mg/ml. Data precision: 5–25%. Images adapted from Ref. 102.

terminated with tetra(ethylene glycol). The adhesion and activation of platelets were not observed on the polySBMA grafted surface. Their results suggested that zwitterionic polySBMA is an effective and stable nonbiofouling material to provide a surface for uses in human blood and biomedical implants.

Bernards and Jiang *et al* described the development of a non-fouling polymer brushes formed from a surface-initiated, two-component ATRP.[108] The polymer brushes were composed of varying mixtures of positively (2-(methacryloyloxy)ethyl]trimethylammonium chloride) and negatively charged (3-sulfopropyl methacrylate potassium salt) monomers depending on the polymerization conditions. The non-specific adsorption of fibrinogen, lysozyme, and bovine serum albumin were measured using a surface plasmon resonance biosensor. It was found that the statistical copolymer surfaces made up of two charged components was protein resistant. The best nonfouling surface coating was found

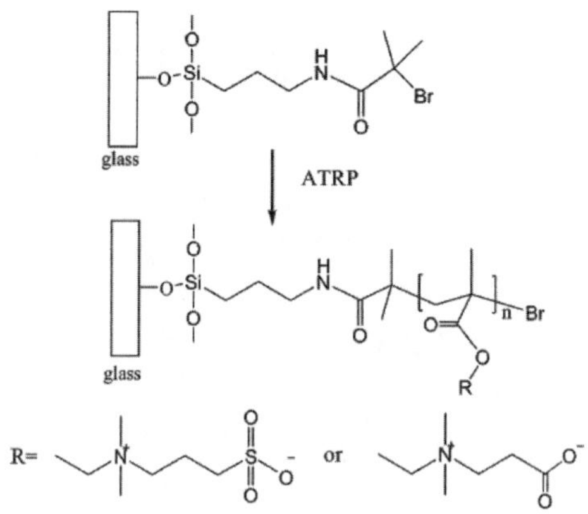

Figure 8.18. Illustration for the process of surface grafting via ATRP from the silanized glass surface covered with initiators. Images adapted from Ref. 103.

for the statistical copolymer brushes formed from a 1:1 homogeneous reaction mixture of the two oppositely charged monomers.

8.4 POLY(METH)ACRYLATE BASED NON-FOULING SURFACES

Genzer *et al.* used assemblies of surface-bound poly(2-hydroxyethyl methacrylate) (PHEMA) brushes with gradual variation of molecular weight and/or grafting density to vary the amount of adsorbed fibrinogen.[109] They achieved a gradient of adsorbed fibrinogen on the surface by changing the molecular weight and graft density with least fibrinogen adsorption found at maximum PHEMA coverage. Cell culture studies using MC-3T3 E1 cells on these surfaces showed that cell morphology and density can be varied as function of fibrinogen (and hence PHEMA) coverage (Fig. 8.20).

Ignatova and Jérôme synthesized copolymer brushes by a two-step "grafting from" method that consisted of a) electrografting of poly(2-phenyl-2-(2,2,6,6-tetramethyl-piperidin-1-yloxy)-ethylacrylate) onto stainless steel b) the nitroxide-mediated radical polymerization of 2-(dimethylamino ethyl)acrylate and styrene or *n*-butyl acrylate initiated from the electrografted polyacrylate chains.[110] The grafted copolymer brushes were quaternized to make them anti-bacterial. Quartz crystal microbalance experiments have proven that fibrinogen adhesion was lower on the hydrophilic quaternized films compared to the nonionic counterpart. Also, the quaternized copolymer surfaces exhibited significant antibacterial activity against the gram-positive bacteria *S. aureus* and the gram-negative bacteria *E. coli*

8.5 POLY(METH)ACRYLAMIDE BASED NON-FOULING SURFACES

Poly(*N*-isopropylacrylamide) (PNIPAAm) is a water-soluble thermo-responsive polymer. PNIPAAm has a lower critical solution temperature (LCST) at about

(a)

(b)

(c)

Figure 8.19. Cell culture on different surfaces: (a) control glass, (b) glass grafted with polySBMA, and (c) glass grafted with polyCBMA. Images adapted from Ref. 104.

32°C in an aqueous solution. It assumes a random coil structure (hydrophilic state) below the LCST and a collapsed globular structure above the LCST (hydrophobic state).[111–115] Because of this unique property, surfaces grafted with PNIPAAm were used for controlled cell adhesion studies. Kang *et al.* synthesized well-defined functional polymer-Si hybrids, consisting of covalently tethered brushes

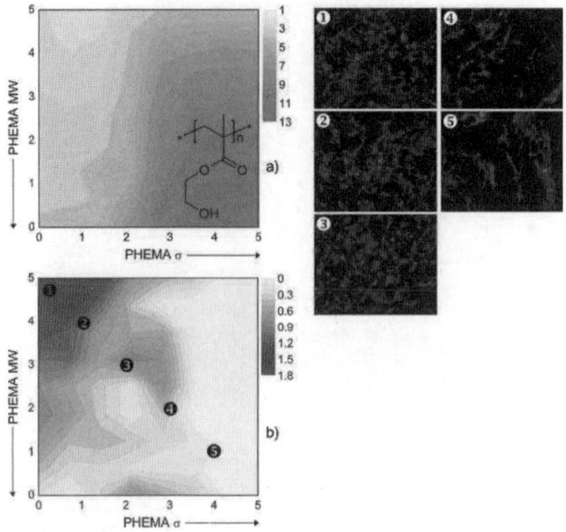

Figure 8.20. Left: contour plots of a) dry thickness of PHEMA in a MW-σ orthogonal PHEMA gradient (scale in nm); b) dry FN thickness in a MW-σ orthogonal PHEMA/FN gradient (scale in nm), the scales depicting the position on the substrate in parts (a) and (b) are in cm. Right: fluorescence microscopy (10x) images of fluorescently labeled MC3T3-E1 cells (nucleus: DAPI blue, cytoskeleton/actin: phalloidin red) cultured on PHEMA/FN gradient substrates. Images were taken at positions on the PHEMA/FN gradient sample marked with the numbers in part (b). The error bars associated with the thickness measurements ($\approx$ 0.1 nm) are smaller than the thickness increments in the contour plots. Images adapted from Ref. 109. For color reference, see page 271.

of POEGMA, PNIPAAm and poly (NIPAAm-OEGMA) copolymers and block copolymers on silicon surfaces via surface-initiated ATRP.[116] Cell adhesion was evaluated by culturing the cell line 3T3-Swiss albino on the hybrid surface. As to PNIPAAm grafted silicon surface, at 30°C (above the lower critical solution temperature of NIPAAm), the seeded cells adhered, spread, and proliferated on the NIPAAm graft polymerized silicon surface. Below the LCST, the cells detached from these surfaces spontaneously. Incorporation of POEGMA units into the NIPAAm chains of the Si-*g*-PNIPAAm surface via copolymerization resulted in more rapid cell detachment during the temperature transition.

Our group utilized aqueous ATRP to graft a series of cleavable hydrophilic PNIPAAm homopolymers and block copolymers of substituted acrylamides from polystyrene latex to give brushes of controlled MW and surface density.[117] Exposure to radiolabeled single proteins or to whole plasma and subsequent analysis by SDS-PAGE showed that PNIPAAm brushes decreased protein adsorption relative to the latex surface or other substituted polyacrylamides. The PNIPAAm brushes exhibited a second-order phase transition around 30°C as reflected by a decrease in the hydrodynamic thickness of the brush at higher temperatures. Total plasma protein adsorption was increased at 40°C compared to 20°C but there was significant differential adsorption behavior among the proteins detected by gel-electrophoresis analysis.

8.6 HYPERBRANCHED POLYGLYCIDOL BASED NON-FOULING SURFACES

Hyperbranched polyglycerols (HPGs) which presented a high density of glycol groups due to its highly branched architecture is another good candidate for making non-fouling surfaces.[118–120]

Siegers and Haag prepared highly protein-resistant, self-assembled monolayers of dendritic disulfide-functionalized polyglycerols (PGs) on gold.[118] Dendritic PGs combined the characteristic structural features of highly protein-resistant surfaces: a highly flexible aliphatic polyether, hydrophilic surface groups and a highly branched architecture. While PG monolayers were as protein resistant as PEG SAMs, they were significantly more active than dextran-coated surfaces. Due to the higher thermal and oxidative stability of the bulk PG as compared to PEG and the easy accessibility of these materials, it is a promising candidate for the development of non-fouling surface coatings.

Recently, our group synthesized monothiol-terminated hyperbranched polyglycerols (HPGs) by ring-opening polymerization of glycidol from partially deprotonated 2,2′-dihydroxyethane disulfide as the initiator and subsequent reduction of the disulfide group (Fig. 8.21).[120] Monothiol-functionalized HPGs readily adsorbed to a gold surface and formed highly uniform thin films on the surface. It was found that the graft density of the HPG layer decreased with an increase in the molecular weight of the polymer. A comparison of protein adsorption to HPG coated surfaces and PEG coated surfaces were also presented. Incubation of a polymer-coated surface (HPG thiols and PEG thiols) with bovine serum albumin and immunoglobulin showed that the high molecular weight hyperbranched polyglycerol was more resistant to protein adsorption than linear PEG of similar

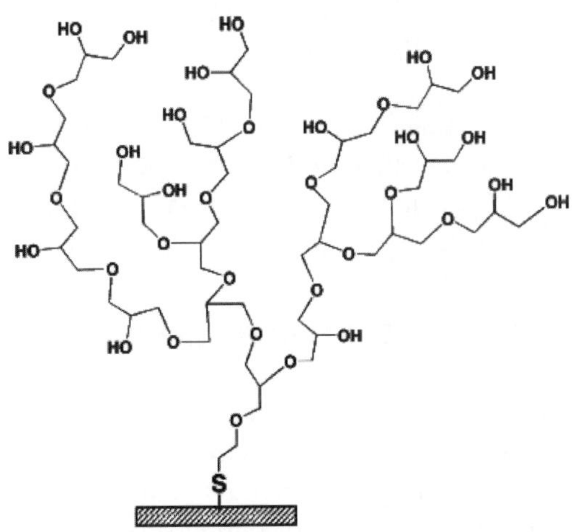

Figure 8.21. Representation of the structures of polymeric films on a gold surface: (a) linear mPEG monothiol and (b) HPG monothiol. Images adapted from Ref. 120.

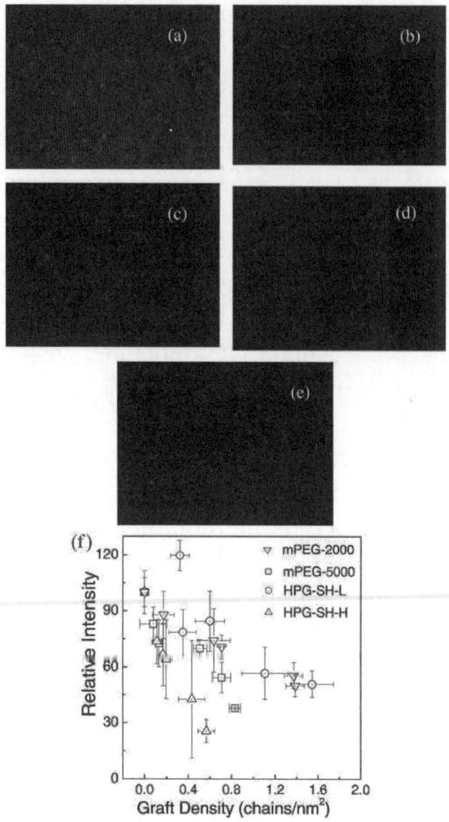

Figure 8.22. Fluorescence photographs of BSA-adsorbed (a) bare gold and (b) linear mPEG-2000-, (c) linear mPEG-5000-, (d) low molecular weight HPG-SH-L-, and (e) high molecular weight HPG-SH-H grafted surfaces. Polymer films were produced by incubating the gold surface in polymer solution at 6 g/L for 16 h. (f) Effect of the graft density on the BSA adsorption of mPEG-, HPG-SH-L-,and HPG-SH-H-grafted surfaces. Images adapted from Ref. 120. For color reference, see page 272.

molecular weight or lower molecular weight HPG. With increasing graft density of the HPG chains on the surface, the protein adsorption was decreased(Fig. 8.22). Our results show that HPG could be a good alternative to PEG in the development of nonfouling functional surfaces.

8.7 PEPTIDE AND PROTEIN GRAFTED POLYMERIC SURFACES

There are lots of interests in recent years in the development of surfaces with specific interaction to a certain biological system.[121] Short chain peptides or protein coupled surfaces were used for this purpose.[122] Although there are several techniques reported for the immobilization of peptides or proteins on the surface,[123] we will confine our discussion on the use of non-fouling polymer surfaces synthesized by surface initiated polymerization in this section.

Table 8.2 Commonly Available Functional Groups in Proteins and Functionalities of the Required Surfaces.[124]

Side groups	Amino acids	Surfaces
$-NH_2$	Lys.hydroxy-Lys	carboxylic acid active ester (NHS) epoxy aldehyde
$-SH$	Cys	maleimide pyridyil disulfide vinyl sulfone
$-COOH$	Asp, Glu	amine
$-OH$	Ser,Thr	epoxy

Most often proteins or peptides were covalently bound to the surfaces through accessible functional groups of exposed amino acids by irreversible binding to the functional groups on the substrate. Table 2 gives the name of functional groups typically used for the peptide/protein immobilization on the support surface.[124] Kang and coworkers grafted the Collagen I onto the titanium surface. Initially the titanium surface was functionalized with an ATRP initiator through a silane coupling agent. PHEMA chains were tethered on the Ti substrate via surface-initiated ATRP. Antibiotics (gentamicin and penicillin) and collagen molecules were covalently immobilized to the grafted polymer brushes after converting the hydroxyl groups of P(HEMA) to carboxyl groups (Fig. 8.23).[125]

Collard and coworkers used polyPOEGMA brushes for the coupling of helical peptide sequence ((GGYGGGPC(GPP)5(GFOGER)(GPP)5GPC onto the titanium surface. POEGMA brushes were prepared by surface initiated ATRP and the peptides were coupled using 4-nitrophenyl chloroformate as coupling reagent

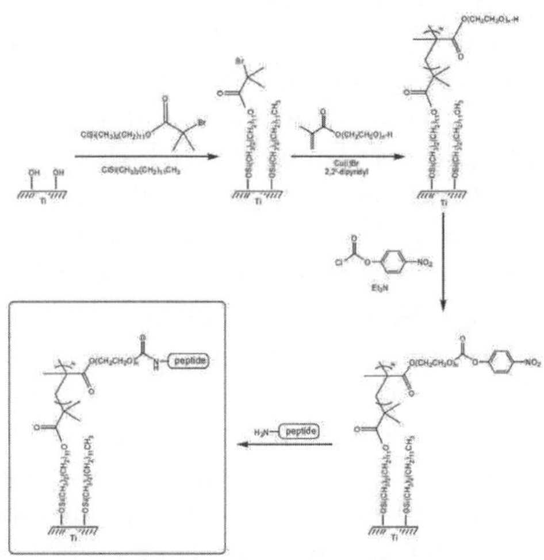

Figure 8.23. Schematic Diagram Illustrating the Route for the Immobilization of Collagen on the Ti$-g-$P(HEMA) Surface. Images adapted from Ref. 125.

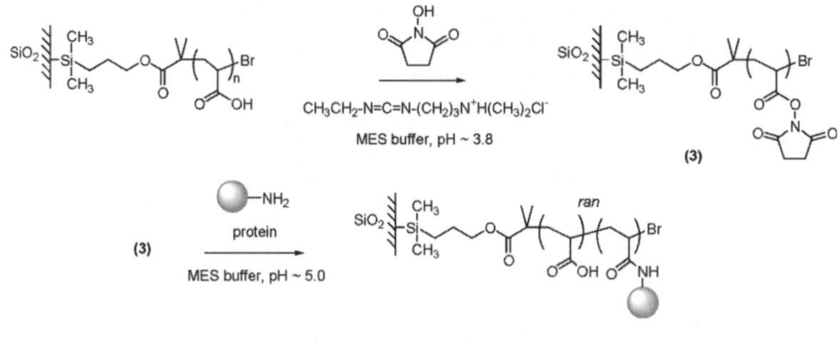

Figure 8.24. Formation of a peptide-modified poly(OEGMA) brush on titanium. Images adapted from Ref. 126. For color reference, see page 272.

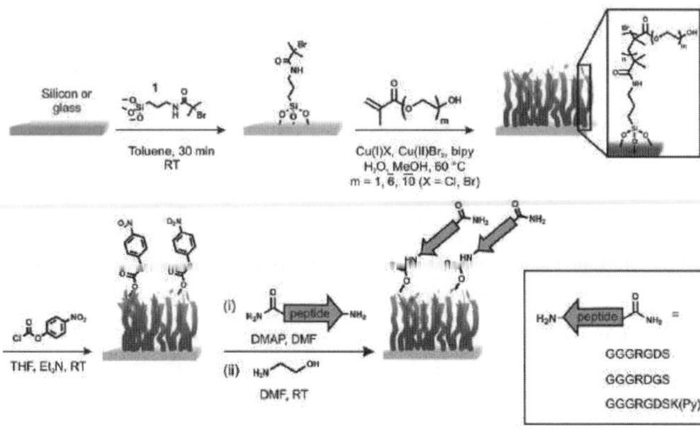

Figure 8.25. RGD functionalized PHEMA and PPEGMA brushes. Images adapted from Ref. 128. For color reference, see page 273.

(Fig. 8.24).[126] In another work Ober and coworkers used poly(acrylic acid) (PAA) brushes for the attachment of proteins. PAA brushes acted as robust templates for protein immobilization by providing a high density of –COOH groups on the surface.[127] They used both bovine serum albumin and avidin for immobilization.

In a related work, Tugulu *et al.* demonstrated the feasibility of surface-initiated atom transfer radical polymerization to prepare thin polymer brushes that can be functionalized with short peptide ligands which could be used as coatings to promote endothelialization of blood-contacting biomaterials. The grafted polymer layer was composed of PHEMA or POEGMA. A protocol was developed to allow the functionalization of the brushes with RGD containing peptide ligands resulting in surface concentrations ranging from ~0.5-12 pmol/cm^2 (Fig. 8.25).[128] They also investigated the use of polymer brushes as platforms for the covalent immobilization of fusion proteins. The AGT-mediated immobilization allowed selective functionalization of polymer brushes with proteins in a defined orientation and the surface density. All of these features make protein-modified polymer brushes very attractive candidates for the development of protein microarrays (Fig. 8.26).[129]

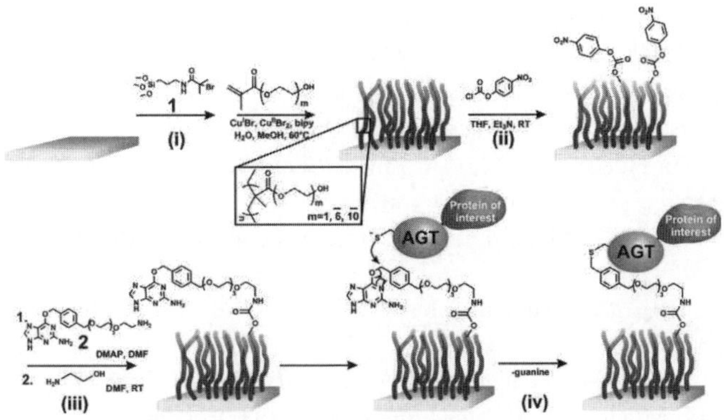

Figure 8.26. Fabrication of protein-functionalized polymer brushes: (i) grafting of ATRP initiator **1** and surface-initiated ATRP; (ii) activation of hydroxyl groups with p-nitrophenyl chloroformate (NPC); (iii) functionalization with benzylguanine derivative **2**and quenching of residual NPC groups; (iv) immobilization of AGT fusion proteins on benzylguanine-displaying surfaces. Images adapted from ref.[129] For color reference, see page 273.

8.8 CONCLUSIONS

We have made attempt to discuss the different approaches used in the development of non-fouling surfaces based on polymeric surface modification. We have also discussed the advantages and disadvantages of different methods used for making non-fouling surfaces. Among the polymeric systems described, poly(ethylene glycol) accounts for the majority of the non-fouling surfaces reported literature. Covalent grafting of polymers especially surface initiated polymerization is gaining popularity due to the excellent stability of grafted layers and non-fouling properties. There is lot of interest in the functionalization of non-fouling surfaces especially for applications in immunoassays, biosensors and microarrays.

ACKNOWLEDGEMENTS

J.N.K. acknowledges the financial support from University of British Columbia, Canadian Blood Services, Canadian Institutes of Health Research and Natural Science and Engineering Research Council of Canada (NSERC) and Canada Foundation for Innovation (CFI). The is the recipient of a Canadian Institutes of Health Research (CIHR)/Canadian Blood Services (CBS) new investigator award in transfusion science KY acknowledges a postdoctoral fellowship from CBS/CIHR.

References

1 M. B. Gorbet and M. V. Sefton, *Biomaterials* **25**, 5681 (2004).

2 B. D. Ratner and S. J. Bryant, *Annu. Rev. Biomed. Eng.* **6**, 41 (2004).

[3] K. J. Kitching, V. Pan and B. D. Ratner, In *Plasma Polymer Films*; Biederman, H., Eds.; Imperial College Press: London, pp 325-377 (2004).

[4] N. Nath, J. Hyun, H. Ma and A. Chilkoti, *Surf. Sci.* **570**, 98 (2004).

[5] K. F. Bohringer, *J. Micromech. Microeng.* **13**, S1 (2003).

[6] C. Mao, Y. Qiu, H. Sang, H. Mei, A. Zhu, J. Shen and S. Lin, *Adv. Colloid Interface Sci.* **110**, 5 (2004).

[7] K. Kottke-Merchant, J. M. Anderson, Y. Umemura and R. E. Merchant, *Biomaterials* **10**, 147 (1989).

[8] A. A. A. DeQueiroz, E. R. Barrak, H. A. C. Gil and O. Z. Higa, *J. Biomater. Sci. Polym. Ed.* **8**, 667 (1997).

[9] W. E. Hennink, L. Dost, J. Feijen and S. W. Kim, *Trans ASAIO* **29**, 200 (1983).

[10] E. Brynda and M. Houska, *J. Colloid Interface Sci.* **183**, 18 (1996).

[11] E. Brynda, M. Housla, M. Jirouslova and J. E. Dyr, *J. Biomed. Mater. Res.* **51**, 249 (2000).

[12] G. W. Bos, N. M. Scharenborg, A. A. Poot, G. H. Engbers, T. Beugeling, W. A. Aken and J. Feijen, *J. Biomed. Mater. Res.* **47**, 279 (1999).

[13] K. D. Park, A. Z. Piao, H. Jacobs, T. Okano and S. W. Kim, *J. Polym. Sci.* **29**, 1725 (1991).

[14] R. Barbucci, A. Magnani, A. Albanese and F. Tempesti, *Int. J. Artif. Organs* **14**, 499 (1991).

[15] Y. Ito, M. Sisido and Y. Imanishi, *J. Biomed. Mater. Res.* **20**, 1157 (1986).

[16] Y. J. Kim, I. K. Kang, M. W. Huh and S. C. Yoon, *Biomaterials* **21**, 121 (2000).

[17] P. Olsson, J. Sanchez, T. E. Mollnes and J. J. Riesenfeld, *J. Biomater. Sci. Polym. Ed.* **11**, 1261 (2000).

[18] T. Matsud and T. Magoshi, *Biomacromolecules* **2**, 1169 (2001).

[19] W. Senaratne, L. Andruzzi and C. K. Ober, *Biomacromolecules* **6**, 2427 (2005).

[20] S. Sharma, R. W. Johnson and T. A. Desai, *Biosens. Bioelectron.* **20**, 227 (2004).

[21] A. L. Klibanov, V. P. Torchilin and S. Zalipsky, In *Liposomes*, 2nd ed.; V. P. Torchilin, V. Weissig, Eds.; Oxford University Press: Oxford, UK, pp 231-265 (2003).

[22] I. Tsujino, J. Ako, Y. Honda and P. J. Fitzgerald, *Exp. Opin. Drug Deliv.* **4**, 287 (2007).

[23] H. M. Burt and W. L. Hunter, *Adv. Drug Deliv. Rev.* **58**, 350 (2006).

[24] D. Kim, S. Park, J. H. Lee, Y. Y. Jeong and S. Jon, *J. Am. Chem. Soc.* **129**, 7661 (2007).

[25] P. Caliceti and F. Veronese, M. *Adv. Drug Deliv. Rev.* **55**, 1261 (2003).

[26] E. van Vlerken Lilian, K. V. Tushar and M. A. Mansoor, *Pharm. Res.* **24**, 1405 (2007).

[27] L. D. Unsworth, H. Sheardown and J. L. Brash, *Biomaterials* **26**, 5927 (2005).

[28] B. Zdyrko, S. K. Varshney and I. Luzinov, *Langmuir* **20**, 6727 (2004).

[29] C. E. McNamee, S.Yamamoto and K. Higashitani, *Biophys. J.* **93**, 324 (2007).

[30] R. Bartucci, M. Pantusa, D. Marsh and L. Sportelli, *Biochim. Biophys. Acta-Biomembranes* **1564**, 237 (2002).

[31] B. M. Steels, J. Koska and C. A. Haynes, *J. Chromatogr., B: Biomed. Sci. Appl.* **743**, 41 (2000).

[32] N. Singh, X. Cui, T. Boland and S. M. Husson, *Biomaterials* **28**, 763 (2007).

[33] M. A. Carignano, I. Szleifer, *Colloids Surf., B* **18**, 169 (2000).

[34] J. Satulovsky, M. A. Carignano and I. Szleifer, *Proc. Natl. Acad. Sci. USA* **97**, 9037 (2000).

[35] A. Halperin, *Langmuir* **15**, 2525 (1999).

[36] Harris, J. M. *Poly(ethylene glycol) Chemistry: Biotechnical and Biomedical Applications* (Plenum Press, New York, 1992).

[37] J. H. Lee, H. B. Lee and J. D. Andrade, *Prog. Polym. Sci.* **20**, 1043 (1995).

[38] Ft. Hesselin, A. Vrij and J. T. Overbeek, *J. Phys. Chem.* **75**, 2094 (1971).

[39] D. H. Napper, *J. Colloid Interf. Sci.* **58**, 390 (1977).

[40] H. Thissen, J. P. Hayes, P. Kingshott, G. Johnson, E. C. Harvey and H. J. Griesser, *Smart Mater. Struct.* **11**, 792 (2002).

[41] J. T. Li and K. D. Caldwell, *Colloids Surf B: Biointer.* **7**, 9 (1996).

[42] J. A. Neff, P. A. Tresco and K. D. Caldwell. *Biomaterials* **20**, 2377 (1999).

[43] M. M. Amiji and K. Park, *J. Appl. Polym. Sci.* **52**, 539 (1994).

[44] M. Amiji and K. Park, *Biomaterials* **13**, 682 (1992).

[45] V. A. Liu, W. E. Jastromb and S. Nn Bhatia, *J. Biomed. Mater. Res.* **60**, 126 (2002).

[46] G. L. Kenausis, J. Voros, D. L. Elbert, N. Huang, R. Hofer, L. Ruiz-Taylor, M. Textor, J. A. Hubbell and N. D. Spencer, *J. Phys. Chem. B* **104**, 3298 (2000).

[47] N.-P. Huang, R. Michel, J. Voros, M. Textor, R. Hofer, A. Rossi, D. L. Elbert, J. A. Hubbell and N. D. Spencer, *Langmuir* **17**, 489 (2001).

[48] N. D. Winblade, I. D. Nikolic, A. S. Hoffman and J. A. Hubbell, *Biomacromolecules* **1**, 523 (2000).

[49] K. L. Prime and G. M. Whitesides, *Science* **252**, 1164 (1991).

[50] C. D. Bain and G. M. Whitesides, *Angew. Chem. Int. Ed.* **28**, 506 (1989).

[51] K. L. Prime and G. M. Whitesides, *J. Am. Chem. Soc.* **115**, 10714 (1993).

[52] M. Mrksich, G. S. Sigal and G. M. Whitesides, *Langmuir* **11** 4383 (1995).

[53] G. S. Sigal, M. Mrksich and G. M. Whitesides, *J. Am. Chem. Soc.* **120**, 3464 (1998).

[54] S. Herrwerth, W. Eck, S. Reinhardt and M. Grunze, *J. Am. Chem. Soc.* **125**, 9359 (2003).

[55] R. L. C. Wang, H. J. Kreuzer and M. Grunze, *J. Phys. Chem. B* **101**, 9767 (1997).

[56] H. J. Kreuzer and M. Grunze, *Europhys. Lett.* **55**, 640 (2001).

[57] D. J. Vanderah, H. La, J. Naff, V. Silin and K. A. Rubinson, *J. Am. Chem. Soc.* **126**, 13639 (2004).

[58] L. Li, S. Chen, J. Zheng, B. D. Ratner and S. Jiang, *J. Phys. Chem. B.* **109**, 2934 (2005).

[59] M. Riepl, M. Ostblom, I. Lundstrom, S. C. T.Svensson, A. W. Denier van der Gon, M. Schaferling and B. Liedberg, *Langmuir* **21**, 1042 (2005).

[60] Y. T. Kim and A. J. Bard, *Langmuir* **8**, 1096 (1992).

[61] C. Schonenberger, J. A. M. Sondaghuethorst, J. Jorritsma and L. G. J. Fokkink, *Langmuir* **10**, 611 (1994).

[62] X. M. Zhao, J. L. Wilbur and G. M. Whitesides, *Langmuir* **12**, 3257 (1996).

[63] M. H. Schoenfisch and J. E. Pemberton, *J. Am. Chem. Soc.* **120**, 4502 (1998).

[64] M. J. Tarlov and J. G. Newman, *Langmuir* **8**, 1398 (1992).

[65] J. P. Bearinger, S. Terrettaz, R. Michel, N. Tirelli, H. Vogel, M. Textor and J. A. Hubbell, *Nat. Mater.* **2**, 259 (2003).

[66] Y.-Y. Luk, M. Kato and M. Mrksich, *Langmuir* **16**, 9604 (2000).

[67] N. Xia, Y. Hu, D. W. Grainger and D. G. Castner, *Langmuir* **18**, 3255 (2002).

[68] J. L. Dalsin, B. H. Hu, B. P. Lee and P. B. Messersmith, *J. Am.Chem. Soc.* **125**, 4253 (2003).

[69] J. L. Dalsin, L. Lin, S. Tosatti, J. Voros, M. Textor and P. B. Messersmith, *Langmuir* **21**, 640 (2005).

[70] G. L. Kenausis, J. Voros, D. L. Elbert, N. Huang, R. Hofer, L. Ruiz-Taylor, M. Textor, J. A. Hubbell and N. D. Spencer, *J. Phys. Chem. B* **104**, 3298 (2000).

[71] J. P. Bearinger, D. G. Castner, S. L. Golledge, S. Hubchak and K. E. Healy, *Langmuir* **13**, 5175 (1997).

[72] L. D. Unsworth, H. Sheardown and J. L. Brash, *Langmuir* **21**, 1036 (2005).

73 S. J. Sofia, V. Premnath and E. W. Merrill, *Macromolecules* **31**, 5059 (1998).
74 X.-Y. Zhu, Y. Jun, D. R. Staarup, R. C. Major, S. Danielson, V. Boiadjiev, W. L. Gladfelter, B. C. Bunker and A. Guo, *Langmuir* **17**, 7798 (2001).
75 B. Dong, H. Jiang, S. Manolache, A. C. L. Wong and F. S. Denes, *Langmuir* **23**, 7306 (2007).
76 S. L. Tao, K. C. Popat, J. J. Norman and T. A. Desai, *Langmuir* **24**, 2631 (2008).
77 J. L. Dalsin, B.-H. Hu, B. P. Lee and P. B. Messersmith, *J. Am. Chem. Soc.* **125**, 4253 (2003).
78 J. L. Dalsin, L. Lin, S. Tosatti, J. Voros, M. Textor and P. B. Messersmith, *Langmuir* **21**, 640 (2005).
79 K. Emoto, J. M. Harris and J. M. Van Alstine, *Anal. Chem.* **68**, 3751 (1996).
80 Martin Malmsten, Ken Emoto and James M. Van Alstine *J. Colloid Interface Sci.* **202**, 507 (1998).
81 K. Mougin, M. B. Lawrence, E. J. Fernandez and A. C. Hillier, *Langmuir* **20**, 4302 (2004).
82 H. Ma, J. Hyun, P. Stiller and A. Chilkoti, *Adv. Mater.* **16**, 338 (2004).
83 H. Ma, M. Wells, T. P. Beebe Jr. and A. Chilkoti, *Adv. Funct. Mater.* **16**, 640 (2004).
84 X. Fan, L. Lin and P. B. Messersmith, *Biomacromolecules* **7**, 2443 (2006).
85 F. J. Xu, Y. L. Li, E. T. Kang and K. G. Neoh, *Biomacromolecules* **6**, 1759 (2005).
86 L. Andruzzi, W. Senaratne, A. Hexemer, E. D. Sheets, B. Ilic, E. J. Kramer, B. Baird and C. K. Ober, *Langmuir* **21**, 2495 (2005).
87 C. Perruchot, M. A. Khan, A. Kamistsi, S. P. Armes, T. von Werne and T. E. Patten, *Langmuir* **17**, 4479 (2001).
88 D. Bontempo, N. Tirelli, K. Feldman, G. Masci, V. Grescenzi and J. A. Hubbell, *Adv. Mater.* **14**, 1239 (2004).
89 F. J. Xu, S. P. Zhong, L. Y. L. Yung, E. T. Kang and K. G. Neoh, *Biomacromolecules* **5**, 2392 (2004).
90 X. Fan, L. Lin, J. L. Dalsin and P. B. Messersmith, *J. Am. Chem. Soc.* **127**, 15843 (2005).
91 R. C. Advincula, W. J. Brittain, K. C. Caster and J. Ruehe, Eds. *Polymer Brushes*, Wiley-VCH: Weinheim, (2004).
92 J. N. Kizhakkedathu, , Y. Jayachandran N. Le, , J. Janzen, R. K. Kainthan and D. E. Brooks, *Polymer Preprints* **46**, 48 (2005).
93 R. E. Holmlin, X. Chen, R. G. Chapman, S. Takayama and G. M. Whitesides, *Langmuir* **17**, 2841 (2001).
94 S. F. Chen, J. Zheng, L. Y. Li and S. Y. Jiang, *J. Am. Chem. Soc.* **127**, 14473 (2005).
95 A. L. Lewis *Colloids Surf B: Biointer.* **18**, 261 (2000).
96 A. L. Lewis, Z. L. Cumming, H. H. Goreish, L. C. Kirkwood, L. A. Tolhurst and P. W. Stratford, *Biomaterials* **22**, 99 (2001).
97 W. Feng, S. Zhu, K. Ishihara and J. L. Brash, *Langmuir* **21**, 5980 (2005).
98 J. D. Patel, Y. Iwasaki, K. Ishihara, J. M. Anderson, *J. Biomed. Mater. Res.* **73A**, 359 (2005).
99 T. Hasegawa, Y. Iwasaki and K. Ishihara, *Biomaterials* **22**, 243 (2001).
100 K. Ishihara, H. Nomura, T. Mihara, K. Kurita, Y. Wasaki and N. akabayashi, *J. Biomed. Mater. Res.* **39**, 323 (1998).
101 E. F. Murphy, J. R. Lu, A. L.Lewis, J. Brewer, J. Russell and P. Stratford, *Macromolecules* **33**, 4545 (2000).
102 W. Feng, J. L. Brash and S. P. Zhu, *Biomaterials* **27**, 847, (2006).
103 W. K. Cho, B. Y. Kong and I. S. Choi, *Langmuir* **23**, 5678 (2007).
104 Z. Zhang, T. Chao, S. F. Chen and S. Y. Jiang, *Langmuir* **22**, 10072 (2006).

105 Y. Chang, S. F. Chen, Z. Zhang and S. Y. Jiang, *Langmuir* **22**, 2222 (2006).

106 Y. Chang, S.-C. Liao, A. Higuchi, R.-C. Ruaan, C.-W.Chu and W.-Y. Chen, *Langmuir* **24**, 5453 (2008).

107 Z. Zhang, S. F. Chen, Y. Chang and S. Y. Jiang, *J. Phys. Chem. B* **110**, 10799 (2006).

108 M. T. Bernards, G. Cheng, Z. Zhang, S. Chen and S. Jiang, *Macromolecules* **41**, 4216 (2008).

109 R. R. Bhat, B. N. Chaney, J. Rowley, A. Liebmann-Vinson and J. Genzer *Adv. Mater.* **17**, 2802 (2004).

110 M. Ignatova, S. Voccia, B. Gilbert, N. Markova, P. S. Mercuri, M. Galleni, V. Sciannamea, S. Lenoir, D. Cossement, R. Gouttebaron, R. Jérôme and C. Jérôme, *Langmuir* **20**, 10718 (2004).

111 D. J. Kim, J. Y. Heo, K. S. Kim and I. S. Choi, *Macromol. Rap. Commun.* **24**, 517 (2003).

112 S. Balamurugan, S. Mendez, S. S. Balamurugan, M. J. O'Brien, II and G. P. Lopez, *Langmuir* **19**, 2545 (2003).

113 J. N. Kizhakkedathu, N. J. Raymond and D. E. Brooks, *Macromolecules* **37**, 734 (2004).

114 J. Gao and C. Wu, *Macromolecules* **30**, 6873 (1997).

115 T. G. Park and A. S. Hoffman, *Macromolecules* **26**, 5045 (1993).

116 F. J. Xu, S. P. Zhong, L. Y. L. Yung, E. T. Kang and K. G. Neoh, *Biomacromolecules* **5**, 2392 (2004).

117 J. Janzen, Y. Le, J. N. Kizhakkedathu and D. E. Brooks, *J. Biomater. Sci. Polym. Ed.* **15**, 1121 (2004).

118 C. Siegers, M. Biesalski and R. Haag, *Chem. Eur. J.* **10**, 2831 (2004).

119 A. Sunder, R. Mulhaupt, R. Haag and H. Frey, *Macromolecules* **33**, 253 (2000).

120 P.-Y. J. Yeh, R. K. Kainthan, Y. Zou, M. Chiao and J. N. Kizhakkedathu, *Langmuir* **24**, 4907 (2008).

121 D. Aili, K. Enander, J. Rydberg, I. Nesterenko, F. Björefors, L. Baltzer and B. Liedberg, *J. Am. Chem. Soc.* **130**, 5780 (2008).

122 U. Hersel, C. Dahmen and H. Kessler, *Biomaterials* **24**, 4385 (2003).

123 J. A. Camarero *Biopolymers* **90**, 450 (2007).

124 F. Rusmini, Z. Zhong and J. Feijen, *Biomacromolecules* **8**, 1775 (2007).

125 F. Zhang, Z. L. Shi, P. H. Chua, E. T. Kang and K. G. Neoh, *Ind. Eng. Chem. Res.* **46**, 9077 (2007).

126 J. E. Raynor, T. A. Petrie, A. J. García and D. M. Collard, *Adv. Mater.* **19**, 1724 (2007).

127 R. Dong, S. Krishnan, B. A. Baird, M. Lindau and C. K. Ober, *Biomacromolecules* **8**, 3082 (2007).

128 S. Tugulu, P. Silacci, N. Stergiopulos and H. Klok, *Biomaterials* **28**, 2536 (2007).

129 S. Tugulu, A. Arnold, I. Sielaff, K. Johnsson and H.-A. Klok, *Biomacromolecules* **6**, 1602 (2005).

Chapter Nine

Vibration-based Anti-Biofouling of Implants

Po Ying Jacob Yeh[*,‡], **Jayachandran N. Kizhakkedathu**[†] **and Mu Chiao**[*,§]

[*] *Department of Mechanical Engineering, 2054-6250 Applied Science Lane, University of British Columbia, Vancouver, B.C., Canada.*
[‡]*E-mails: ypy0816@interchange.ubc.ca,* [§]*muchiao@mech.ubc.ca*
[†] *Centre for Blood Research and Department of Pathology and Laboratory Medicine, 4th Floor, Life Sciences Centre, 2350 Health Sciences Mall, University of British Columbia, Vancouver, B.C., Canada.*
E-mail: jay@pathology.ubc.ca

The interfacial behavior of proteins at surfaces has attracted much attention over the years because of the problem of fouling, which is encountered in implants, microfiltration membrane, medical implantable devices, drug delivery system and biosensors. A novel anti-biofouling mechanism based on the combined effects of electric field and shear stress is reported. The mechanism was observed in mm scale piezoelectric plates coated with different metal materials and MEMs (MicroElectroMechanical system) fabricated membrane based device. Experimental observation of quantity of proteins adsorption and theoretical calculations of protein-surface interactions (van der Waals (VDW), electrostatic, and hydrophobic) and shear stress reveal the mechanism for protein desorption from surface. This anti-fouling mechanism is also able to implant to micromachined MEMs membrane. The dimension of membrane (Si(2 μm)/SiO_2(1 μm)) is 2000 μm × 500 μm × 3 μm. The membrane vibrates in a flexural plate wave (FPW), which is further verified by finite element simulation. The surface charges on the membrane and the fluid shear stress contribute in attenuating the protein adsorption on the membrane surface. Potentially, a microelectromechanical system (MEMS)-based vibrating membrane could be a tool to minimize biofouling of *in vivo* implants.

9.1 INTRODUCTION

The success of implantable device technologies, such as artificial joint, vascular graft, cardiopacemaker, cardiovascular, drug delivery system, and biosensor,

Biomaterials for MEMS, Edited by M. Chiao and J.-C. Chiao

www.panstanford.com
978-981-4241-46-5

had improved the patients' life quality. However, with the advances of these technologies, the view of biocompatibility had extended from the "the host tolerates the device" to "the device tolerates the host and vice versa".[1,2] For example, a subcutaneous glucose sensor comprises of immobilized enzymes, electrodes and a semi-permeable membrane. The semi-permeable membrane acts as a protecting layer that keeps the enzyme and electrodes from the attack of the immune system and it allows glucose from interstitial space to diffuse through. In order to be successful, the semi-permeable membrane must not induce pathogenic reactions to the host. On the other hand, the glucose permeability of the membrane must be maintained at all times. Hence, the problem of biofouling needs to be reduced as much as possible.

Biofouling is a process started immediately after a foreign object comes into contact with biological fluids because of the immune response system. And it is considered to be one of the greatest challenges of medical implantation.[2–5] The immediate response is to flood the injured area with blood followed by adsorption of blood proteins onto the implant surface and activation of the molecular and cellular defense systems.[6,7] The coagulation, complement, immune and inflammatory pathways can all be initiated by this process as can platelet and white cell activation. The extent to which these responses occur depends on the nature of the protein adsorption, specific binding or rejection reactions that occur at the impalnt-blood interface. Usually an adsorbed protein layer of thickness from 0.5 to 9 µm accumulates over time. Approximately 21 days after implantation, an avascular fibrous capsule of up to 100 µm will form around the implant and may reduce its performance.[8] Prevention of protein adsorption unto implants is critical for increasing their longevity. Some methods had been provided so far, such as polymer-based coatings,[9–19] plasma surface functionalization,[20] surface charges,[21] detergent coatings,[22] nano-structured surfaces,[8] surface roughness and topography,[23,24] mechanical vibration,[8,25–27] and flow flushing upon the surface,[28,29] etc. Coatings, surface roughness, surface topography are catalogued as passive methods, while physical means, such as mechanical vibration and flow flushing are considered as active methods of anti-biofouling.

Numerous studies have indicated that the surface chemistry of and microstructure on the implant surface can modulate protein adsorption.[3,4,8,30] And among those passive methods, polymer coating is attractive for certain applications, such as long-circulating sterically protected liposome,[31] biosensor,[32] and protein chip.[33] Those polymer candidates usually are composed by three parts, anchor to surface, monomer repeated backbone, and end functional group.

The most popular polymer used is probably the hydrophilic poly(ethylene glycol) (PEG). When there is sufficient high PEG graft density, the coiled PEG with one end anchored to the surface will form brushes by extending the coils away from surface, and so that to create a steric barrier from protein adsorption.[34,35] However, polymer coatings are successful for short term and often fail in the case of a highly complex mixture of proteins, for example, blood plasma,[9,13] or under *in vivo* conditions.[5] As an alternative to polymer coating, researchers have also fabricated or assembled nano-structures on the implant surface, such as nanotextures and nanowires, to tailor the surface chemistry for reducing proteins adsorption.[8] It is also possible to add enzymes into carbon nanotubes, such as proteases, which can break down the accumulation of proteins, to clean the surface.[5]

On the another hand, the effects of an active surface by physical means for reducing biofouling had also been investigated.[25–29,36] It had been shown that attenuation of protein adsorption can be achieved if the surface vibrates in specific frequency, amplitude, and mode.[25,26] For example, in oceanographic study, the biofouling on the sensor surfaces can be reduced by attaching a vibrating polyvinylidene fluoride (PVDF) actuator, which supplies the mechanical vibration energy to sensor surface.[25] Furthermore, non-specific adsorption of proteins was reduced by a high-frequency (~MHz) macro scale quartz resonators (thickness-shear mode), which supplies the shear waves. Interestingly, the biosensor based on the site protein–ligand specific binding on the quartz surface for detect protein of interest. By controlling the output energy of quartz, one could reduce the background because of the protein adsorption while keeping the site protein–ligand specific binding so that to lower down background noise and increase the detection limit of biosensor.[26]

Another energy supply, such as periodic magnetic field, was also reported to be able to vibrate magnetic nanowires assembled on the surface and to attenuate the adsorption of proteins onto surface.[8] Another physical mean, flow, can be used to sweep the surface for the protein adsorption attenuation.[28,29,36] For instance, a glucose sensor implanted inside the rat to *in vivo* measure the body glucose level had been designed to incorporate a continuous and external supplied saline flow to flush the sensor surface.[28,29] It was shown that the glucose sensor with this saline flushing over the surface had a longer life because that the sensor reading drifts less over time. However, the difficulty to integrate both flow generator and biosensor in *in vivo* limits its application.

The ultrasound-induced acoustic flow can be considered as the *in vivo* flow generator.[37–39] The flow is generated by the surface acoustic wave or the flexural plate wave and was shown to capable move μm-sized particles close the surface[36] if the adhesion interaction between particle and surface can be overcome. Similarly, it was reported that the acoustic radiation force can concentrate and collect biomaterials such as proteins and DNA in the ethanol.[40] Biomaterials can be gathered on the pressure node in the ultrasonic standing wave field. However, since the interactions between biomaterial with the scale of proteins and surface are quite different with μm-sized particles, to understand the involved interactions are critical to modulate protein behaviors over the surfaces.

In this chapter, the vibration-based antibiofouling mechanism is investigated. The vibration is activated by a piezoelectric material (lead zirconium titanate, PZT). The antifouling mechanism is based on overcoming adsorption interactions between protein and surface and moving proteins by acoustic flow. Theoretical calculation of those surface interactions and experimental results of protein adsorption are discussed. The antifouling performance of the vibration-based

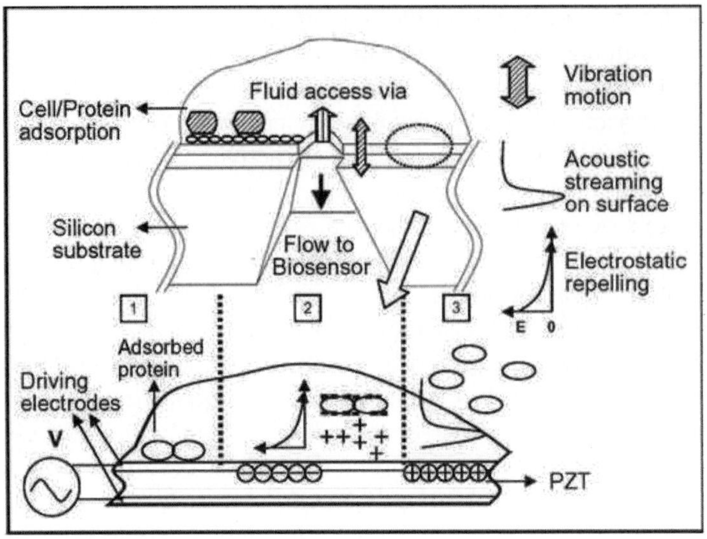

Figure 9.1. Schematic diagram of the proposed piezoelectric membrane as an implantable sensor coating. The membrane is composed of two driving electrodes with a piezoelectric material in between. Adsorbed protein can be desorbed by an electric field and carried away by acoustic flow generated by the vibration. Reproduced with the permission from Yeh *et al.* (2007).

MEMs fabricated device is evaluated by the criteria of protein quantity on the surface. Figure 9.1 shows the schematic diagram of an antibiofouling mechanism which is able to integrate an implant, such as, biosensor. Three steps of this mechanism shown in the Fig. 9.1 are: (1) protein adsorption, (2) protein desorption by surface charges, and (3) the desorbed proteins swept by the flow.

9.2 PROTEINS VISUALIZATION

There are many methods to identify or visualize the protein adsorbed on the surface, to name a few, silver stain, radioactive labeling, reverse stain, fluorescence, chemiluminescence,[41] AC impedance, quartz crystal nanobalance,[42] and so on. Mainly the fluorescence detection and the electrochemical impedance measurement are used to investigate the adsorption of proteins In this chapter. Fluorescence detection can determine the relative amount of adsorbed proteins,[10,43,44] the advantage of this method is its linearity (intensity vs quantity) and sensitivity. It is a relatively convenient method to detect proteins. By attaching fluorescence dyes of emitting different colors, it is easy to identify different proteins on the surfaces.[45] But it is not the best method to quantify adsorbed proteins especially if the substrate has a fluorescence background. The electrochemical impedance spectroscopy (EIS), on the other hand, is sensitive to the adsorption on interface. It also can be used to investigate the dynamic adsorption behavior of protein and the thickness of adsorbed proteins.[42,46–48]

For the former method, many fluorescence dyes with different emitting colors are commercialized already and can be chosen to attach onto proteins before or after adsorption. They usually are excited by a specific wavelength of light and emit a specific longer wavelength of light, for example, Fluorescein IsoThio-Cyanate (FITC) can be excited by blue light and emit green light. The fluorescence photos of surfaces (PZT) adsorbed by BSA-FITC and the relation between intensity and BSA concentration are shown in Fig. 9.2(a). The fluorescence intensity is linearly depends on the quantity of adsorbed BSA. In addition, from Fig. 9.2(a), the intensity is a function of BSA solution concentration and incubation time and the function can be portrayed well by the Langmuir adsorption isotherm.

The EIS measures the impedance between two electrodes. The impedance (Z) is like resistance, and follows the Ohm's law; and is not like resistance because that it depends on the frequency of signal and voltage and current of signal is unnecessarily in phase. The measurement is usually a three electrodes setup: working, counter, and reference electrodes. The surface of interest, such as protein adsorption surface, is as working electrode, the much larger inert material, such as platinum, is as counter electrode. Two plots, Nyquist and Bode plots, are used to interpret the measured data.[46] Nyquist plot separates Z into real and imaginary parts, resulted from the equivalent electrical circuit, can represent the parameters, such as resistance and capacitance, between working and counter electrodes. Bode plots shows Z or phase with the function of frequency so that reveals the frequency information. The Nyquist plot of BSA adsorbed surface is shown in Fig. 9.2(b). The inset is the equivalent circuit of the plot, where R_{el} represents the electrolyte resistance between working and reference electrodes, R_{ct} and constant-phase element (CPE) are charge-transfer resistance and electric double layer capacitance between electrolyte and working electrode. If there is protein adsorption on the surface, the R_{ct} and CPE will change. For example, as shown in Fig. 9.2(b), the R_{ct} decreases 90% when the surface was incubated in BSA solution for 1 hour because that BSA can help charge transfer from the electrolyte to the electrode. In addition, the surface after vibration shows increasing resistance which means less proteins adsorbed on the surface. The vibration on the attenuation of protein adsorption will discuss in the later section.

9.3 INTERACTIONS BETWEEN PROTEINS AND SURFACE

In the scale of protein, some interactions between protein and surface are critical: van der Waals (VDW), electrostatic double-layer (EDL), and hydrophobic interactions.[49,50] Since all the experiments were performed on a piezoelectric plate, all parameters related to the surface were those of piezoelectric plate. In this chapter, we use PZT as the material.

Van der Waals force includes momentary attractions between molecules, diatomic free elements, and individual atoms and other than those due to covalent bonds or to the electrostatic interaction of ions with one another or with neutral

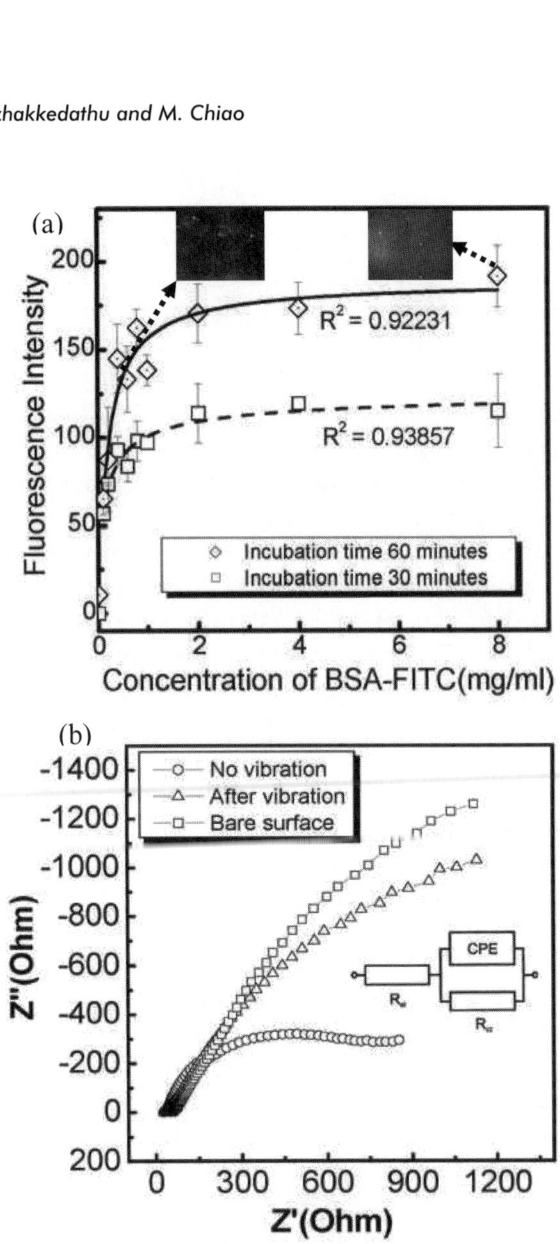

Figure 9.2. (a) BSA adsorption isotherms on a piezoelectric (PZT) surface incubated at 37°C for 1 hour and 30 minutes. The fluorescence intensity was transfer from tri-color to 256 gray scale by Adobe Photoshop 6.0. The intensity is linearly to the quantity of adsorbed BSA, and is a function, which follows the Langmuir isotherm, of BSA concentration. Two inserted photos are fluorescence photos in pointed experimental condition. (b) Nyquist plot of impedance between working electrodes (PZT) and platinum counter electrode. The voltage data (in Z measurement) is corresponding to the reference voltage. The inset is the corresponding equivalent circuit. Reproduced with the permission from Yeh *et al.* (2007). For color reference, see page 274.

molecules. And it comes from the temporary fluctuating dipoles of molecules and is correlated as the molecules come closer to originate an attractive force. There is always a VDW force between two molecules or surfaces, and has a power-law dependence on the separation between protein and surface. The VDW force,

F_V(N), between a sphere (radius R) and a flat surface can be written as:[49,50]

$$F_V = -HR/6D^2 \; (N) \tag{9.1}$$

where H is the Hamaker constant, which is 10^{-20} (J) for protein interacted with water or salt solution,[51] and D (nm) is the separation between sphere and the surface.

The VDW interaction energy(J), E_V, between a sphere (radius R) and a flat surface is hence as:[49,50]

$$E_V = -HR/6D \; (J) \tag{9.2}$$

In addition to VDW, the other major force that exists between two molecules in aqueous solution is electrostatic double layer force. VDW and EDL are actually the forces described in so-called DLVO theory, which was developed by the Derjaguin, Landau,[52] Verwey, and Overbeek[53] for colloid stability. This electrostatic force between two similarly charged molecules is repulsive and is exponential decaying with the separation. The electrostatic double-layer force $F_e(N)$ between a sphere (radius R) and a flat surface can be written as:[49,50]

$$F_e = -\kappa R Z e^{-\kappa D} \; (N) \tag{9.3}$$

where Z is a constant derived from the Gouy-Chapman electrostatic double-layer model. And $1/\kappa$ is the Debye length, for the NaCl-based electrolyte buffer at 37°C, it can be written as

$$\frac{1}{\kappa} = \frac{0.3}{[NaCl]^{0.5}} \; (nm) \tag{9.4}$$

For example, for a molecule in a phosphate buffer solution (0.15M NaCl, 0.0067M PO4) 37°C, $1/\kappa$ is 0.8 nm. The constant Z in Eq. (9.3) at 37°C can be written as:[49,50]

$$Z = 64\pi\varepsilon\varepsilon_o (kT/e)^2 \tanh^2(ze\psi_o/4kT) \; (J/m) \tag{9.5}$$

where ε_o is vacuum permittivity, ε is the buffer permittivity, k is Boltzmann constant, T is the temperature in Kevin, z is the electrolyte valence, e = -1.6*10^{-19} C, and ψ_o is the surface potential. The surface potential ψ_o is related to surface charge density σ given by the Graham equation for 1:1 electrolytes such as NaCl at 37°C:[49,50]

$$\sigma = 0.116 \tanh(\psi_o/53.4)[NaCl]^{0.5} \; (C/m^2) \tag{9.6}$$

Experimentally, σ of a vibrating piezoelectric plate can be calculated by Eq. (9.7):[49,50]

$$\sigma = Q/A = CV/A \; (C/m^2) \tag{9.7}$$

were Q is surface charges, V is the surface voltage, C is the capacitance of piezoelectric material in air at resonance frequency, and A is the plate area. By measuring V, by a voltmeter, and C, by an impedance spectroscopy, and implanting in the Eq. (9.7), the surfaces charges, σ, can be calculated. By applying AV (frequency is from $10^2 \sim 10^6$ Hz) on a PZT plate, the impedance of the plate is changing with the frequency (f). The section (cover the sweeping frequencies in the following experiment) of linear dependence between log Z and log f can deduce the capacitance of PZT plate in this range of frequency. The surface potential, ψ_o, is estimated by Eq. (9.6) after getting σ. The constant Z is computed from eq. (9.5) after calculation of ψ_o. Finally, EDL force and can be calculated by Eq. (9.3). The EDL energy is written as:[49,50]

$$E_e = -RZe^{-\kappa D}(J) \tag{9.8}$$

For only DLVO interactions, the energy summation is Eq. (9.2)+(9.8).

Hydrophobic interaction energy is an attractive interaction in water between inert, non-polar molecules or surfaces, and it is much larger than VDW force between non-polar molecules or surfaces. The hydrophobic interaction energy between a sphere and a flat surface is as the following:[54]

$$E(D) = -1.903^*kT^*Re^{D/14}(J) \tag{9.9}$$

The hydrophobic interaction energy will be affected by the hydrophobicity of surface, which is related to the surface tension of the liquid-surface. The surface tension can be evaluated by measuring water contact angle on the surface. The following Young-Dupré equation depicts that surface tension is a function of contact angle:[49]

$$\gamma_i = \gamma_{SL} = \gamma_S - \gamma_L cos\theta \tag{9.10}$$

where γ_i, γ_S, γ_L are surface energies of surface-liquid, surface-air, and liquis-air, respectively. θ is the contact angle, which is the angle between interfaces of liquid-air and liquid-surface. The larger the contact angle, the more hydrophobicity is. For example, for the partially hydrophobic surfaces, the contact angle is in the range of $75{\sim}90°$, γ_i is 6-30 mJ/m^2, and for the fully or strongly hydrophobic surfaces, the contact angle is $100{\sim}115°$ γ_i is 40–55 mJ/m^2.

For evaluating these three interactions between a molecule and the surface, BSA in a PBS buffer at 37°C adsorbed on PZT was taken as the example. The dimension of a BSA molecule is 14 nm × 4 nm × 4 nm[55] and is approximating to a sphere with 3.04 nm ($\sim 0.5(14^*4^*4)^{1/3}$) geometric mean radius (R). The contact angle of PZT surface 77.8 °. The capacitance of PZT is evaluated from impedance measurement, and surface voltage (−1V, in the following calculation) is measured by the voltage probe. Hence, these three interaction energies with the functions of separation can be calculated. VDW, EDL, hydrophobic, and the total energy with separation are shown in the Fig. 9.3. The variables used in Eq. (9.1)–(9.9) are listed in Table 9.1, the D in the Table 9.1 is the cut-off distance, which represents the effective separation between two surfaces or particles in contact and is 0.3 nm for most of the biological case.[50] From Fig. 9.3, when a BSA approaches the surface, it is first attracted because that the total of VDW and hydrophobic interactions is much larger than EDL. And then it stops at the distance of ~1.2 nm. For the

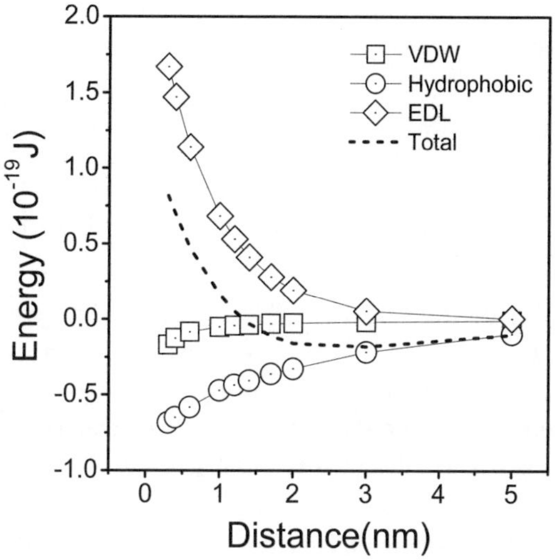

Figure 9.3. Calculated protein-surface interaction energy between a BSA molecule and a negatively charged PZT surface (pH = 7.4). In the proximity of the surface, the total attractive energy of VDW and hydrophobic interaction can be compensated by the repelling electrostatic interaction.

Table 9.1 Variables used in eqs. (9.1)–(9.10). Reproduced with the permission from Yeh *et al.* (2007)

		Electrosttic Interaction		Hydrophobic Interaction		VDW Interaction		Total
1/k (nm)	0.8	ψ_o (mV)	-175	R (nm)	3.04	R(nm)	3.04	
σ (C/m^2)	0.5962	Z (J/m)	$8.06{*}10^{-11}$	k	$1.38{*}10^{-23}$	D (nm)	0.3	
C (farad)	$1.6{*}10^{-5}$	R (nm)	3.04	T (K)	310	H (J)	10^{-20}	
V (volt)	-1	D(nm)	0.3	D (nm)	0.3			
E_e= $1.67{*}10^{-19}$				E_h= $-6.82{*}10^{-20}$		E_V= $-1.69{*}10^{-20}$		E_T= $8.18{*}10^{-20}$

1. D is the distance from surface when the BSA is adsorbed on the surface. We used a cut-off distance of 0.3∼0.4 nm.[50]
2. The calculated magnitude of attracted protein-protein-surface forces is compared with the report in the literature.[56] Blomberg *et al.* measured a force-distance profile of human serum albumin (HSA) and a mica surface by a protein-surface force apparatus. The average hydrophobic energy between a HSA molecule and a mica surface is around $3.6{*}10^{-19}$ (J), which is comparable to the $6.82{*}10^{-20}$ (J) listed in Table 9.1.

BSA adsorbed on the surface without conformation change, it is repelled from the surface to that separation. From Fig. 9.3, VDW and hydrophobic attractions are possibly overcome by EDL repulsion.

Table 9.2 Protein adsorption on positively and negatively charged surfaces. Reproduced with the permission from Yeh *et al.* (2007).

Voltage(V)	Fluorescence Intensity		Normalized
−1	58.92 ± 4.7	n = 3	35.2 ± 2.8%
+1	165.38 ± 10.34	n = 3	100 ± 6.3%

The calculation is verified by the experiment. A DC voltage is applied on the PZT surface in BSA solution and the results are shown in Table 9.2. A −1V potential level surface has much less BSA adsorption than +1V surface. In pH 7.4 solution environment, BSA is negatively charged (isoelectric point of BSA is 4.9), so BSA prefers to adsorbs on the positive charged surface than negative charged surface.

The surface charges control the quantity of protein adsorption. Striemer *et al.* fabricated a 15 nm nano crystalline silicon membrane with pore size 5-25 nm. By applying different charges on the membrane, they had proved the separation of similarly sized molecules carrying different charges (molecules carry different charges with the surface can not pass through the pores, while with similarity of charges can pass the pores).[57]

Pasche *et al.* also shown the importance of electrostatic forces in protein adsorption.[21] In their experiment, three small model proteins (Lysozyme, Myoglobin and R-lactalbumin) at different ionic strengths on the five types of surfaces were observed. Lysozyme is similarly to a hard, positively charged particle, and is repelled by positively charged surfaces. The hard protein owns the similarity to a solid sphere and can be described well by above equations. The BSA used in Fig. 9.3 and Table 9.1 belongs to such kind of proteins. The theoretical calculation of electrostatic interaction was investigated. One example is from Sivasanar *et al.*[58] They used surface force apparatus to determine the pH dependent electrostatic charge density of a single binding face of streptavidin. And this direct approach to probe the electrostatic features of proteins is applicable to investigations of other kind of proteins with different charges.

9.4 SHEAR STRESS ON THE PROTEIN

The vibration of piezoelectric material in the liquid generates an acoustic flow over the surface. And the effect of shear stress on a BSA is evaluated by the rotating plate system since there is no surface charges accumulation so that the effect mentioned in Sec. 3 can be ruled out. The rotating plate setup is shown in Fig. 9.4(a), which consists of a rotor connected to a sample Cu plate coated with the same material of the electrodes of piezoelectric plate. The fluid wall shear stress (τ_ω,dyn/cm^2) at the surface of the rotating plate is:[59]

$$\tau_\omega = 0.8\eta r(\omega^3/v)^{1/2} \qquad (9.11)$$

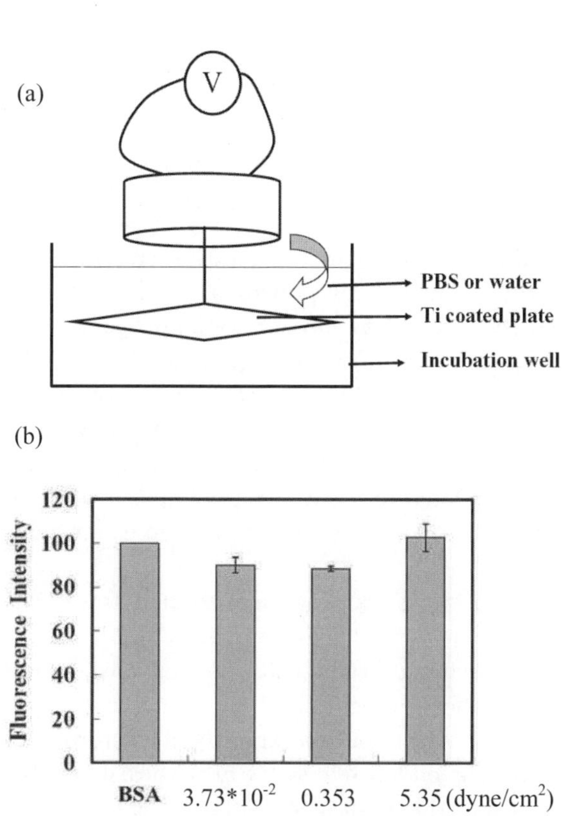

Figure 9.4. (a) Schematic diagram of rotating plate system. (b) Fluorescence intensity of different rotating plate surfaces with different shear stresses applied on one BSA molecule. Reproduced with the permission from Yeh *et al.* (2007).

where η(cPoise) is the viscosity of the medium, r(cm) is the radial distance from the center of the disk, ω(rpm) is the radial velocity of the rotating disk (which can be measured by laser doppler vibrometer (LDV)), and v(Stokes) is the kinematic viscosity of the medium. Properties of buffer solution for experiment are close to those of water at 37°C, in which, η and v are 0.89 (cPoise) and 0.008583 (Stokes), respectively. The shear stress is a function of radius, so the average shear stress over the whole plate is:[59]

$$\int 0.8\eta(\omega^3/v)^{1/2} r 2\pi r dr / \pi r^2 \tag{9.12}$$

The wall shear stress on an individual BSA molecule can then be estimated by knowing the surface area of BSA, which is simplified as a sphere with radius of 3.04 nm. The shear stress on one BSA molecule at different rotating speeds is listed in Table 9.3; for example, for a rotating speed of 40.79 Hz, the average shear stress is $0.267*0.8*8.9*10^{-3}*((40.79)^3/8.583*10^{-3})^{0.5} = 5.35$ dyne/cm^2 on the circular area of plate from radius 0.1 to 0.38 cm. Now, consider the shear stress from acoustic flow induced by vibration on one BSA molecule. The shear stress is defined as $\tau = \eta\partial u/\partial z$, where η is the viscosity of the medium, u is the flow velocity and y is the distance from the surface. The flow velocity is related to the vibration amplitude of the surface as the following:[60]

$$u = \frac{5}{4}\frac{(fAmp)^2}{v_p}\left(\frac{\kappa_S}{\alpha}\right)^3\left(\frac{\alpha}{\kappa_S}+\frac{\kappa_S}{\beta}\right)\left(1-\frac{d\alpha}{2}\right)^2 \tag{9.13}$$

where u is the maximum velocity of the acoustic stream; f is the frequency of vibration; Amp is the z-directed component of membrane vibration amplitude; v_p is the phase velocity of the plate wave; κ_s is the propagation constant, equal to $2\pi/\lambda$; λ is the wavelength of FPW; $\alpha = \kappa s(1\text{-}(v_P/v_F)^2)^{1/2}$; v_F is the sound velocity in PBS; $\beta^2 = \rho f/2\eta$, which relates to the viscosity effect of PBS; ρ is the density of PBS; η is the dynamic viscosity of PBS; and d is the membrane thickness. From Eq. (9.13), the flow velocity is proportional to the square of the vibration amplitude, which is linearly related to the driving voltage. This relationship between vibration amplitude and flow velocity has also been experimentally observed in the literature.[60,61]

The velocity of acoustic streaming is calculated from Eq. (9.13), which is $8.84*10^{-6}$ m/s (10 V_{pp}) and is the maximum velocity parallel to the plate surface. The flow on the wall is assumed to be no-slip, hence the velocity at the wall is zero. The velocity of flow reaches a maximum at the evanescent decay length, which is the reciprocal of α in Eq. (9.13). For our experimental setup, $1/\alpha$ equals $3.26*10^{-3}$ m. Assuming that the dependence of velocity with distance from wall is linear, the shear rate of acoustic streaming is $8.84*10^{-6}/3.26*10^{-3} = 2.71*10^{-3}$ 1/s. The shear stress, $\tau = \eta\partial u/\partial z$, is $8.9*10^{-4}$ (kg/m.s) * $2.71*10^{-3}$ (1/s) = $2.41*10^{-6}$ N/m^2 = $2.41*10^{-5}$ dyne/cm^2 where η is $8.9*10^{-4}$(kg/m.s). Parameters for calculating shear stress from acoustic streaming are also listed in Table 9.3.

The fluorescence intensities for plates rotating at different speeds for 5 min were compared with plates without rotation (Fig. 9.4(b)) and no significant difference occurs between different speeds. Since shear stresses induced by a rotating

Table 9.3 Shear stress applied on a BSA molecule. Reproduced with the permission from Yeh *et al.* (2007)

	Applied Voltage ($v_p p$)	Rotating Speed (Hz)	Shear Stress (dyne/cm^2)	
Rotating Plate	0.7	1.49	3.73×10^{-2}	
	1	6.66	3.53×10^{-1}	
	3	40.79	5.35×10^{0}	
Vibrating PZT	Applied Voltage ($V_p p$)	Max Vibration Amplitude (nm)	Shear Rate (1/s)	Shear Stress (dyne/cm^2)
	10	500	2.71×10^{-3}	2.41×10^{-5}

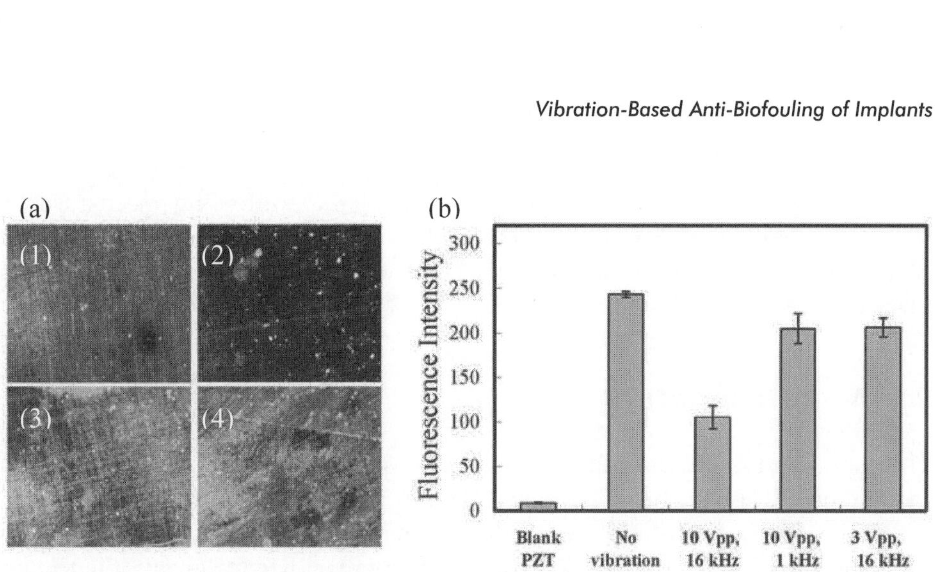

Figure 9.5. ((a) Fluorescence microscope images of PZT plate surfaces with four vibrating conditions: (9.1) No vibration, (9.2) 10 V_{pp} at 16 kHz, (9.3) 10 V_{pp} at 1 kHz, and (9.4) 3 V_{pp} at 16 kHz for 5 minutes. (b) BSA adsorption measurement based on Fig. 9.2(a). Reproduced with the permission from Yeh *et al.* (2007).

plate are much larger than those of acoustic streaming, we conclude that the acoustic streaming shear stress itself cannot desorb BSA from a PZT plate surface.

For the DC application only, the surface charges can be produced and play a role on protein desorption. For the flow application only, the shear stress seems not important to protein desorption. When an AC sigmoid is applied on a piezoelectric plate, the plate can vibrate in different modes at resonance frequencies. With the vibration, the flow is generated and applies shear stress on the BSA. At the same time, the charges accumulate on the surface. The relative quantity of BSA adsorption at different vibration conditions (PZT plates) are shown in Fig. 9.5. For the PZT plate with dimension 1cm $\times$ 0.5 cm $\times$ 0.5 mm, the resonance frequency is 16 kHz in the PBS buffer at 1st blending mode. Interestingly, the strong attenuation of BSA adsorption is seen under that resonance frequency of PZT in buffer (Fig.9.5(a(9.1)&(9.2)), the photos had been transferred to gray scale). However, for the surfaces under vibration at lower AC amplitude or off-resonance, the less attenuation is observed (Fig. 9.5(a(9.3)&(9.4))). The former is related to less surface charges and shear stress, and the later is related to less shear stress. For AC sigmoid application, the attenuation is also shear stress modulated.

The surface charges accumulation under an AC sigmoid is estimated to compare with the case of DC application. One half period of the 16 kHz AC sigmoid is $(1/16\text{kHz})/2 = 3.125*10^{-2}$ms. The resistance of the whole system in PBS is 2~5 Ω, and the capacitance of the PZT plate at 16kHz is $1.6*10^{-5}$ F. The time constant (resistance $\times$ capacitance) of the plate is $3.2\sim8*10^{-2}$ms. The average voltage in one half AC sigmoid of 10 V_{pp} is $2*5/\pi = 3.18$ V. Hence, within one half period, the PZT plate can be charged to the surface voltage of 3.18*(1-exp(−3.125/8))~3.18*(1-exp(−3.125/3.2)) = 1.03 to 1.98V.[62] The amount of surface voltage is comparable to the DC voltage listed in Table 9.2. Therefore, for the above AC sigmoid, the accumulated surface charges are larger enough to attenuate

BSA adsorption. So under the AC application, the antifouling mechanism is two steps, firstly, the BSA is desorbed by the surface charges, and secondly, the shear stress prevents the re-adsorption of BSA when the surface charges becomes positive because of the AC application.

9.5 MICROFABRICATION

The vibration-based antifouling mechanism is applied on the micromachined membrane.[63] The scheme of setup is shown in Fig. 9.6(a). The device starts from the SOI wafer (Si(2 μm)-SiO_2(1 μm)-Si(400 μm)) as substrate. The process is briefly described as following. Firstly, SOI wafer is thermal oxidized to grow SiO_2 around SOI wafer as protection layer. The cavity (2000 μm × 500 μm) within the substrate is anisotropic etched from the 400 μm silicon side by 20% Tetra Methyl Ammonium Hydroxide (TMAH) at 85°C. During the etching, areas except cavity opening are protected by SiO_2. The TMAH etching stops on SiO_2 of SOI wafer. After TMAH etching, the protective thermal SiO_2 was removed and the silicon/SiO_2 membrane is then remained. The diced PZT plate (3000 μm × 1000 μm × 500 μm) is attached on the silicon/SiO_2 membrane by silver epoxy. The SEM photos of the front view and back view of the device are shown in Fig. 9.6(b) and 9.6(c), respectively. The cavity in Fig. 9.6(c) can be filled with protein solution for protein adsorption and PBS for washing, only back side of the membrane is contacting the solution. During the vibration and protein incubation, the plastic cover is shed on the cavity.

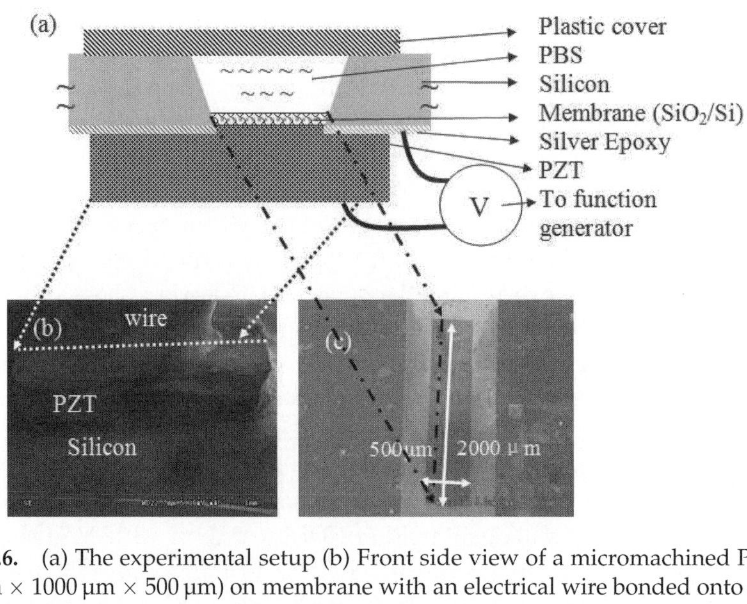

Figure 9.6. (a) The experimental setup (b) Front side view of a micromachined PZT plate (3000 μm × 1000 μm × 500 μm) on membrane with an electrical wire bonded onto the plate surface (c) The back side view of the Si/SiO_2 membrane. In the following experiments, the proteins were adsorbed onto the SiO_2 side of the Si/SiO_2 membrane. Reproduced with the permission from Yeh *et al.* (2008).

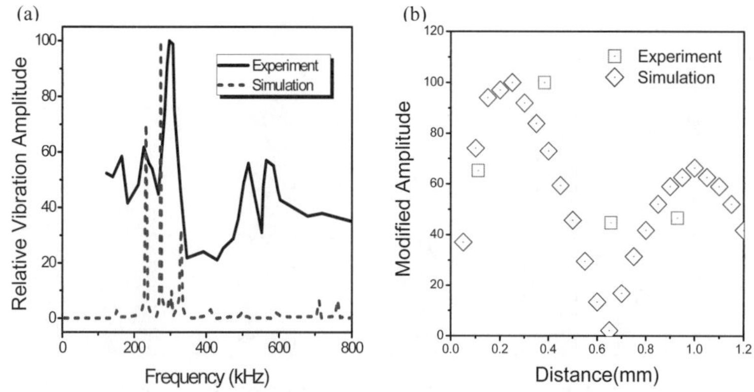

Figure 9.7. (a) The simulation and experimental vibration spectrum of the PZT plate/siliconmembrane. The vibration at 308, 320, 500, and 575 kHz correspond to 1.5, 2.5, 3.5 wavelength bending mode, and longitudinal mode, respectively. (b) The simulation and experimental vibration amplitude distribution across the Si/SiO_2 membrane along the membrane center line. The x axis is given as the distance from the center of the membrane to the left periphery. The half length of the membrane is 1000 μm. Reproduced with the permission from Yeh *et al.* (2008). For color reference, see page 274.

The vibration of membrane is measured by LDV and simulated by finite element software (ANSYS 8.0). Fig. 9.7 (a) shows the vibration spectrum at the center of a Si/SiO_2 membrane, with cavity filled with PBS solution. The PZT was driven by a 10 V_{pp} sinusoid waveform from 120~900 kHz to find the resonance frequency. The observed resonance frequency with largest vibration amplitude is 308 kHz, and the out-of-plane vibration amplitude at 308 kHz is 105 nm. Simulation result matches well with the experiment (other minor resonance frequencies and major resonance frequency, which is(269 kHz, compared to measured 308 kHz). Figure 9.7(b) shows the displacement at different locations of the membrane at the major resonance frequency (269 kHz). The displacement is defined as the distance measured from the center of the membrane to left periphery. Both simulation and experiment show that the vibration of membrane is in a FPW mode with the wavelength of 4000 µm at major resonance frequency. The vibration at 308, 320, 500, and 575 kHz in Fig. 9.7(a) correspond to 1.5, 2.5, 3.5 wavelength bending mode, and longitudinal mode, respectively.

The FPW mode happens when the thickness of plate is much less than the wavelength of FPW.[64,65] If contacted with fluid, the advantage of the FPW mode is that the phase velocity is lower than the wave velocity in fluid. Hence, the energy can be restricted within an so-called evanescent decay length and can be transferred to generate acoustic flow near the surfaces.

The attenuation of proteins adsorption by the membrane is tested on BSA, IgG, and blood plasma. Figure 9.8 shows the results of relative protein desorption due to vibration. About 57 $\pm$ 10% of BSA and 47 $\pm$ 13% of IgG were removed from the membrane surface after the vibration was applied (10 V_{pp} at major resonance frequency). This shows the same mechanism works on both PZT plate and

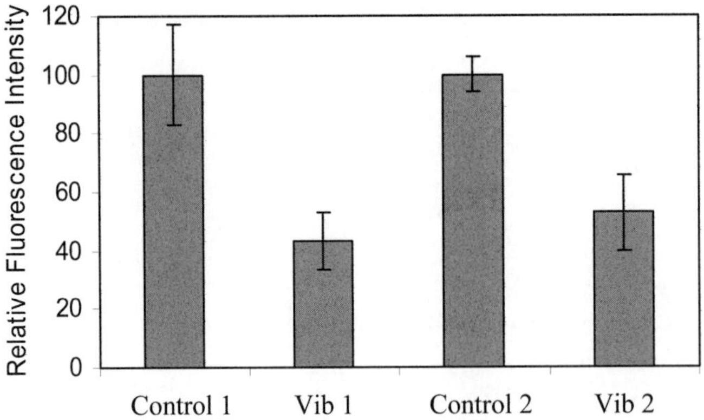

Figure 9.8. Comparison of protein adsorption with and without PZT vibration. The vibration was initiated by a 10 V_{pp} AC signal at 308 kHz for 5 minutes. The proteins were incubated in 2 mg/ml BSA and 0.1 mg/ml IgG for 1 hour. Control 1 and 2 are the surfaces incubated in BSA and IgG solution, respectively. For Vib 1 and Vib 2, BSA and IgG adsorbed surfaces, respectively, were used. The fluorescence intensity at every condition was measured after washing the surface after that condition 5 times by PBS. Reproduced with the permission from Yeh *et al.* (2008).

micromachined membrane. The BSA adsorption in two other vibration conditions: (9.1) lower AC voltage input (1 V_{pp}) at 308 kHz (resonance frequency) and (9.2) 10 V_{pp} at 400 kHz (off-resonance frequency) are also tested. Both of the surfaces under such a vibration condition have less attenuation of BSA adsorption (data is not shown). Under control of the same mechanism, for micromachined vibrating being able to reduce protein adsorption, it is need sufficient large surface charges and strong shear stress.

The surface charge comes from the applied voltage on a dielectric material. The surface voltage was measured on the SiO_2 side of the membrane in PBS when a DC or AC was applied on the PZT plate. Only by DC application, the attenuation of BSA adsorption is also observed data is not shown. By AC application, the period for the surface to accumulate negative charges was (1/308 kHz)/2 = $1.623*10^{-3}$ms The average voltage of a 6.5 V_{pp} sinusoid (voltage measured with the cavity filled with buffer if 10 V_{pp} is applied) is $2*3.25/\pi = 2.1$ V. In addition, the time constant, 2~5 Ω (system resistance) multiplied by $2.3*10^{-11}$ F (membrane capacitance), is $4.6{\sim}11.5*10^{-8}$ ms. The much longer period of $1.623*10^{-3}$ms than time constant ($4.6{\sim}11.5*10^{-8}$ ms) means that surface can be fully charged to 2.1 V.

Similar experiments had also been tested on the blood plasma. Plasma consists of mainly albumin (50~60%), immunoglobulins (15~35%), fibrinogen (~3%), and other regulatory proteins.[66] About 55.3 ± 5% of plasma proteins were desorbed from the surface due to vibration (Fig. 9.9(a)). The vibrated surface (protein desorbed) was further incubated with plasma for another 20 minutes and vibrated again to remove the adsorbed plasma proteins. The second cycle of adsorption and the vibraton cycle resulted in the removal of 56.4 ± 12% of adsorbed proteins,

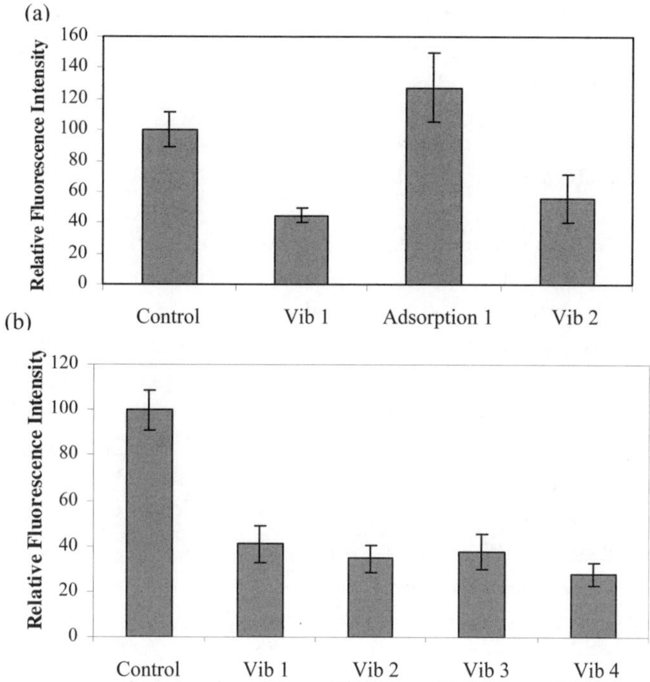

Figure 9.9. Comparisons of (a) plasma protein adsorption at different times without and with PZT vibration, (b) plasma protein adsorption at 4 hours without and with PZT vibration in different media and time intervals. The conditions in (a) are Control incubation in plasma protein solution for 1 hour; **Vib 1** vibration application in PBS for 5 minutes after incubation; **Adsorption 1** second plasma protein incubation for 20 minutes after **Vib 1** on the sample sample; **Vib 2** vibration application in PBS for 5 minutes after Adsorption 1 on the sample sample. The conditions in (b) are **Control** Incubation in plasma protein solution for 4 hours; **Vib 1** vibration application in plasma prtein solution for 5 minutes; **Vib 2** vibration application in PBS for 5 minutes after **Vib 1**; **Vib 3** vibration application in PBS for 10 minutes after Vib 2; **Vib 4** vibration application in PBS for 20 minutes after Vib 3. All vibrations were applied by a 10 V_{pp} AC signal at 308 kHz. The plasma protein concentration was 0.3 mg/ml. The fluorescence intensity at every condition was measured after washing the surface after that condition 5 times by PBS. Reproduced with the permission from Yeh *et al.* (2008).

which shows that continuous protein desorption can be achieved by appropriate vibration strategies.

Another experiment where tested to show the long term effect, the membrane surface was exposed to plasma proteins for 4 hours and vibrated at different conditions (Fig. 9.9(b)). Results show that ~ 59.2 to 72.1 % of the adsorbed proteins can be removed by vibration at various conditions. The reason no statistical difference between conditions in Vib 1, 2, 3, 4 in Fig. 9.8(b) is that the loosely bounded proteins were effectively removed in Vib 1. It seems to increase vibration time does not help to further remove protein from surface. A comparision of the relative fluorescence intensities of the control samples at 1-hour incubation

(Fig. 9.8(a)) and 4-hour incubation (Fig. 9.8(b)) shows that the 4-hour incubated sample has ~69.5 % higher protein adsorption than the 1-hour incubated sample. Neverthless, vibration removed most of the adsorbed

9.6 CONCLUSIONS AND OUTLOOK

A novel anti-fouling mechanism based on the combined effects of electric field and shear stress is reported. The mechanism was observed in mm scale piezoelectric plates coated with different metal materials and MEMs (MicroElectroMechanical system) fabricated membrane based device. The piezoelectric material, lead zirconate titanate (PZT) composite, coated with electrodes after application of voltage (DC or AC) can (1) accumulates the surface charges, which generates an electric field against proteins adsorption (with the similar charge of proteins); and (2) induced an acoustic streaming shear stress, which attenuates the desorption of proteins. In vitro characterization showed that 58 ± 5.5% of adsorbed bovine serum albumin (BSA) can be effectively removed from PZT. Theoretical calculations of protein-surface interactions (van der Waals (VDW), electrostatic, and hydrophobic) and shear stress reveal the mechanism of attenuation of protein (BSA) adsorption.

On the other hand, a micromachined vibrating membrane is proved to be able to attenuate proteins adsorption as well. The (2000 μm x 500 μm x 3 μm) silicon membrane attached by a lead zirconate titanate (PZT) composite (3 mm × 1 mm × 0.5 mm) can vibrate in a FPW, which is also verified by FEM simulation. The surface charges on the membrane and the fluid shear stress contribute in reducing protein adsorption on the membrane surface. In vitro characterization shows that 57±10% of the adsorbed BSA, 47±13% of IgG, and 55.3~59.2±8% of the blood plasma proteins are effectively removed from the vibrating membrane. Potentially, a microelectromechanical system (MEMS)-based vibrating membrane could be a tool to minimize biofouling of *in vivo* implants.

ACKNOWLEDGMENTS

This work was supported by the Natural Sciences and Engineering Research Council of Canada (NSERC), Canada Foundation for Innovation (CFI). I would like to thank Dr. M. Chiao, and Dr. J. N. Kizhakkedathu for the valuable comments on the early drafts.

References

1 M. Schlosser and M. Ziegler, in *Biosensors in the Body: Continuous in vivo Monitoring*, Ed. D. M. Fraser (John Wiley & Sons, Chichester, 1997).

2 D. M. Fraser, in *Biosensors in the Body: Continuous in vivo Monitoring*, Ed. D. M. Fraser (John Wiley & Sons, Chichester, 1997).

3 N. Wisniewski, F. Moussy and W. M. Reichert, Fresenius, *J. Anal. Chem.* 366, 611 (2000).

4 N. Wisniewski and M. Reichert, *Colloids Surf., B, Biointerfaces* 18, 197 (2000).

5 P. B. Messersmith and M. Textor, *Nat. Nanotechnol.* 2, 138 (2007).

6 B. D. Ratner and S. J. Bryant, *Annu. Rev. Biomed. Eng.* 6, 41 (2004).

7 J. M. Anderson, *Cardiovasc. Pathol.* 2, S33 (1993).

8 K. M. Ainslie, G. Sharma, M. A. Dyer, C. A. Grimes and M. V. Pishko, *Nano Lett.* 5, 1852 (2005).

9 R. Gref, M. Lück, P. Quellec, M. Marchand, E. Dellacherie, S. Harnisch, T. Blunk and R. H. Müller, *Colloids Surf., B, Biointerfaces* 18, 301 (2000).

10 T. T. Huang, J. Sturgis, R. Gomez, T. Geng, R. Bashir, A. K. Bhunia, J. P. Robinson and M. R. Ladisch, *Biotechnol. Bioeng.* 81, 618 (2003).

11 C. M. Nolan, C. D. Reyes, J. D. Debord, A. J. Garcia and L. A. Lyon, *Biomacromolecules* 6, 2032 (2005).

12 R. G. Nuzzo, *Nat. Mater.* 2, 207 (2003).

13 K. L. Prime and G. M. Whitesides, *J. Am. Chem. Soc.* 115, 10714 (1993).

14 C. Allen, N. D. Santos, R. Gallagher, G. N. C. Chiu, Y. Shu, W. M. Li, S. A. Johnstone, A. S. Janoff, L. D. Mayer, M. S. Webb and M. B. Bally, *Biosci. Rep.* 22, 225 (2002).

15 E. A. S. Doherty, R. J. Meagher, M. N. Albarghouthi and A. E. Barron, *Electrophoresis* 24, 34 (2003).

16 H. Jacobs, D. Grainger, T. Okano and S. W. Kim, *Artif. Organs* 12, 506, (1988).

17 H. Makamba, J. H. Kim, K. Lim, N. Park and J. H. Hahn, *Electrophoresis* 24, 3607 (2003).

18 A. Whelan and F. Regan, *J. Environ. Monit.* 8, 880 (2006).

19 S. Chen, L. Liu, J. Zhou and S. Jiang, *Langmuir* 19, 2859 (2003).

20 G. B. Sigal, M. Mrksich and G. M. Whitesides, *J. Am. Chem. Soc.* 120, 3464 (1998).

21 S. Pasche, J. Vörös, H. J. Griesser, N. D. Spencer and M. Textor, *J. Phys. Chem.* B 109, 17545 (2005).

22 M. Zhang and M. Ferrari, *Biotech. Bioeng.* 56, 618 (1997).

23 A. Sharkawy, B. Klitzman, G. Truskey and W. Reichert, *J. Biomed. Mater. Res.* 40, 586 (1998).

24 C. E. Campbel, A. F. von Recum, *J. Invest Surg.* 2, 51 (1989).

25 G. D. Meyer, J. M. Morán-Mirabal, D. W. Branch and H. G. Craighead, *IEEE Sens. J.* 6, 254 (2006).

26 M. Rahmoune and M. Latour, *Smart Mater. Struct.* 4, 195 (1995).

27 P. Y. Yeh, J. N. Kizhakkedathu, J. D. Madden and M. Chiao, *Colloids Surf., B, Biointerfaces* 59, 67 (2007).

28 G. P. Rigby, P. W. Crump and P. Vadgama, *Analyst* 121, 871 (1996).

29 G. P. Rigby, P. W. Crump and P. Vadgama, *Med. Biol. Eng. Comput.* 33, 231 (1995).

30 A. L. Klibanov, V. P. Torchilin, S. Zalipsky, In *Liposomes*, Ed. V. P. Torchilin and V. Weissig (Oxford University Press, Oxford, 2003), p. 231–265.

31 S. Sharma, R. W. Johnson and T. A. Desai, *Biosens. Bioelectron* 20, 227 (2004).

32 H. Zhu and M. Snyder, *Curr. Opin. Chem. Biol.* 7, 55 (2003).

33 C. E. McNamee, S. Yamamoto, and K. Higashitani, *Biophys. J.* 93, 324 (2007).

34 A. Halperin, *Langmuir* 15, 2525 (1999).

35 P. Roach, D. Farrar and C. C. Perry, *J. Am. Chem. Soc.* 128, 3939 (2006).

36 Q. Qi and G. J. Brereton, *IEEE T. Ultrason. Ferr.* 42, 619 (1995).

37 D. S. Ballantine, R. M. White, S. I. Martin, A. I. Ricco, E. T. Zellers, G. C. Frye, and H. Wohltjen, in *Acoustic Wave Sensors: Theory, Design, and Physico-Chemical Applications* (Academic Press, San Diego, 1997).

38 N. T. Nguyen and R. M. White, *Sensor. Actuat. A-Phys.* 77, 229 (1999).

39 N. T. Nguyen and R. M. White, *IEEE Trans Ultrason. Ferroelectr. Freq. Control* 47, 1463 (2000).

40 K. Yasuda, M. Kiyama, S. Umemura and K. Takeda, *J. Acoust. Soc. Am.* 99, 1248 (1996).

41 W. F. Patton, *J. Chromatogr.* B 771, 3 (2002).

42 J. E. I. Wright, N. P. Cosman, K. Fatih, S. Omanovic, S. G. Roscoe *J. Electroanal. Chem.* 564 185 (2004).

43 R. Bashir, *Adv. Drug Deliv. Rev.* 56, 1565 (2004).

44 F. G. Tseng, S. C. Lin, D. J. Yao, H. Huang and C. C. Chieng, in *Protein Microarrays*, Ed. M. Schena (Jones and Bartlett Publishers, Sudbury, 2004) p. 360-362.

45 P. N. Prasad, in *Introduction to Biophotonics* (John Wiley & Sons, Hoboken, 2003) p. 372–378.

46 S. Omanovic and S. G. Roscoe, *Langmuir* 15, 8315 (1999).

47 J. E. I. Wright, K. Fatih, C. L. Brosseau, S. Omanovic and S. G. Roscoe, *J. Electroanal. Chem.* 550/551, 41 (2003).

48 D. Bizzotto, Y. Yang, J. L. Shepherd, R. Stoodley, J. Agak, V. Stauffer, M. Lathuilli, A. S. Akhtar and E. Chung, *J. Electroanal. Chem.* 574, 167 (2004).

49 D. Leckband and J. Q. Israelachvili, *Rev. Biophys.* 34, 105 (2001).

50 J. Israelachvili, *Intermolecular & Protein-surface Forces*, 2nd ed., (Academic Press London, 1991), pp 176–212.

51 J. J. Marra, *Colloid Interf. Sci.* 107, 446 (1985).

52 B. V. Derjaguin and L. Landau, *Acta Physicochim.* (URSS) 14 633 (1941).

53 E. J. W. Verwey and J. Th. G. Overbeek, in *Theory of Stability of Lyophobic Colloids* (Elsevier, Amsterdam, 1948).

54 S. I. Jeon and J. D. Anderade, *J. Colloid and Interface Sci.* 142, 159 (1991).

55 M. Zhang, T. Desai and M. Ferrari, *Biomaterials* 19, 953 (1998).

56 E. Blomberg, P. M. Claesson and J. C. Froberg, *Biomaterials* 19, 371 (1998).

57 C. C. Striemer, T. R.. Gaborski, J. L. MaGrath, and P. Fauchet, *Nature* 445, 749 (2007).

58 S. Sivasankar, S. Subramaniam and D. Leckband, *Proc. Natl. Acad. Sci.* 95, 12961 (1998).

59 W. Kao, *Biomaterials* 21, 2295 (2000).

60 R. M. Moroney, R. M. White and R. T. Howe, *Appl. Phys. Lett.* 59, 774 (1991).

61 N. T. Nguyen, A. H. Meng, J. Black and R. M. White, *Sensor. Actuat. A-Phys.* 79, 115 (2000).

62 J. R. Cogdell, *Foundations of Electrical Engineering*, (Prentice Hall, Englewood Cliffs, 1990).

63 P. Y. Yeh, Y. Le, J. N. Kizhakkedathu, M. Chiao, *Biomed. Microdevices* published on line Apr 22 2008.

64 M. S. Weinberg, B. T. Cunningham and C. W. Clapp, *J. Microelectromech.* S. 9, 370 (2000).

65 M. S. Weinberg, C. E. Dubé, A. Petrovich and A. M. Zapata *J. Microelectromech.* S. 12, 6645 (2003).

66 B. Blombäck and L. A. Hanson, in *Plasma Proteins*, (The Pitman Press, Bath, 1975), p. 17–21.

Chapter Ten

Characterization of Biomaterials

Haishan Zeng

Cancer Imaging Department, British Columbia Cancer Research Centre,
675 West 10th Avenue, Vancouver, BC, V5Z 1L3, Canada
E-mail: hzeng@bccrc.ca

This chapter starts with an introduction of the issues related to biomaterials characterization and the overall strategies of performing three levels of tests: initial tests for defining the general properties of a biomaterial and for quality assurance; level II tests for defining the properties closely related to the end-use applications including testing of retrieved materials after*in vitro* or *in vivo* exposure to the host biological environment; level III tests for evaluating the biological responses of blood or tissue to the implanted biomaterial. The methods for biomaterial bulk property analysis are then reviewed with discussions focused on two new technologies: X-ray micro-computed tomography(μCT) and X-ray microdiffraction. The rest of this chapter provides detailed reviews on selected surface analysis techniques including *microscopy methods*: transmission electron microscopy (TEM), scanning electron microscopy (SEM), scanning tunneling microscopy (STM), atomic force microscopy (AFM), confocal laser scanning microscopy (CLSM); *spectroscopy methods*: X-ray photoelectron spectroscopy (XPS), second ion mass spectrometry (SIMS), Infrared spectroscopy; *microspectroscopy and spectral imaging methods*: IR microspectroscopic imaging, Raman microspectroscopic imaging; *thermodynamic methods*: contact angle analysis; and *emerging optical methods for in vivo analysis*: rapid Raman spectroscopy, multi-photon excitation (MPE) microscopy, optical coherence tomography (OCT). The chapter ends by emphasizing the importance of combing multiple analysis methods for a comprehensive solution of biomaterial characterization for specific applications.

10.1 INTRODUCTION

The rigorous characterization of biomaterials offers the baseline information upon which the performance of an MEMS device in the biological host setting can

Biomaterials for MEMS, Edited by M. Chiao and J.-C. Chiao

www.panstanford.com
978-981-4241-46-5

be related. This includes characterization of both the bulk and the surface properties of biomaterials. Lyman has summarized the required characterization into three different levels.[1] The first level (initial) tests are used to define the general properties of a candidate biomaterial and can also be used for quality assurance on new materials. These include determination of chemical composition and contaminants as well as tests of basic mechanical properties, such as tensile strength, elongation, elastic modulus, hardness, density, surface roughness and surface energy. The results obtained from these tests should be used in guiding the preliminary selection of potential biomaterials.

The second level of characterization defines the properties that are directly related to the intended applications of the biomaterial, and thus can help to refine the selection of candidate materials. Tests should be done on materials which have been processed in a manner similar to that used in the fabrication of the final MEMS device (including sterilization) since these processings could alter both the bulk and the surface properties of the biomaterial. Example tests include bulk measurements of porosity, creep, stress relaxation, fatigue and wear testing, as well as surface measurements using techniques such as contact angle, X-ray photoelectron spectroscopy (XPS), second ion mass spectrometry (SIMS), attenuated total reflectance Fourier transform infrared spectroscopy (ATR-FTIR), and scanning electron microscopy (SEM). If the biomaterial is melt formed polymer or is subjected to heat in any processing or sterilization steps, then thermal property measurements, such as differential thermal analysis (DTA) and dynamic-mechanical-thermal analysis (DMTA), will also provide useful structural information since the morphology and the crystallinity of a polymer material may be affected during these steps. Additionally, information on protein adsorption and biomaterial responses to solvents (water, saline, lipids, etc.), which may affect the biomaterial surface and even the bulk structure), may also be important in assessing the interaction of the biomaterial with its final biological environment. Retrieved materials should also be retested after *in vitro* and/or *in vivo* exposure to the host biological environment to determine if any changes have occurred in the material that may adversely affect its long-term performances.

The third level of testing involves measuring the histological response of blood and tissue to the presence of the implanted biomaterial using both standard histological test methods for specific applications (soft and hard tissue, blood, etc.) as well as molecular biology analyses for the presence of specific cell types and molecules such as matrix proteins, growth factors. In this regard, the emerging optical techniques such as rapid Raman spectroscopy, optical coherence tomography (OCT), and multi-photon excitation (MPE) microscopy may play a more importatnt role and eventually replace some of the histolgy tests with non-invasive *in vivo* imaging and analyses. These test results help in understanding the mechanisms of blood and tissue compatibility, aid in development of new biomaterials, and assist in the modification of implant design. Level III testing can also involve more dedicated techniques, such as STM (scanning tunneling microscopy) and AFM (atomic force microscopy). These surface characterizations, which allow three-dimensional imaging of surfaces from micron to nanometer molecular scales, may assist the biomaterials investigator to more accurately determine the surface properties of his/her biomaterial and characterize molecular level interactions that might occur at the interface of the biomaterial with tissue and biofluids.

10.2 BULK ANALYSIS METHODS

Bulk analyses of biomaterials include chemical composition analysis, measurements of porosity, creep, stress relaxation, and fatigue testing, assessment of mechanical properties such as tensile strength, elongation, and elastic modulus, determination of hardness and density, and thermal property measurements. Most bulk analysis methods have been well developed. For example, infrared absorption spectroscopy is widely used for chemical composition analysis, X-ray diffraction for 3-D structure determination, and thermal properties can be determined by differential scanning calorimetry (DSC), differential thermal analysis (DTA), thermal gravimetric analysis (TGA), and dynamic-mechanical-thermal analysis (DMTA). The details of these methods as well as the mechanical property analysis methods have been well documented in numerous textbooks[1–3] and will not be discussed in this chapter. We will focus our discussions on two recently developed new technologies for obtaining 3D information of biomaterials at micron and nanometer resolutions respectively.

10.2.1 X-ray Micro-computed Tomography

X-ray micro-computed tomography (also called microCT or μCT) is similar to conventional CT systems used in medical diagnoses and research. Unlike these systems, which typically have a maximum spatial resolution of about 0.5 mm, advanced μCT is capable of achieving a spatial resolution up to 0.3 μm i.e. about three orders of magnitude lower.[4] To image in such a high resolution, the size of the specimen has to be reduced to about a few cubic millimeters. Most μCTs are performed with synchrotron radiations that allow quantitative high spatial resolution images to be generated with high signal-to-noise ratio.[4–7] The super performance of μCT is resulted from the unique advantages of synchrotron X-rays: (1) a high photon flux permits measurements at high spatial resolution; (2) the tunable capability allows measurements at different energies; (3) the highly monochromatic synchrotron radiation eliminates beam hardening effects; and (4) parallel beam acquisition allows the use of exact tomographic reconstruction algorithms.[4]

The spatial resolution of a CT image is dependent on the number of parallel beam projections and the number of data points in each projection. In μCT, hundreds of two-dimensional (2D) projection radiographs are taken of the specimen at many different angles. The information contained in each radiograph is a projection of the absorption density distribution in the sample along the direction of X-ray beam onto the plane perpendicular to the direction of the X-ray propagation. 3D (volume) information on the sample structure can be obtained by using computer reconstruction algorithms. 3D renderings of obtained data after reconstruction are then made by electronically stacking up the slices. Quantitative architectural parameters from the tomographic images can then be extracted.

MicroCT has become a powerful tool for biomaterials research. It is a non-destructive and noninvasive procedure and suitable for *in vivo* applications. For example, a few recent studies have looked into the imaging and quantification of scaffold-bone interactions *in vivo*.[5,8,9] Direct μCT-based image analysis allows to accurately quantify scaffold porosity, surface area, and 3D measures such as pore size, pore distribution, and strut thickness.[10] Furthermore, it allows for a precise measurement of bone growth into the scaffold and onto its surface (Fig. 10.1).[4] This methodology is useful for quality control of scaffold fabrication processes, to assess scaffold degradation kinetics, and to assess bone tissue response.

10.2.2 X-ray Microdiffraction Technique

X-ray diffraction is the most commonly used technique for studying the 3-D structures of solid materials. Its spatial resolution is limited by the X-ray beam cross-section. In X-ray microdiffraction technique, x-rays are focused to a spot size of one micron or less to perform x-ray diffraction experiments with very high spatial resolution. X-ray microdiffraction is providing new insights in the fields of materials science.[4] Materials properties such as strength and fatigue resistance are highly dependent on microstructure. X-ray microdiffraction can measure local variations in stress, orientation, and plastic deformation between grains and within individual grains, helping us understand mechanical properties

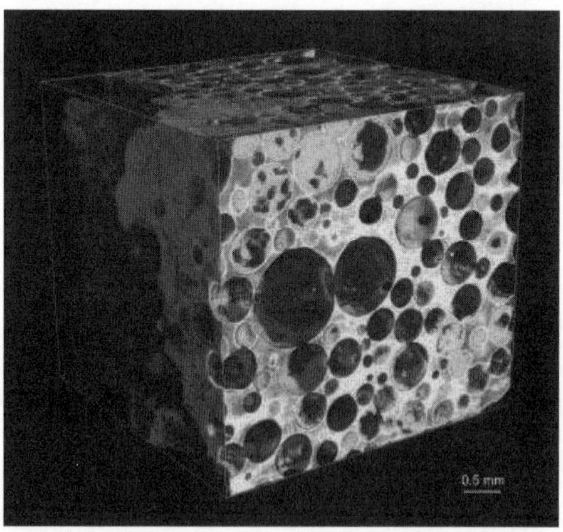

Figure 10.1. 3D microCT image of hydroxyapatite scaffold (Engipore) loaded with sheep BMSC and implanted in an immunodeficient mouse, showing newly formed bone (green) onto the inner surface of the scaffold (yellow and red); the organic phase is blue. With permission from.[4] For color reference, see page 275.

at this critical length scale. It has been used to study the hierarchical structure of a biological material such as a bone. In particular it plays a crucial role in studying the lowest hierarchical level, within the nanometer scale, where the distribution of organic fibers (collagen) and crystalline mineral nano-particles has a major influence for mechanical optimization of bone.[4]

10.3 SURFACE ANALYSIS METHODS

The surface structure characterization of a biomaterial presents different analytical challenges since the composition and spacial arrangement of the molecular segments on the surface may not be representative of the bulk structure. In addition, the surface is the interface where the biomaterial meets and interacts with the molecular constituents of the biological host setting (e.g., bone, soft tissue, blood, cells, body fluids). These interactions can modify the material and result in macroscopic events in the biological media which ultimately determine whether or not an MEMS device will have successful performance in the host biological environment. Therefore, surface characterization will be discussed in more detail in this chapter.

The properties that are of interest in the characterization of biomaterial surfaces include the chemical structure, the hydrophilicity or hydrophobicity, the presence of ionic groups, the morphology and topography.[1,3] Varying degrees of information about these properties can be obtained using different analysis methods including microscopic, spectroscopic and thermodynamic and other methods. We aim to provide an overview on some of the specific techniques frequently used in the characterization of biomaterial surfaces. In addition, discussions and references are made to emerging methods such as rapid Raman spectroscopy, optical coherence tomography (OCT), and multi-photon excitation microscopy that have recently come to the forefront of biomedical research and possess great potentials for studying the interactions of biomaterials with real biological environment *in vivo*.

10.3.1 Microscopic Methods

Microscopy techniques often used in biomaterial analysis include scanning electron microscopy (SEM), transmission electron spectroscopy (TEM), scanning tunneling microscopy (STM) and atomic force microscopy (AFM).[11] Recent developments in light microscopy include confocal laser scanning microscopy (CLSM) and optical near-field microscopy.[12]

10.3.1.1 Electron Microscopy

An electron microscope is a type of microscope that uses electrons to illuminate a specimen in a vacuum environment and create an enlarged image. Electron microscopes have much greater resolving power than light microscopes. Electron microscopy can be performed in either transmission or secondary electron modes. In transmission electron microscopy (TEM),[13] images are formed by diffraction or phase contrast mechanisms. The detectors are phosphor screens coupled with digital CCD cameras. Unlike the TEM, where electrons of the high energy beam form the image of the specimen, the scanning electron microscope (SEM)[14] produces images by detecting low energy secondary electrons which are emitted from the surface of the specimen due to excitation by the primary electron beam. In the SEM, the electron beam is rastered across the sample, with detectors building up an image by mapping the detected signals with beam positions. SEM and TEM are commonly used for studying both the surface morphology of and the cellular response to biomaterials.[15,16]

SEM sample preparation involves fixation (when proteins, cells, or tissues are present), followed by drying, attachment to a metallic stub, and then coating with a metal prior to data collection. The thin metallic coating is typically 20 to 30 nm in thickness. Common conductive metals used include gold, platinum, or gold/palladium alloy. It should be noted that the drying and metal coating processes used in the preparation of some polymeric materials might alter surface morphology, particularly those surfaces that may undergo changes in a hydrated environment. In addition to imaging the surface morphology of biomaterials, the SEM can be combined with other analysis methods such as energy dispersive X-ray analysis to determine elemental distribution[15] and IR and Raman spectroscopy to monitor surface modification procedures.[17] Figure 10.2 shows an example application of using SEM to assess the effect of sterilization on biomaterial surface morphology.[18]

TEM sample preparation involves fixation (if biological materials are present), processing, embedding and sectioning. Samples must be very, very thin (50 nm–90 nm). This is achieved by embedding samples in hard plastic resins and using very sharp diamond knife ultramicrotomes. TEM provides extraordinary resolution of organelle and membrane structures in single cells. There is probably

Figure 10.2. SEM micrograph of as-deposited surface finish of sputter deposition elemental titanium (left side) compared with SEM micrograph of surface finish of autoclaved sputter deposition elemental titanium (right side). Pictures were taken from a candidate biomaterial for developing implantable MEMS devices. With permission from.[18]

no other instrument that provides a higher resolution image of the intimate cell membrane-biomaterial interface. Unfortunately the challenge nature and limitations of TEM sample preparation mean that this technique is underutilized in biomaterial characterization. It has been noted that the embedding and sectioning processes used in the preparation of some polymeric materials may alter the polymeric material itself.[19]

10.3.1.2 *Scanning Tunneling Microscopy*

Scanning tunneling microscopy (STM) is a powerful technique for viewing surface topography at the atomic level.[20] The STM is based on the concept of quantum tunneling. When a conducting tip is brought very near to a metallic or semiconducting surface, a bias between the two can allow electrons to tunnel through the space between them. Variations in current as the probe passes over the surface are translated into an image. STM can achieve a lateral resolution of 0.1 nm and a depth resolution of 0.01 nm, not possible with electron microscopes. Compared to SEM and TEM, another advantage of STM is that no special sample preparation procedures are required. STM can be used not only in ultra high vacuum but also in air and various other liquid or gas ambients. However, the application of STM in the characterization of polymeric biomaterials is somewhat limited in that most polymers are not sufficiently conductive to allow STM images to be generated.[19]

10.3.1.3 *Atomic Force Microscopy*

Like STM, atomic force microscopy (AFM) or scanning force microscopy (SFM) is a very high-resolution type of scanning probe microscope, with demonstrated resolution of fractions of a nanometer.[21,22] However, AFM overcomes the problem of the required sample conductivity needed for STM, and is therefore, more widely used in biomaterial characterization. A 3-D image of the surface is created by scanning a nanometer size tip attached to the end of a soft cantilever across the surface and sensing the forces between the tip and the sample surface.[22] To this end, the sample is mounted on a piezoelectric scanner, which allows 3-D positioning with sub-nanometer accuracy while the force is monitored with pico-newton sensitivity by measure the deflection of the cantilever. Typically, the deflection is measured using a laser beam reflected from the top of the cantilever into an array of photodiodes.

A number of different AFM imaging modalities are available, which differ mainly in the way the tip is moving over the sample. In the so-called *contact mode*, the AFM tip is raster scanned over the sample while the cantilever deflection, thus the force applied to the tip, is kept constant using feedback control. In *tapping mode*, also called dynamic or intermittent mode, an oscillating tip is scanned over the surface and the amplitude and phase of the cantilever are monitored near its resonance frequency. Because lateral forces during imaging are greatly reduced, this mode may be advantageous for imaging soft samples. More detailed information on the different imaging modalities can be found in the literature.[21–23]

Figure 10.3 shows that tapping mode phase imaging and amplitude imaging can be performed simultaneously, so the position of the surface features observed in a phase image can be correlated directly with the surface topography.[24] This combination can therefore be used to determine the size, shape, and spacing of different material domains that could not otherwise be discerned from height alone.[25,26]

Unlike electron microscopes, a significant advantage of AFM is that sample topographies, as well as surfaces roughness values, can be obtained without surface treatment or coating which may damage or alter the material surface under investigation.[23,25] Furthermore, AFM images can be acquired under vacuum, air, or liquid conditions. The ability to image biomaterials within an aqueous environment is extremely useful, as it allows for the examination of the surface of biomaterials in an environment similar to one that would be found in an implant situation, thus allowing for the examination of dynamic processes such as erosion, hydration, and adsorption at interfaces. For example, it is possible to visualize individual plasma protein molecules under aqueous environments using tapping mode phase imaging AFM.[27–29] A more recent development is to use AFM to conduct force spectroscopy, the measurement of force-distance curves. It has been used to measure nanoscale contacts, atomic bonding and perform single molecule analyses.[30] We anticipate that these techniques will be increasingly used in biomaterials research for probing cell-cell and cell-material interactions.[30]

10.3.1.4 Confocal Laser Scanning Microscopy

Confocal laser scanning microscopy (CLSM, also referred to as CSLM, confocal scanning laser microscopy) has been established as a valuable tool for obtaining high resolution images and 3-D reconstructions of a variety of biomaterial samples

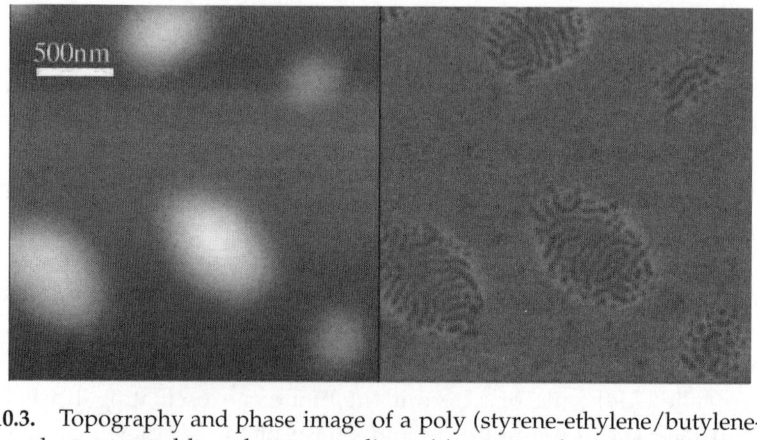

Figure 10.3. Topography and phase image of a poly (styrene-ethylene/butylene-styrene) (SEBS) copolymer treated by solvent annealing. (a) topography image, (b) phase image. The phase image in (b) shows unambiguous resolution of the two different phases of the copolymer. The phase angle contrast of two different copolymers results from the different probe–sample interaction forces. With permission from.[24] For color reference, see page 275.

and their interfaces with biological environments.[31] The basic principle of a CLSM is shown in Fig. 10.4. A beam of laser light is focused onto the specimen by a microscope objective lens. The back-scattered light or laser induced fluorescence light from the sample is then collected by the same microscope objective and recorded by a photodetector. The optical system is designed so that the laser beam's focal point in the sample is imaged exactly on the detection pinhole in front of the photodetector (i.e., confocal). By this confocal arrangement, any photons originated from the point of laser focus will be focused on the detection pinhole and be detected by the photodetector, and any photons originated from points other than the point of laser focus will be out of focus at the detection pinhole and be rejected. Thus, the detected fluorescence or back-scattered light can be limited to a region very close to the point of focus of the laser beam. The smaller the aperture is, the higher the resolution will be. In CLSM, the focal point of the laser spot is stepped across the sample in a raster (x-y) pattern, always maintaining the confocal nature of the image at the detection pinhole. The detected optical signal is accumulated on a point-by-point basis with a digital processing system, and an optical cross section of the sample at the focal plane is obtained. By stepping the focus vertically (z), multiple slices can be used to build up a full three dimensional image. With non-opaque samples, the interior structure can be clearly seen. By scanning in the x-z direction, a vertical cross section can be obtained.

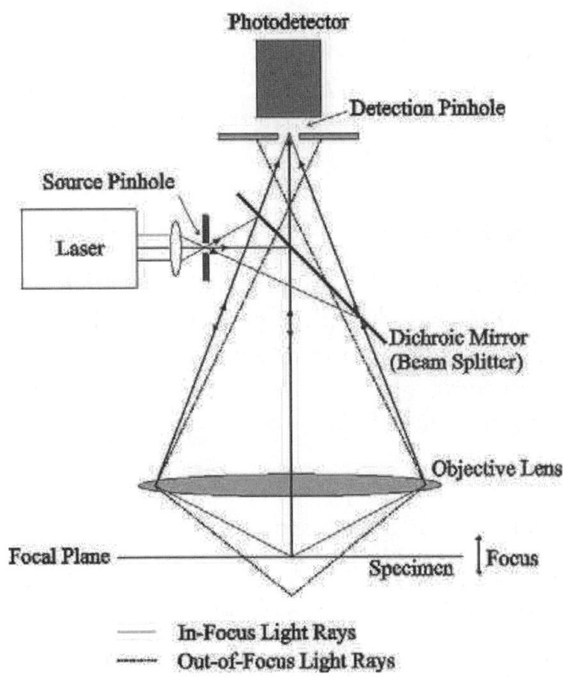

Figure 10.4. Schematic illustration of the principle of a CLSM — confocal laser scanning microscope.

The spatial resolution of CLSM is diffraction limited, i.e. at the order of the wavelength of light used for imaging. Thus, micron and sub-micron resolutions are readily achievable with CLSM.[31] The actual useful depth of CLSM is limited to 100–200 μm, even with the most penetrating near IR illumination. Different imaging modalities are available with CLSM depending on the types of optical signals utilized for imaging formation. For example, reflectance confocal imaging mode uses elastic backscattering light, while fluorescence confocal imaging utilizes sample fluorescence emission. Confocal Raman imaging can be performed by utilizing inelastic (Raman) scattering.

CLSM applications in biomaterials research is expanding with advantages such as convenience of use, in that no special sample preparation is required as compared to electron microscopy, and high spatial resolution than conventional light microscopy. Confocal microscopy can resolve details at interfaces that could previously only be seen using electron microscopy of specially processed specimens.[32] To assess the biocompatibility of poly-β-hydroxy butyric acid (PHB) polymer, CLSM has been used to study the structural organization of cellular molecules involved in adhesion using osteoblastic and epithelial cell lines.[33] Figure 10.5 shows example confocal images from this study demonstrating that for cells cultured on PHB polymer, the organization of the actin cytoskeleton was impaired. To devise better strategies for developing biomaterials optimized for particular therapies and treatments, a novel, CLSM based 3-D screening method for analyzing interactions between cells and new biomaterials has been developed to cut initial search times by more than half.[34] Recent technology advancement has resulted in a reflectance mode CLSM that is capable of acquiring video rate cellular resolution images from human skin *in vivo*.[35] This could presumably be adopted to study the interactions of MEMS biomaterials with biological tissues *in vivo* in animal models.

10.3.2 Spectroscopy Methods

Spectroscopy methods are routinely used to analyze the constituent elements and chemical structures of biomaterial surfaces. In this section, we will discuss the most commonly used three types of traditional spectroscopy techniques, namely, X-ray photoelectron spectroscopy (XPS), secondary ion mass spectroscopy (SIMS), and attenuated total reflection Fourier transform infrared spectroscopy (ATR-FTIR).

10.3.2.1 X-ray Photoelectron Spectroscopy

X-ray photoelectron spectroscopy (XPS) also known as ESCA (electron spectroscopy for chemical analysis) is the most extensively used technique for the characterization of biomaterial surfaces,[1,3] It is used to determine the surface chemistry of a biomaterial in its "as received" state or after surface modifications, e.g. adsorption and retention of chemicals such as antibiotics and bonding agents,[15] for the detection of immobilized proteins,[36] understanding the chemistry

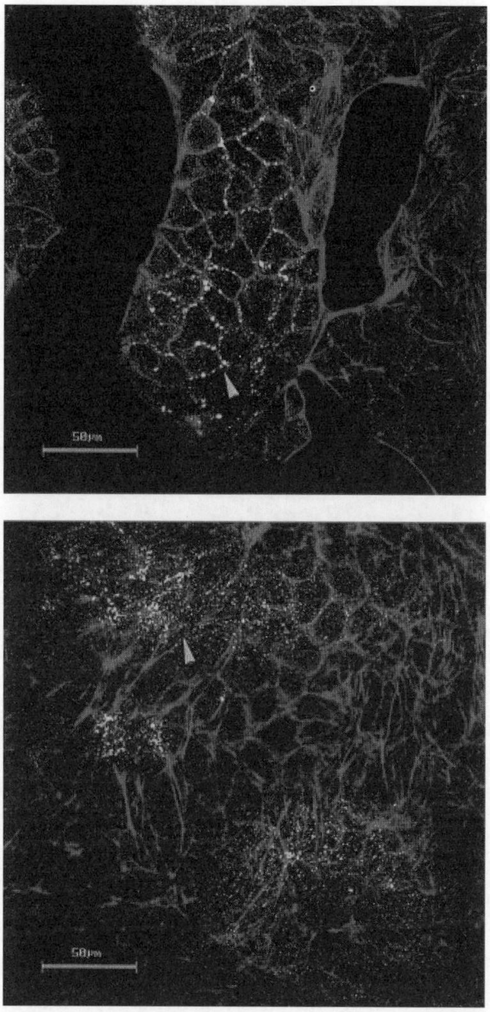

Figure 10.5. Fibronectin formation (green) and actin filament formation (red) in epithelial mHepR1 cells cultured on the biomaterial PHB (below) compared to control cultures (above). Note that in control cultures the actin filaments are structured in fibrils showing colocalization with fibronectin (arrow head) in contrast to PHB cultured cells, where the organization of actin filaments is more irregular and the colocalization of fibronectin with actin is lost (arrow head) (CLSM images taken with LSM 410, Carl Zeiss Jena). With permission from.[33] For color reference, see page 276.

of the structure, formation and stability of plasma treated surfaces,[19] as well as for the characterization of the steps of formation of thin coatings.[37]

XPS spectra are obtained by irradiating a material with a beam of aluminium or magnesium X-rays while simultaneously measuring the kinetic energy (KE) and number of electrons that escape from the top 1 to 10 nm of the material being analyzed.[38] XPS requires ultra-high vacuum (UHV) conditions. The kinetic energy of the emitted photoelectrons is unique for different elements and is sensitive to the

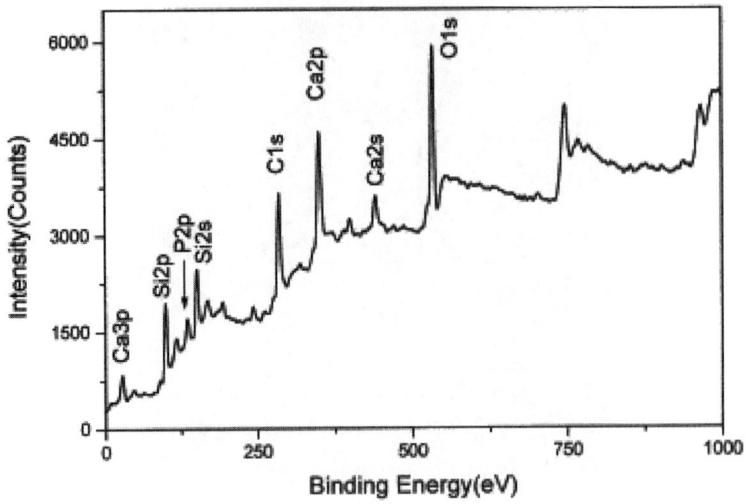

Figure 10.6. XPS spectra of survey scans for bioactive hydroxyapatite (HA) layer deposited onto pretreated hydrophilic Si (111) from supersaturated solutions containing Ca^{2+} and PO_4^{3-} ions. The objective is to develop a biocompatible and protective coating for silicon based implantable MEMS device. With permission from.[39]

chemical state of the atoms. XPS detects all elements with an atomic number (Z) between those of lithium (Z=3) and lawrencium (Z=103). This limitation means that it cannot detect hydrogen or helium. Detection limits for most of the elements are in the parts per thousand range. Detection limits of parts per million (ppm) are possible, but require special conditions: concentration at top surface or very long collection time (overnight).

XPS survey scans (see Fig. 10.6) are about 1000 eV wide and are often used to identify and quantify the elements present at the investigated biomaterial surface, in this example, a hydroxyapatite layer is deposited on hydrophilic silicon MEMS device for biocompatability and protection purposes.[39] High resolution scans (see Fig. 10.7) are typically 20 to 30 eV wide and are used to obtain information about chemical shifts and derive the chemical states (Table 10.1) of the detected elements.[19,40–41] With these two types of scans, XPS can characterize both the elemental and chemical structure of the surface.

10.3.2.2 Secondary Ion Mass Spectrometry

Bombardment of a sample surface with a primary ion beam followed by mass spectrometry analysis of the emitted secondary ions constitutes secondary ion mass spectrometry (SIMS).[42] The elemental, isotopic, or molecular composition of the specimen surface can be obtained. The primary ion beam can be focused to less than 1 μm in diameter. A SIMS analysis has a relatively high lateral resolution, capable of microanalysis, the measurement of the lateral distribution of elements on a microscopic scale.

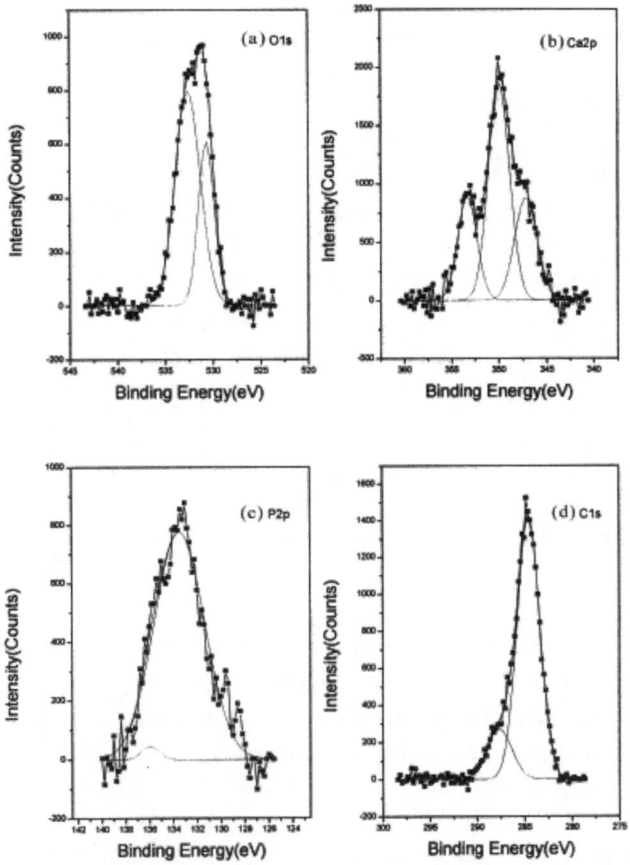

Figure 10.7. XPS spectra of the same specimen for Fig. 10.6; (a)O1s; (b)Ca2p; (c)P2p and (d)C1s. With permission from[39]

SIMS can be used for practically all elements of the periodic table, including hydrogen. Only the noble gases are difficult to measure because they do not ionize easily. SIMS is the most sensitive surface analysis technique, being able to detect trace elements present in the parts per billion range.

During SIMS analysis, the sample surface is slowly sputtered away. Continuous analysis while sputtering produces information as a function of depth, called a depth profile. When the sputtering rate is extremly slow, the entire analysis can be completed while consuming less than a tenth of an atomic monolayer. This slow sputtering mode is called static SIMS in contrast to dynamic SIMS used for depth profile.

Three different mass analyzers are used in SIMS: sector, quadrupole and time-of-flight. A sector field mass spectrometer uses a combination of an electrostatic analyzer and a magnetic analyzer to separate the secondary ions by their mass to charge ratio. A quadrupole mass analyzer separates the masses by resonant electric fields, where only masses of choice are able to pass. The time of flight

Table 10.1 XPS profile parameters for C1s, $Ca2p_{1/2}$, $Ca2p_{3/2}$, P2p and O1s states derived from Fig. 10.7. With permission from.[39]

Binding energy(eV)	FWHM of deconvolution	core level	Chemical state	Binding energy(eV)
284.6	2.35	C1s	Adventious C	–
287.7	2.61		CO_3 ion	–
347.3	2.33	Ca2p3/2	Ionic Ca^{2+}	347.2
350.0	2.20	Ca2p1/2	Ionic Ca^{2+}	350.7
353.4	2.05	Ca2p1/2	Ionic Ca^{2+}	–
133.4	4.20	p2p	FO_4^{3-}	133.5
135.9	1.51	p2p	FO_4^{3-}	–
530.7	1.68	O1s	O^{2-}	531.1
532.6	2.65		O^{2-}	–

mass analyzer separates the ions in a field-free drift path according to their kinetic energy. It requires pulsed secondary ion generation using either a pulsed primary ion gun or a pulsed secondary ion extraction. It is the only analyzer type able to detect all generated secondary ions simultaneously and is the standard analyzer for static SIMS instruments.

SIMS has been increasly used in biomaterial analysis in recent years. The technique can be used to obtain direct proof of covalent binding of molecules to a surface, as well as to predict various surface properties including wettability, adhesiveness, and biological reactivity.[19] It has also been used to correlate surface chemistry with cell growth[43] and for monitoring the degradation kinetics of biodegradable polymers.[44,45]

10.3.2.3 Infrared Spectroscopy

Infrared (IR) spectroscopy is one of the most common spectroscopic techniques used for molecular analysis. It is the absorption measurement of different IR frequencies by a sample positioned in the path of an IR beam [46]. The main goal of IR spectroscopic analysis is to determine the chemical functional groups in the sample. Different functional groups absorb characteristic frequencies of IR radiation. The most frequently used wavelength range is mid-IR, where light absorption corresponds with molecular vibrations. Using various sampling accessories, IR spectrometers can accept a wide range of sample types such as gases, liquids, and solids. Thus, IR spectroscopy is an important and popular tool for structural elucidation and compound identification.

Fourier transform infrared (FTIR) spectroscopy is a measurement technique for collecting IR spectra. Instead of recording the amount of energy absorbed when the frequency of the IR light is varied (by monochromator), the IR light after passing the sample is guided through an interferometer. The measured signal is the interferogram. Performing a mathematical Fourier transform on this signal results in a spectrum identical to that from conventional (dispersive) infrared spectroscopy. FTIR spectrometers are cheaper than conventional spectrometers because building of interferometers is easier than the fabrication of a monochromator in the IR wavelength range. In addition, measurement of a single spectrum is faster for the FTIR technique because the information at all frequencies is collected simultaneously. Because of its various advantages, virtually all modern IR spectrometers are FTIR instruments.

A special technique called attenuated total reflection Fourier transform infrared spectroscopy (ATR-FTIR) is usually employed to characterize biomaterial surfaces without using any distructive sample preparation procedures. ATR-FTIR couples IR spectroscopy analysis with the physical phenomena of total internal reflection to restrict the analyzed volume to within a few micronmeter depth of the sample surface. For this technique, the incident electromagnetic waves (IR light) are entirely reflected back into the initial medium. The electromagnetic field is established in the second medium as an evanescent wave.[47] In attenuated total reflectance (ATR) sampling mode the second medium is the material to be studied and has higher refractive index than the initial medium. Information about the molecular structure of the biomaterial, inter- and intra-molecular interactions, crystallinity, conformation (e.g. proteins) and orientation of molecules can be obtained through analysis of the detected infrared spectra.[19,48] It has also been used to study the relationship of surface characteristics to cellular attachement.[49] Figure 10.8 shows an example of IR spectroscopy applications in characterizing SU-8 surface modifications, aimed to functionize SU-8 for immobilization of biomolecules on a Bio-MEMS structure.[50]

10.3.3 Microspectroscopy and Spectral Imaging Methods

There is a clear need to understand not only what chemical and biological species are present in a biomaterial but also how those species are spatially distributed. In principle, if we perform any spectroscopy analysis discussed in Section 10.3.2 in a point by point fashion over an area of the sample surface we will get a spectral image, in which a spectrum curve is available for each pixel. This huge amount of data set is often referred as a spectral cube. Analysis of the spectral cube can generate 2-D distributions/images of the chemical constituents and structures of the sample surface. In fact, XPS imaging and mass spectrometry imaging (SIMS imaging and MALDI — matrix-assisted laser desorption ionization imaging) has recently been realized in biomaterial researches.[51,52] However, widespread pratical applications of spectral imaging techniques are often hamppered by the length data acquisition time of point by point spectral mapping and the achievable spatial resolution of the spectral measurments. Therefore, success of a spectral imaging mode often depends or is pending on technical advancements to improve the

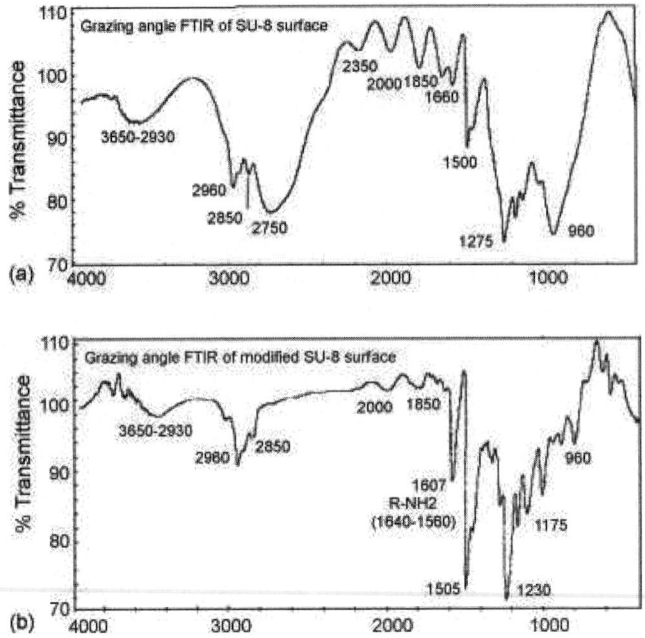

Figure 10.8. ATR-FTIR of SU-8 surface (a) before and (b) after surface modification showing additional R–NH_2 group at 1607 cm^{-1}. The other peaks are 960–1275 cm^{-1}: phenyl in plane bending; 1500–1600 cm^{-1}: phenyl nucleus; 1660–2000 cm^{-1}: aromatic overtone of dCH; 2350 cm^{-1}:CO_2; 2750 cm^{-1}: chelation intermolecular bond with C=O; 2850 cm^{-1}:$-CH_2-$; 2960 cm^{-1}:$-CH_3-$; 2930–3650 cm^{-1}: water. With permission from.[50]

detection sensitivity and shorten the data acquisition time. In the optics field, advances on light sources/lasers, spectrographs, detectors, and fast computer data processing have enabled the successful applications of IR spectral imaging and Raman spectral imaging in biomaterials reseach. These two techiques will be discussed in more details below.

10.3.3.1 Infrared Microspetroscopic Imaging

Infrared microspectroscopy measurement was first realized in the 1940s by coupling an FTIR spectrometer to a light microscope.[53] Thus, it is also called FTIR microspectroscopy. In early instruments, IR spectral imaging is acquired through point-by-point mapping utilizing mechanical scanning of the sample. These early instruments has been used for bone characterization and widely applied to polymer research.[53] Infrared microspectroscopy has also been used to study tissues and individual cells.[54,55] as well as to identify pathologic inclusions and soft tissue calcifications.[56] Data collection using point-by-point mapping is relatively slow and tedious. A technical improvement, the coupling of the FTIR microscope with an array detector for FIIR imaging changed the way data was collected and processed.[57,58] The entire field of view of the sample is analyzed simultaneously by Fourier transform of the 2-D data series acquired by the 2-D array detector. No mechanical scanning of the sample is required. The spectral imaging data cube can be obtained and analyzed in a much short period of time. The spatial resolution of IR microspectroscopy imaging is low compared to other light microscopy techniques because mid-IR light is used, which has longer wavelengths than visible and near IR light. The diffraction limited spatial resolution is approximately 1.7 μm (at 4000 cm^{-1}) to 13 μm (at 500 cm^{-1}).[59] ATR-FTIR based spectral imaging technique has also been successfully developed recently.[60] The improved instrumentation has led to increasing applications of FTIR microspectroscopy in biomaterials research such as identification and mapping of contamination, surface imperfections, identification of layers in multilayered structures, identification of polymer additives, and microanalysis of multicomponent phases in polymers.[61] Figure. 10.9 shows FTIR imaging of post repair cartilage tissues.[53]

10.3.3.2 Raman Microspectroscopic Imaging

Raman spectroscopy is a noninvasive, nondestructive analytical method capable of obtaining chemical constituents and molecular structures of a sample based on molecular vibrations. When monochromatic light strikes a sample, almost all the observed light is scattered elastically (Rayleigh scattering) with no change in energy (or frequency). A very small portion of the scattered light is inelastically scattered (Raman scattering) with a corresponding change in frequency. The difference between the incident and scattered frequencies corresponds to an excitation of the molecular system, most often excitation of vibrational modes. By measuring

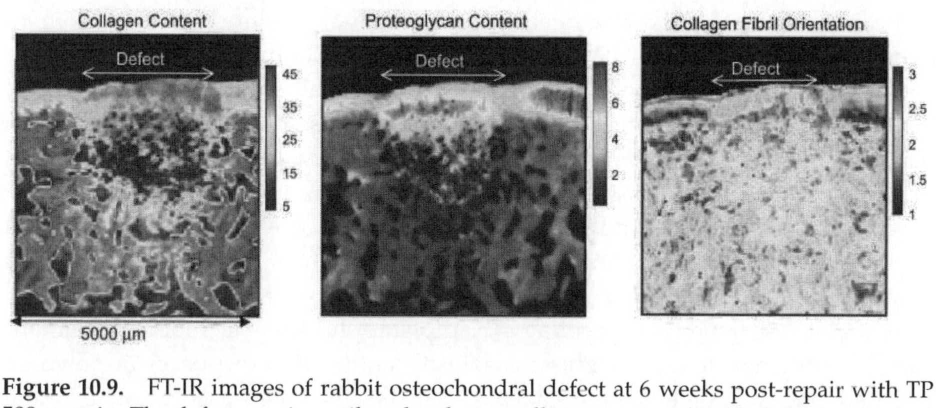

Figure 10.9. FT-IR images of rabbit osteochondral defect at 6 weeks post-repair with TP 508 protein. The defect repair cartilage has lower collagen content (amide I area) compared to the adjacent articular cartilage, but also has regions with PG content (PG sugar ring C-O absorbance (985–1140 cm^{-1}) of a similar magnitude to that in the native cartilage. There is some orientation present in the repair cartilage, as evidenced by the layer of fibrils parallel to the surface. For the orientation data, the boundaries on the color bar indicate the three collagen fibril orientation categories with respect to the articular surface, corresponding to amide I/amide II peak area ratios$>$2.7 (parallel fibrils), between 2.7 and 1.7 (random fibrils), and $<$1.7 (perpendicular fibrils), respectively. With permission from.[53] For color reference, see page 276.

the intensity of the scattered photons as a function of the frequency difference, a Raman spectrum is obtained. A laser is usually used as the excitation source for Raman spectroscopy. Raman peaks are typically narrow and in many cases can be attributed to the vibration of specific chemical bonds (or normal mode dominated by the vibration of a functional group) in a molecule. As such, it is a "fingerprint" for the presence of various molecular species and can be used for both qualitative identification and quantitative determination.[62]

Raman spectroscopy has a rich history of applications in the characterization of biomaterials such as polymers.[61] Raman spectroscopy is complimetary to IR spectroscopy for measuring molecular vibrations, but has several advantages over IR spectroscopy: (1) Little sample preparation or pretreatment is necessary since scattering from the sample can be measured at the back-reflection configuration; (2) The Raman slection rules are less restrictive than IR, so more vibrational modes are observable; (3) Raman can be used to monitor biomaterial interactions with water rich biological environments such as tissue, blood and other body fluids since water is a weak Raman scatter and Raman siganl is measured at a wavelength range closing to the visible or near IR laser line where water absorption is minimal as well when compared to the mid-IR wavelength range. The major disadvantge of Raman spectroscopy is the weakness of Raman scattering signal that leads to length data acquisition time. But recent technical advancement has significantly improved this situation, a Raman spectrum from biological tissues *in vivo* can be acquired within one second.[63] This is very encouraging for characterizing biomaterial in and its interactions with *in vivo* biological environments.

Raman spectroscopy makes microscopic and mapping analysis simpler than with IR spectroscopy for the following reasons: (1) The laser beam is collimated and can be easily focused to a micro-spot, the spatial resolution of micro-Raman is on the order of micron or even sub-micron; (2) Raman is measured in the visible or near IR wavelength range, so conventional microscope can be used for Raman microscopy, while IR microscopy requires special optics; (3) Raman scattering is inherectly a scattering process, and back scattering at 180° is particularly useful for mapping purposes because positioning of the beam is simple. As a consequence of these advantages, Raman microscopy has been widely used for the study of materials. With recent instrument advances, especially confocal Raman microscopy, the use of micro-Raman spectroscopy and spectral imaging in biomaterial researches is expanding. Example applications include contaminant analysis, polymer degradation assessments, identification of phases in polymer, and pathological tissue sample analysis.[61] It has also been used to analyze the composition of bone and apatitic biomaterials,[64] image load-bearing surfaces in artificial hip joints,[65] and monitor the effects of traetment protocols and subcutaneous implantation on bovine percardium.[66]

10.3.4 Thermodynamic Methods

Thermodynamic methods can be used to obtain surface parameters related to the interfacial free energy. These are the earliest developed methods for solid material surface characterization. They are convenient to use, low cost, yet capable of obtaining preliminary information on the biocompatibility of biomaterials. The most used thermodynamic method for biomaterial surface characterization is wetting or contact angle experiments.

Contact angle analysis characterizes the wettability of a surface by measuring the surface tension of a solvent droplet at its interface with a homogenous surface. The phenomenon of contact angle can be explained as a balance between the force with which the molecules of the liquid (in the drop) are being attracted to each other (a cohesive force) and the atraction of the liquid molecues for the molecues that make up the solid surface. An equilibrium is established between these forces, the energy minimum. Contact angle is one of the most sensitive but inexpensive surface analysis techniques and is capable of measuring ∼3–20 Å deep.

There are five common techniques that can be employed to measure the contact angle: the Wilhelmy plate method, the sessile drop method, the capative bubble method, the capillary rise method, and the tilted-drop measurement. The details can be found in.[3,67,68] The technique chosen depends principally on the geometry and location of the surface or coating to be studied. In all the methods, the contact angle is the angle of the liquid at the interface relative to the plane of the model surface. As application examples, contact angle analysis has been successfully used to predict the performance of vascular grafts and the adhesion of cells to surfaces.

10.3.5 Emerging Optical Methods for *in vivo* Analysis

Over the last two decades, great progress has been made on biomedical optical imaging and spectroscopy technology development. Emerging novel optical techniques enabled the imaging and analysis of biological tissues at micrometer spatial resolution and up to millimeter depth below the tissue surface in spite of the challenge of light scattering by biological tissues. These technologies are definitely useful for monitoring implant biomaterials and MEMS devices in tissue or other biological environment as well as for studying the interactions of biomaterials with *in vivo* and other real biological environments. The most powerful biomedical optical analysis technologies include confocal laser scanning microscopy (CLSM), Raman spectroscopy and imaging, multi-photon excitation (MPE) microscopy, and optical coherence tomography (OCT). CLSM and Raman techniques have been introduced in previous sections, MPE microscopy and OCT will be discussed below.

10.3.5.1 Multi-Photon Excitation Microscopy

Multi-photon excitation microscopy is a powerful tool that combines multi-photon excitation fluorescence and confocal scanning microscopy to create high-resolution, 3-D images of microscopic samples.[69] In traditional fluorescence mode confocal microscopy, a single photon of light is used to excite a molecule from its ground state to an upper excited electronic state to induce fluorescence emission at longer wavelengths that are then used for image formation. Although the interaction probability is greatest for single-photon absorption, if two or more lower energy (longer wavelength) photons arrive simultaneously, there is probability that they can excite the molecule as long as the sum of the multi-photon energies equals to the corresponding single-photon excitation related energy difference. The multi-photon absorption probability is non-linear and increases with the square of light intensity (I^2) for two-photon absorption and as the cube for three-photon absorption. For this reason, ultrafast lasers (hundreds of fs light pulses, often Ti:Sapphire laser) are used to generate instantaneously super high power density at a tightly focused spot in space to facilitate high probability multi-photon absorption for multi-photon excitation microscopy applications.

Multi-photon absorption brings advantages over single-photon absorption. With single-photon absorption, when a laser is focused to a point within a sample, the sample may, because of the linear dependence on the light intensity for single-photon absorption, fluoresce throughout the entire beam path. Using multi-photon absorption, induced fluorescence occurs only at, or near, the focal point of the beam thanks to the non-linear (quadric, cubic, or higher order) dependence on the light intensity. Since the position of the focal point can be precisely determined, multiphoton fluorescence can yield a great deal of information about specific points below the sample surface. Furthermore, longer wavelengths, particularly the near-IR, penetrate deeper in biological materials and are not scattered as much as shorter wavelengths. These lead to a couple of advantages for MPE microscopy over conventional fluorescence confocal microscopy: higher axial resolution, greater sample penetration, reduced fluorescence photobleaching, reduced photodamage and increased tissue and cell viability.

A conventinal CLSM can be modified to perform MPE microscopy with two major changes: (1) the light source is changed to a ultrafast laser (usually Ti:Sapphire) with very high peak power but low average power; (2) The confocal detection aperture is removed because all of the fluorescent light originates from

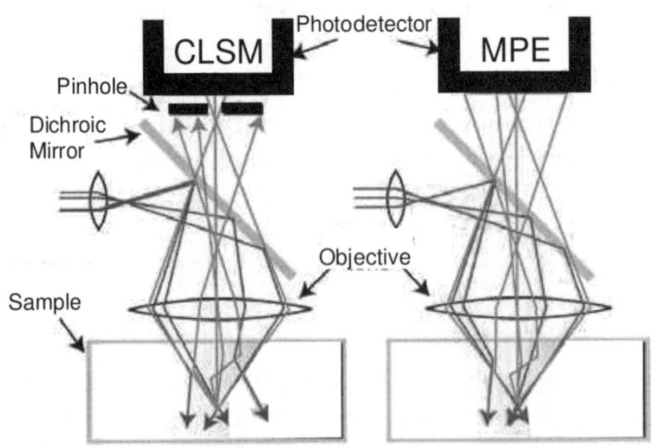

Figure 10.10. Comparison of a conventional confocal microscope (CLSM) with a multi-photon excitation (MPE) microscope. For color reference, see page 277.

the laser focus spot. Figure 10.10 illustrates the differences between the optical configurations of a CLSM and a MPE microscope. In the conventional CLSM case, fluorescence occurs throughout the beam path inside the sample and must be blocked by the detection pinhole aperture. This not only eliminates the fluorescence away from the focal point, but also the scattered (diffusing) fluorescence from the focal point. Only the ballistic (straight line) fluorescence is detected. In the MPE microscopy case, both the ballistic and the diffusing photons are collected. Furthermore, since the excitation light has a longer wavelength, less excitation light is lost to scattering. Combination of these two factors results in great improvement on detection sensitivity and signal-to-noise ratio for MPE microscopy.[31] In fact, *in vivo* tissue autofluorescence based MPE microscopy has been reported,[70] while *in vivo* autofluorescence based conventional single-photon fluorescence microscopy had never been realized due to the low quantum yield of tissue autofluorescence and the low detection sensitivity of single-photon fluorescence confocal microscopy due to the use of a pinhole detection aperture.

The commonly used MPE microscopy in biomedical research is based on two-photon excitation fluorescence (TPEF) and second harmonic generation (SHG). In second harmonic generation, two photons interact via a virtual electronic state of certain molecules to generate a new photon with energy of two times the excitation photons. SHG can be observed in biological tissue from non-centrosymmetric structures, for example, collagen fibers.[71] Combing collagen SHG imaging with elastin TPEF imaging forms a powerful tool for studying the extracellular matrix structures in biological tissues. This non-invasive tool has been used to diagnose extracellular matrix structures and cell resettlement of tissue engineered (TE) heart valves prior to implantation.[72] Figure. 10.11 shows overlay of collagen SGH images (excited by 840 nm, 250 fs, 80 MHz laser light) with elastin TPEF images (excited by 760 nm, 250 fs, 80 MHz laser light) for two TE valve leaflets which were cultured under normal and supra-physiologic different pressure conditions in a bioreactor system. Whereas photographs of TE valves demonstrated macroscopic similar leaflets (Fig. 10.11(C) and (D)), clear differences of extracellular matrix structures were only explicitly verifiable by multi-photon excitation imaging. As seen in Fig. 10.11(B) which represents a typical image out of image stacks taken from different leaflets, supra-physiological pressures (>200 mm Hg systolic) cause detachment and damage of reseeded cells, impaired formation of elastin network and subsequently, decreased mechanical strength. Consequently, a successful valve replacement would be impossible. By contrast, TE valves cultured under fixed frequency (1 Hz) and pressure conditions (3 l/min and 60/40 mm Hg) demonstrated a close network of elastic fibers and collagen matrix components (Fig.10. 11(A)). Concurrently conducted animal studies demonstrated that these valves could be successful implemented into living animals (sheep).

10.3.5.2 Optical Coherence Tomography

Optical Coherence Tomography (OCT) is a technique for obtaining sub-surface images of translucent or opaque materials at a resolution equivalent to a low-power microscope.[73] OCT is analogous to ultrasound B-mode imaging except that

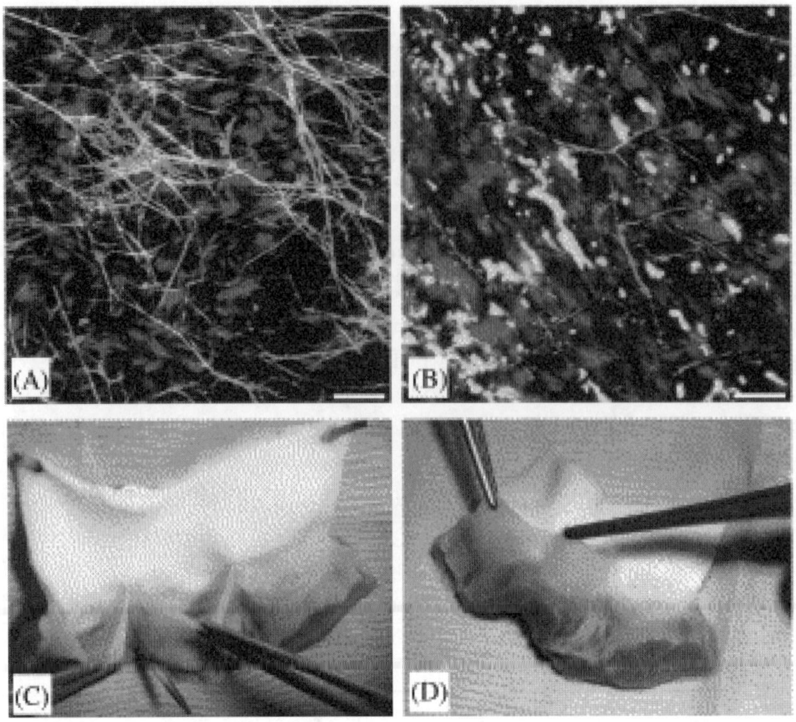

Figure 10.11. Overlay of optical sections showing elastic fibers (green) and collagen (red) of two TE valve leaflets, cultured under different conditions within the bioreactor system. The differences between the tissue states were only explicitly verifiable by the use of the MPE images. (A) TE heart valve, exposed to defined physical signals with a fixed frequency (1 Hz) and pressure conditions (3 l/min and 60/40 mm Hg), show normal extracellular matrix structures (bar: 20 μm). (C) Corresponding macroscopic appearance of the TE valves. (B) TE leaflet, exposed to supra-physiologic (>200 mm Hg) pressures, revealing an impaired matrix formation and a high amount of cell detritus (yellow). (D) Corresponding macroscopic pictures. With permission from.[72] For color reference, see page 277.

OCT uses near infrared light rather than sound. The optical echo delay time (related to depth) and intensity of light backscattered from biological tissues are measured using a technique called low-coherence interferometry. In conventional interferometry with long coherence length (laser interferometry), interference of light occurs over a distance of meters. In OCT, this interference is shortened to a distance of micrometres, thanks to the use of broadband light sources.

Light in an OCT system is broken into two arms — a sample arm and a reference arm (a mirror in the simplest OCT system). The combination of reflected light from the sample arm and reference light from the reference arm gives rise to an interference pattern, but only if light from both arms have the same optical pathlength ("same" meaning a difference of less than a coherence length). By scanning the mirror in the reference arm, a reflectivity profile of the sample can be obtained (time domain OCT). Areas of the sample that reflect back a lot of light will create greater interference than areas that don't. Any light that is outside the short coherence length will not interfere. This reflectivity profile, called an A-scan, contains information about the spatial dimensions and location of structures within the sample. A cross-sectional tomograph (B-scan) may be achieved by laterally combining a series of these axial depth scans (A-scan). An axial resolution of 1–10 μm has been achieved using advanced low-coherence light sources. And imaging depths of 1–2 mm can be achieved in most highly scattering tissues using a near infrared light source in the 800–1300 nm wavelength range. Optical coherence tomography is a very rapid evolving field, for details on verious types of imaging formation methods and practical instrumentations, refer to.[74]

In biomedical researches, OCT is widely used to obtain cross-sectional images of internal microstructures of various biological tissues *in vivo* and in real time. OCT is becoming a clinically viable imaging modality for early detection of diseases, precision guidance of surgical interventions, and repeated monitoring of treatment. OCT has also been used to non-invasively evaluate wound healing processes.[75] As an inexpensive and non-invasive quantitative 3-D imaging technology, OCT has great potential for biomaterials research.[76] It has already been demonstrated for online quality monitoring of engineered tissue for surgical implantation.[77,78] Dunkers *et al*[79] have also demonstrated a functional OCT system for imaging cellular activity within tissue scaffolds, as well as imaging the scaffold itself. In their system, functional information regarding cell activity and structural data is obtained using a combination of confocal fluorescence microscopy and high numerical aperture OCT.

10.4 CONCLUDING REMARKS

We have introduced a handful of methods and techniques for characterization of bulk and surface properties of biomaterials. However, space limitations prevent us from discussing every single analysis method used in biomaterials characterization. One has to remember that each individual analysis method only look at a specific aspect of a biomaterial. A combination of characterization methods (whether listed here or not) is usually required to provide the comprehensive information necessary for correlation with biomaterials performances in the real biological environment. The instrumentation of material analysis steadily advances and new methods and techniques will provide additional and invaluable information about biomaterials and lead to better performance of MEMS based medical devices.

References

[1] D. Lyman., Characterization of biomaterials, in *Integrated Biomaterials Science*, R. Barbucci Ed. New York: Kluwer Academic/Plenum Publishers, 325–337, (2002).

[2] A. F. von Recum Ed., *Handbook of Biomaterials Evaluation*, New York: Macmillan Publishing Company, (1986).

[3] B. D. Ratner, A. S. Hoffman, F. J. Schoen, J. E. Lemons, Eds., *Biomaterials Science: An Introduction to Materials in Medicine*, 2nd ed., San Diego, CA: Elsevier Academic Press, (1996).

[4] R. Cancedda, A. Cedola, A. Giuliani, V. Komlev, S. Lagomarsino, M. Mastrogiacomo, F. Peyrin, F. Rustichelli, Bulk and interface investigations of scaffolds and tissue-engineered bones by X-ray microtomography and X-ray microdiffraction, *Biomaterials*, **28**, 2505–2524, May, (2007).

[5] E. N. Landis, E. N. Nagy, D. T. Keane, Microstructure and fracture in three dimensions, *Eng Fract Mech*, **70**, 911–925, May, (2003).

[6] S. Nuzzo, F. Peyrin, P. Cloetens, J. Baruchel, G. Boivin, Quantification of the degree of mineralization of bone in three dimension using synchrotron radiation microtomography, *Med. Phys.*, **19**, 2672–2681, November, (2002).

[7] M. Salomé, F. Peyrin, P. Cloetens, C. Odet, A. M. Laval-Jeantet, J. Baruchel *et al.*, A synchrotron radiation microtomography system for the analysis of trabecular bone samples, *Med. Phys.*, **26** 2194–2204, October (1999).

[8] G. H. van Lenthe, H. Hagenmüller, M. Bohner, S. J. Hollister, L. Meinel, R. Müller, Nondestructive micro-computed tomography for biological imaging and quantification of scaffold–bone interaction *in vivo*, *Biomaterials*, **28**, 2479–2490,May, (2007).

[9] B. D. Porter, A. S. P. Lin, A. Peister, D. Hutmacher, R. E. Guldberg, Noninvasive image analysis of 3D construct mineralization in a perfusion bioreactor, *Biomaterials*, **28**, 2425–2533, May (2007).

[10] A. C. Jones, C. H. Arns, A. P. Sheppard, D. W. Hutmacher, B. K. Milthorpe, M. A. Knackstedt, Assessment of bone ingrowth into porous biomaterials using MICRO-CT, **28**, 2491–2504, May, (2007).

[11] B. Kasemo, J. Lausmaa, Biomaterials, in *Surface Characterization: A User's Sourcebook*, D. Brune, R. Helborg, O. Hunderi and H. Whitlow Eds. Weinheim: Scandinavian Science Publisher, 660–680, (1997).

[12] J. Hwang, L. K. Tamm, Bohm, T. S. Ramalingam, E. Betzig, and M. Eddin, Nanoscale complexity of phospholipid monolayers investigated by near-field Scanning Optical Microscopy, *Science* **270**, 610–614, October, (1995)

[13] G. I. Goldstein, D. E. Newbury, P. Echlin, D. C. Joy, C. Fiori, E. Lifshin Eds., *Scanning Electron Microscopy and X-ray Microanalysis*, New York: Plenum Press, (1981).

[14] D. K. Racker Ed., Transmission Electron Microscopy: Methods of Application, Springfield, IL: Thomas, (1983).

[15] R. N. S. Sodhi, Application of surface analytical and modification techniques to biomaterial research, *J. Elec. Spectroscopy and Related Phenomena*, **81**, 269–284, September, (1996).

[16] C. Battistioni, M. P. Casaletto, G. M. Ingo, S. Kaciulis, G. Mattogno, L. Pandol., Surface characterization of biocompatible hydroxyapatite coatings, *Surf. Interface Anal.* **29**, 773–781, (2000).

[17] R. Born, D. Scharnweber, S. Robler, M. Stolzel, M. Thieme, C. Wolf, H. Worch, Surface analysis of titanium based biomaterials, *Fresenius J. Anal. Chem.*, **361**, 697–700, August, (1998).

[18] G. Kotzar, M. Freas, P. Abel, A. Fleischman, S. Roy, C. Zorman, J. M. Moran, J. Melzak, Evaluation of MEMS materials of construction for implantable medical devices, *Biomaterials*, **23**, 2737–2750, (2002).

[19] K. Merrett, R. M. Cornelius, W. G. Mcclung, L. D. Unsworth, H. Sheardown, Surface analysis methods for characterizing polymeric biomaterials, *J. Biomater. Sci. Polymer Edn*, **13**, 593–621, June (2002).

[20] C. Bai, *Scanning Tunneling Microscopy and Its Applications*, 2nd ed., New York: Springer Verlag, (1999).

[21] D. Sarid, *Scanning Force Microscopy, Oxford Series in Optical and Imaging Sciences*, New York: Oxford University Press, (1991).

[22] A. D L. Humphris, M. J. Miles, J. K. Hobbs, A mechanical microscope: High-speed atomic force microscopy, *Applied Physics Letters*, **86**, 034106 (1–3), January, (2005).

[23] T. Cunningham, F. M. Serry, L. M. Ge, D. Gotthard, D. J. Dawson, Atomic force profilometry and long scan atomic force microscopy: new techniques for characterisation of surfaces, *Surface Engineering*, **16**, 295–298, August, (2000).

[24] F. Z. Fang, Z. W. Xu, S. Dong, Study on phase images of a carbon nanotube probe in atomic force microscopy, *Measurement Science and Technology*, **19**, 055501 (7pp), (2008).

[25] N. B. Holland, R. E. Marchant, Individual plasma proteins detected on rough biomaterials by phase imaging AFM, *J. Biomed. Mater. Res. 51*, 307–315, September 2000.

[26] W. Stocker, S. N. Magonov, H. J. Cantow, J. C.Wittman, B. Lotz, Contact faces of epitaxially crystallized .alpha.- and .gamma.-phase isotactic polypropylene observed by atomic force microscopy, *Macromolecules*, **26**, 5915–5923, October, 1993.

[27] S. O. Vansteenkiste, M. C. Davies, C. J. Roberts, S. J. B. Tentfler, P. M. Williams, Scanning probe microscopy of biomedical interfaces, *Progress in Surface Science*, **57**, 95–136, February, (1998).

[28] C. Werner, H. J. Jacobasch, Surface characterization of polymers for medical devices, *International Journal of Artificial Organs* **22**, 160–176, March (1999).

[29] R. E. Marchant, M. D. Barb, J. R. Shainoff, S. J. Eppell, D. L. Wilson and C. A. Siedlecki, Three dimensional structure of human fibrinogen under aqueous conditions visualized by atomic force microscopy, *Thromb Haemostasis* **77**, 1048–1051, June (1997).

[30] V. Dupres, C. Verbelen, Y. F. Dufrêne, Probing molecular recognition sites on biosurfaces using AFM, *Biomaterials*, **28**, 2393–2402, May, (2007).

[31] A. Diaspro, Ed., *Confocal and Two-photon Microscopy, Foundations, Applications, and Advances*, New York: Wiley-Liss, Inc., 2002.

[32] T. Appenzeller and C. Norman, New Eyes on Hidden Worlds, *Science*, **276**, 1981, June (1997).

[33] B. Nebe, C. Forster, H. Pommerenke, G. Fulda, D. Behrend, U. Bernewski, K. P. Schmitz, J. Rychly, Structural alterations of adhesion mediating components in cells cultured on poly-β-hydroxy butyric acid, *Biomaterials*, **22**, 2425–2434, September (2001).

[34] Y. Yang, M. L. Becker, D. Bolikal, J. Kohn, D. N. Zeiger, C. G. Simon, Jr., Combinatorial polymer scaffold libraries for screening cell-biomaterial interactions in 3-D. *Advanced Materials*, 20, 1-7, January 2008.

[35] M. Rajadhyaksha, R. R. Anderson, R. H. Webb, Video-rate confocal scanning laser microscope for imaging human tissues *in vivo, Appl. Opt.*, Vol. 38, 2105–2115, April 1999.

[36] B. D. Ratner, Advances in the analysis of surfaces of biomedical interest, *Surf. Interface Anal.*, **23**, 521-, (1995).

[37] L. Dai, H. S. W. St. John, J. Bi, P. Zientek, R. C. Chatelier, H. J. Griesser, *Surf. Interface Anal.*, Biomedical coatings by the covalent immobilization of polysaccharides onto gas-plasma-activated polymer surfaces, **29**, pp46–55, January (2000).

[38] J. T. Grant, D. Briggs, Eds., *Surface Analysis by Auger and X-ray Photoelectron Spectroscopy*, Chichester, UK: IM Publications, (2003).

[39] R. Chung, M. F. Hsieh, R. N. Panda, T. S. Chin, Hydroxyapatite layers deposited from aqueous solutions on hydrophilic silicon substrate, *Surface and Coatings Technology*, **165**, 194–200, February (2003).

40 B. D. Ratner, B. J. Tyler, A. Chilkoti, Analysis of biomedical polymer surfaces: polyurethanes and plasma-deposited thin films, *Clinical Materials,* **13** 71–84, (1993).

41 D. Klee, H. Hocker, Polymers for Biomedical Applications: Improvement of the Interface Compatibility, *Advances in Polymer Science,* **149**, 1–57, September, (1999).

42 A. Benninghoven, F. G. Rüdenauer, and H. W. Werner, *Secondary Ion Mass Spectrometry: Basic Concepts, Instrumental Aspects, Applications, and Trends*, New York: Wiley, (1987).

43 A. Chilkoti, B. Ratner and D. Briggs, Plasma-deposited polymeric films prepared from carbonyl-containing volatile precursors: XPS chemical derivatization and static SIMS surface characterization, *Chem. Mater.* **3**, 51–61, January, (1991).

44 J. Chen and J. A. Gardella, Jr., Time-of-flight secondary ion mass spectrometry studies of *in vitro* hydrolytic degradation of biodegradable polymers, *Macromolecules*, **32**, 7380–7388, October, (1999).

45 J. W. Lee, J. A. Gardella, Jr., *In vitro* hydrolytic surface degradation of poly (glycolic acid): Role of the surface segregated amorphous region in the induction period of bulk erosion, *Macromolecules*, **34**, 3928–3937, May, (2001).

46 M. Diem, *Introduction to Modern Vibrational Spectroscopy*, New York, Wiley, (1993).

47 R. Barbucci, M. Casolaro, A. Magnini, Characterization of biomaterial surfaces: ATR-FTIR, potentiometric and calorimetric analysis, *Clinical Materials* **11**, 37–51, (1992).

48 K. K. Chittur, Surface techniques to examine the biomaterial-host interface: An introduction to the papers, *Biomaterials,* **19**, 301–305, March (1998).

49 M. Kiremitçi, A. Pesmen, M. Pulat, I. Gürhan, Relationship of surface characteristics to cellular attachment in PU and PHEMA, *Journal of Biomaterials Applications*, **7**, 250–264, January, (1993).

50 M. Joshi, N. Kale, R. Lal, V. R. Rao, S. Mukherji, A novel dry method for surface modification of SU-8 for immobilization of biomolecules in Bio-MEMS, *Biosensors and Bioelectronics*, **22**, 2429–2435, May, (2007).

51 B. J. Tyler, G. Rayal, D. G. Castner, Multivariate analysis strategies for processing ToF-SIMS images of biomaterials, *Biomaterials*, **28**, 2412–2423, February (2007).

52 P. Chaurand, J. L. Norris, D. S. Cornett, J. A. Mobley, R. M. Caprioli, New developments in profiling and imaging of proteins from tissue sections by MALDI mass spectrometry, *J. Proteome Res.*, **5**, 2889–900, (2006).

53 A. Boskey, N. P. Camacho, FT-IR imaging of native and tissue-engineered bone and cartilage (review), *Biomaterials*, **28**, 2465–2478, May 2007.

54 R. Manoharan, J. J. Baraga, R. P. Rava, R. R. Dasari, M. Fitzmaurice, M. S. Feld, Biochemical analysis and mapping of atherosclerotic human artery using FT-IR microspectroscopy, *Atherosclerosis*, **103**, 181–93, (1993).

55 P. Lasch, A. Pacifico, M. Diem, Spatially resolved IR microspectroscopy of single cells, *Biopolymers*, **67**, 335–338, (2002).

56 L. M. Pachman, A. Veis, S. Stock, K. Abbott, F. Vicari, P. Patel, *et al*, Composition of calcifications in children with juvenile dermatomyositis: Association with chronic cutaneous inflammation, *Arthritis Rheum.*, **54**, 3345–3350, (2006).

57 E. N. Lewis, P. J. Treado, R. C. Reeder, G. M. Story, A. E. Dowrey, C. Marcott, *et al.*, Fourier transform spectroscopic imaging using an infrared focal-plane array detector, *Anal. Chem.* **67**, 3377–3381, (1995).

58 C. Marcott, R. C. Reeder, E. P. Paschalis, D. N. Tatakis, A. L. Boskey, R. Mendelsohn., Infrared microspectroscopic imaging of biomineralized tissues using a mercury–cadmium–telluride focal-plane array detector, *Cell Mol. Biol.*, **44**, 109–115, (1998).

[59] L. M. Miller and R. J. Smith, Synchrotrons versus globars, point-detectors versus focal plane arrays: selecting the best source and detector for specific infrared microscopy and imaging applications, *Vib. Spec.*, **38**, 237–240, (2005).

[60] K. L. A. Chan and S. G. Kazarian, New opportunities in micro- and macro-attenuated total reflection infrared spectroscopic imaging: Spatial resolution and sampling versatility, *Applied Spectroscopy*, **57**, 381–389, April, (2003).

[61] J. L. Koenig, Microscopic imaging of polymers, Washington, DC: American Chemical Society, (1998).

[62] J. G. Grasselli and B. J. Bulkin, *Analytical Raman Spectroscopy, Chemical Analysis Series*, **114**, New York: John Wiley, (1991).

[63] Z. Huang, H. Zeng, I. Hamzavi, D. I. McLean, H. Lui, A rapid near-infrared Raman spectroscopy system for real-time *in vivo* skin measurements, *Optics Letters*, **26**, 1782–1784, (2001).

[64] G. Penel, C. Delfosse, M. Descamps, G. Leroy, Composition of bone and apatitic biomaterials as revealed by intravital Raman microspectroscopy, *Bone*, **36**, 893–901 (2005).

[65] G. Pezzotti, Raman piezo-spectroscopic analysis of natural and synthetic biomaterials, *Anal. Bioanal. Chem.*, **381**, 577–590, February, (2005).

[66] C. P. Tarnowski, S. Stewart, K. Holder, L. Campbell-Clark, R. J. Thoma, A. K. Adams, M. A. Moore, M. D. Morris, Effects of treatment protocols and subcutaneous implantation on bovine pericardium: A Raman spectroscopy study, *J. Biomed. Opt.*, **8**, 179–184, April (2003).

[67] A. Ulman, *An Introduction to Ultrathin Organic Films: From Langmuir-Blodgett to Self-assembly*. Boston, MA: Academic Press, (1991).

[68] A. W. Neumann and R. J. Godd, Techniques of measuring contact angles. In *Surface and Colloid Science—Experimental Methods*, R. J. Good, R. R. Stromberg, Eds., **11**, New York: Plenum Publishing; 31–61, (1979).

[69] W. Denk, J. Strickler and W. Webb, Two-photon laser scanning fluorescence microscopy, *Science*, **248**, 73–76, (1990).

[70] K. König and I. Riemann, High-resolution multiphoton tomography of human skin with subcellular spatial r–439, (2003).

[71] R. M. Williams, W. R. Zipfel, W. W. Webb, Interpreting second harmonic generation images of collagen I fibrils, *Biophys. J.* **88**, 1377–1386, (2005).

[72] K. König, K. Schenke-Layland, I. Riemann, U. A. Stock, Multiphoton autofluorescence imaging of intratissue elastic fibers, *Biomaterials*, **26**, 495–500, February (2005).

[73] D. Huang, E. A. Swanson, C. P. Lin, J. S. Schuman, W. G. Stinson, W. Chang, M. R. Hee, T. Flotte, K. Gregory, C. A. Puliafito, J. G. Fujimoto, Optical coherence tomography, *Science.* **254**, 1178–1181, November, (1991).

[74] X. Li, J. G. Fujimoto, Optical Coherence Tomography, In *Encyclopedia of Biomaterials and Biomedical Engineering*, 1:1, 1120— 1132, (2004).

[75] M. J. Cobb, Y. Chen, R. A. Underwood, M. L. Usui, J. Olerud, X. Li, Noninvasive assessment of cutaneous wound healing using ultrahigh-resolution optical coherence tomography, *J. Biomed. Opt.* **11**, 064002, November (2006).

[76] P. H. Tomlins, R. K. Wang, Theory, developments and applications of optical coherence tomography, *J. Phys. D: Appl. Phys.*, **38**, 2519–2535, (2005).

[77] C. Mason, J. F. Markusen, M. A. Town, P. Dunnill, R. K. Wang, The potential of optical coherence tomography in the engineering of living tissue, *Phys. Med. Biol.* **49**, 1097–1115, April (2004).

78 C. Mason, J. F. Markusen, M. A. Town, Y. He, P. Dunnill, R. K. Wang, Doppler optical coherence tomography for measuring flow in engineered tissue, *Biosensors Bioelectron.*, **20**, 414–423, October, (2004).

79 J. P. Dunkers, M. T. Cicerone, N. R .Washburn, Collinear optical coherence and confocal fluorescence microscopies for tissue engineering, *Opt. Express,* **11** 3074–3079, November, (2003).

Color Inserts

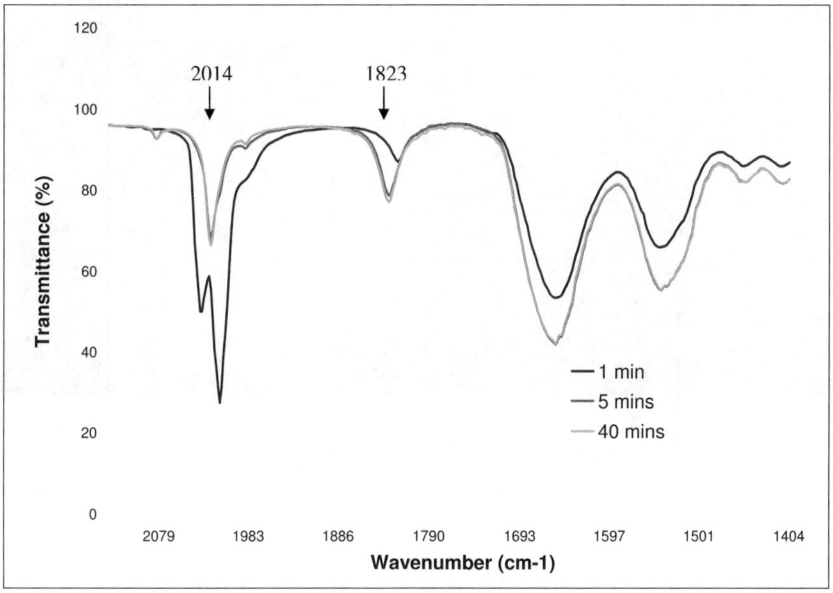

Figure 2.6 FTIR spectrums of iron spider silk sample after different exposure time.

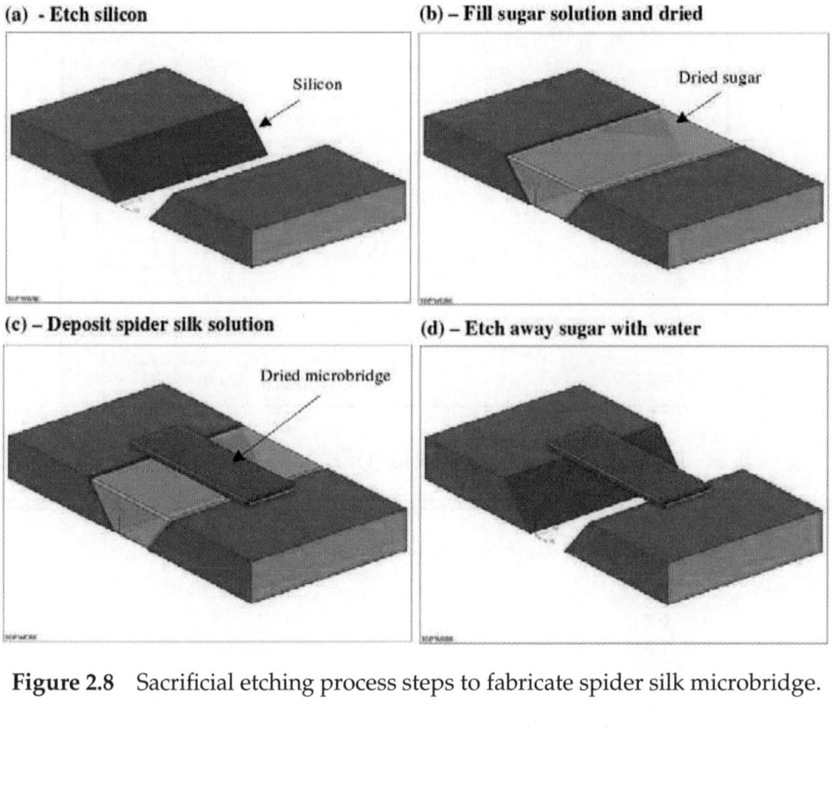

Figure 2.8 Sacrificial etching process steps to fabricate spider silk microbridge.

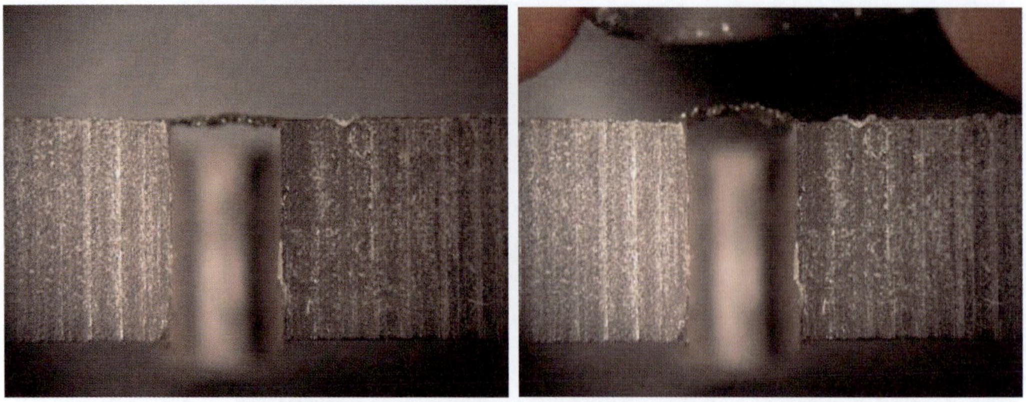

Figure 2.13 Fe/Ni 50%spider silk beam.

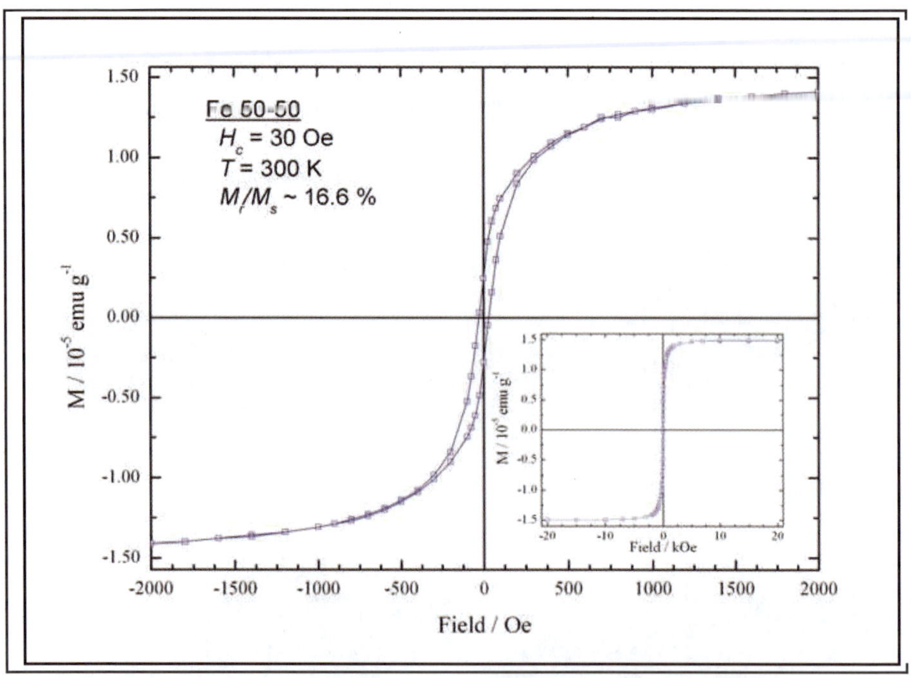

Figure 2.14 Hysteresis curve for 1:1 v/v iron spider silk sample.

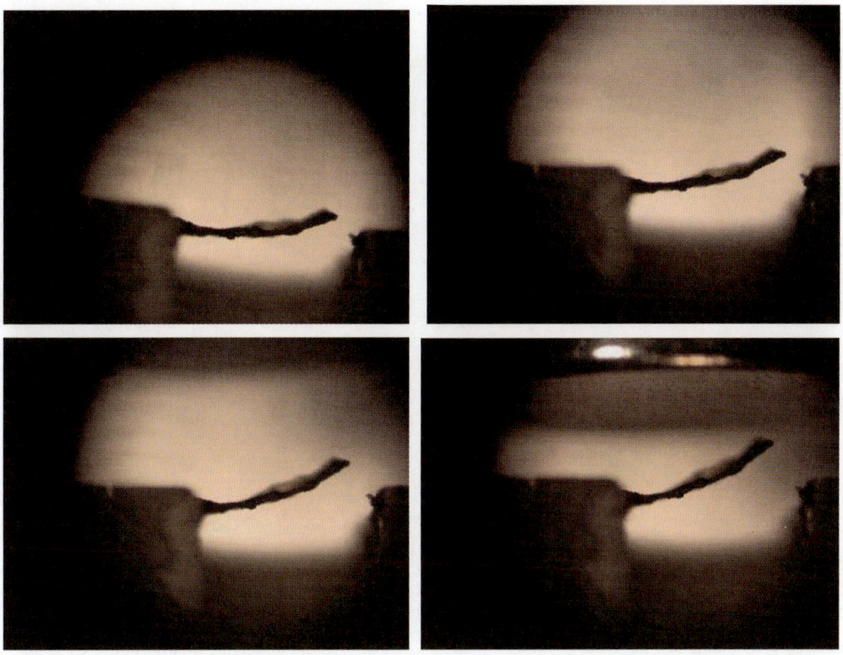

Figure 2.15 Images used to measure bending movement of the beam under increasing magnetic field. Sample size 3.25 × 0.98 × 0.17 mm (*Length* × *Width* × *Thickness*).

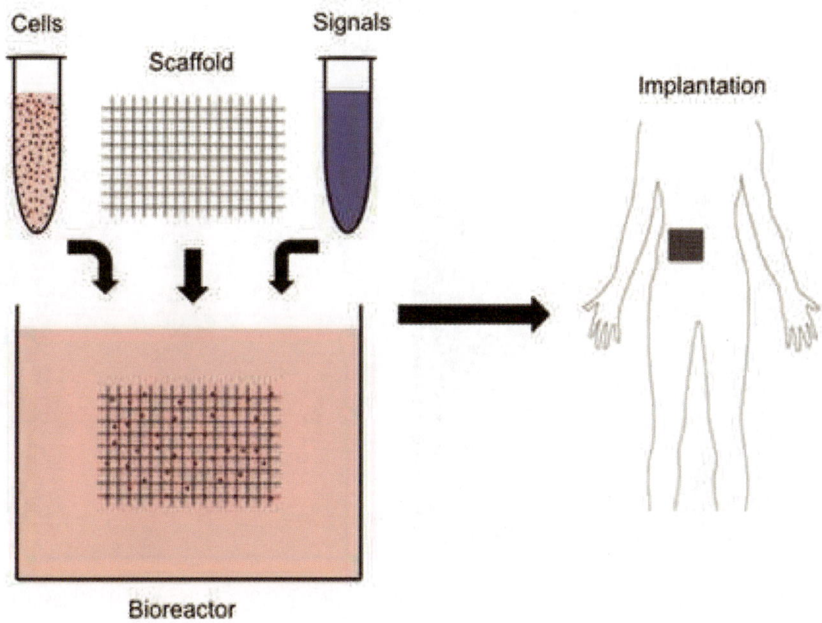

Figure 3.1 The key elements involved in the classic tissue engineering paradigm.

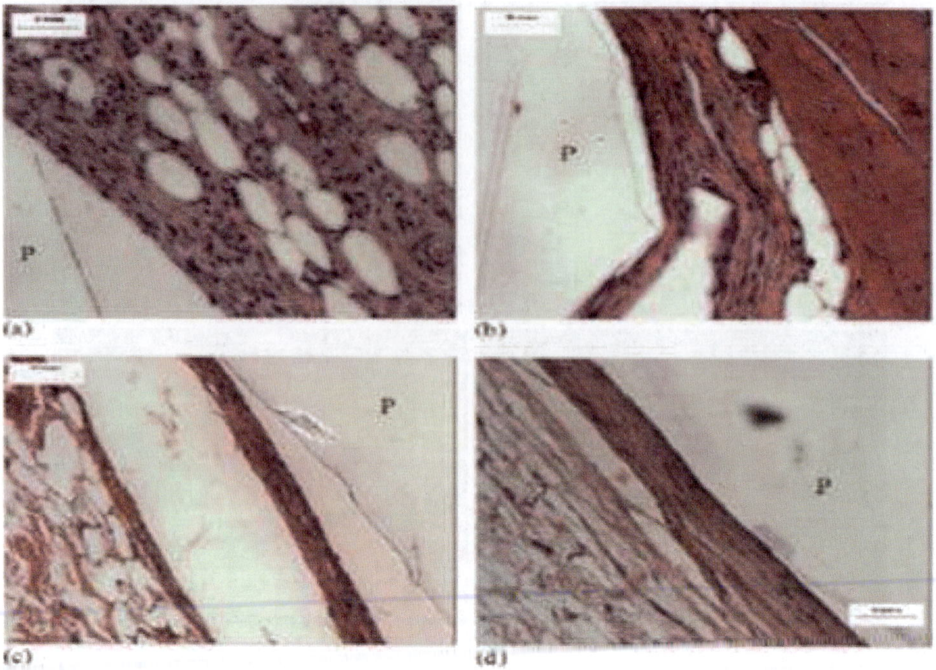

Figure 3.2 Foreign body response of POC (120°C, 2 Pa, 3 d) implanted subcutaneously in female Sprague–Dawley rats (scale bar = 50 μm). Implants and surrounding tissues were harvested after (a) 1 week; (b) 1 month; (c) 2 months; (d) 4 months implantation for H&E staining. "P" represents polymer section. Reprinted from Biomaterials, 27, Jian Yang *et al.*, Synthesis and evaluation of poly(diol citrate) biodegradable elastomers, 1889–1898, 2006, with permission from Elsevier.

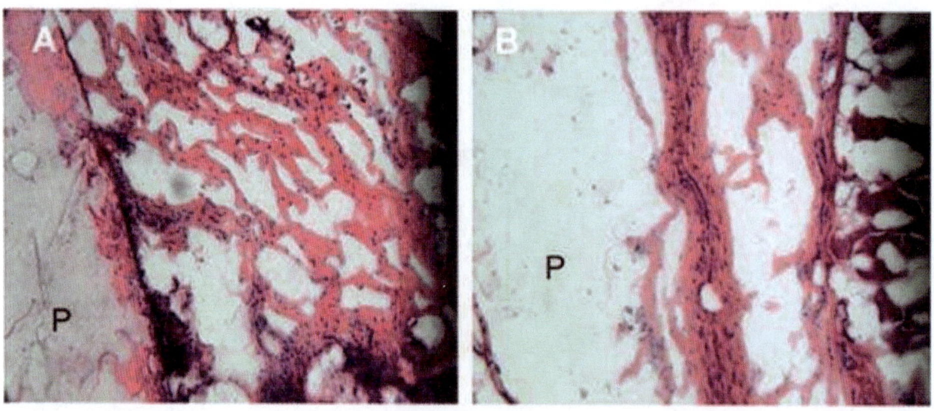

Figure 3.3 Foreign body response of PAMC (POMC 0.8, 20 mins. UV crosslinking) implanted subcutaneously in female Sprague–Dawley rats. Implants and surrounding tissues were harvested after (a) 1 week and (b) 4 weeks. "P" represents polymer section.

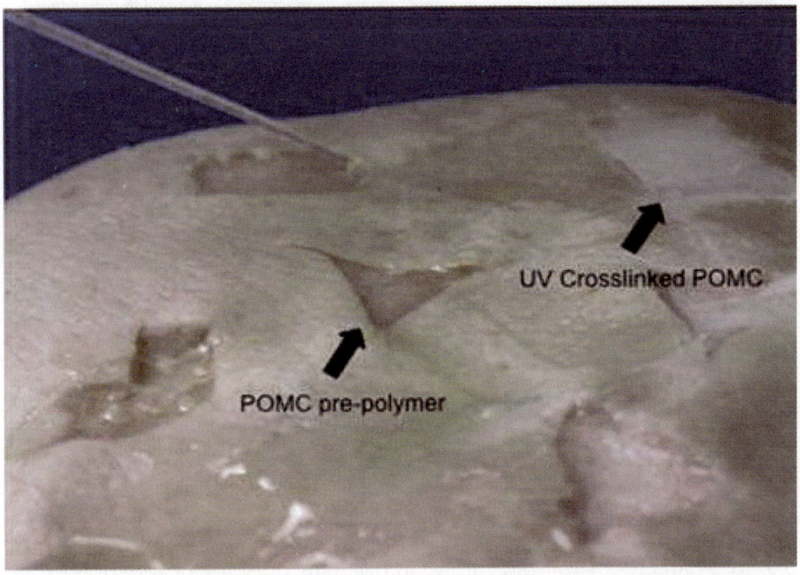

Figure 3.4 Pre-PAMC and ultraviolet crosslinked PAMC on porcine skin.

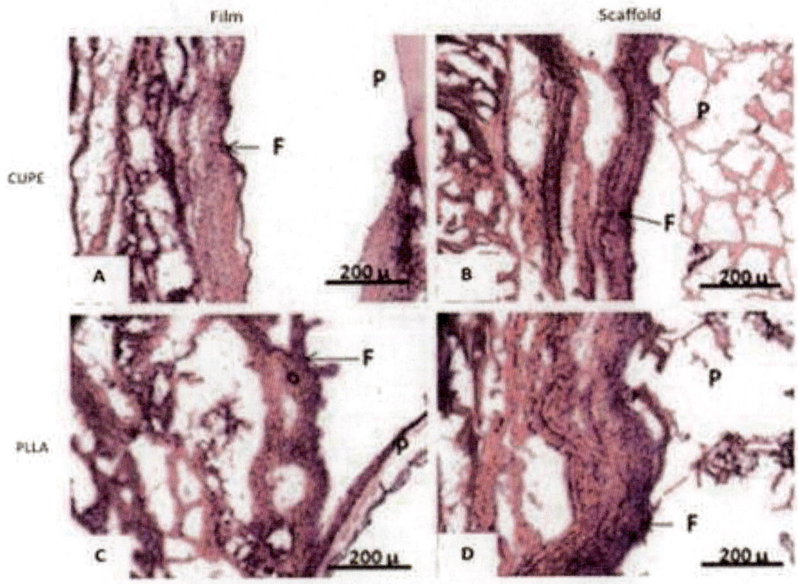

Figure 3.5 Histology of *in vivo* response to CUPE film (A) and scaffold discs (B). PLLA films (C) and scaffolds (D) served as control. P and F are used to indicate the regions of polymer and fibrous capsule respectively. All images were taken at 10x magnification. On the 1 week samples, although all samples were covered by a well defined fibrous capsule, CUPE implants were consistently surrounded by a thinner fibrous capsule as opposed to PLLA implants. In the case of the 4 week implants, overall, all the implants appear to trigger similar extent of tissue responses.

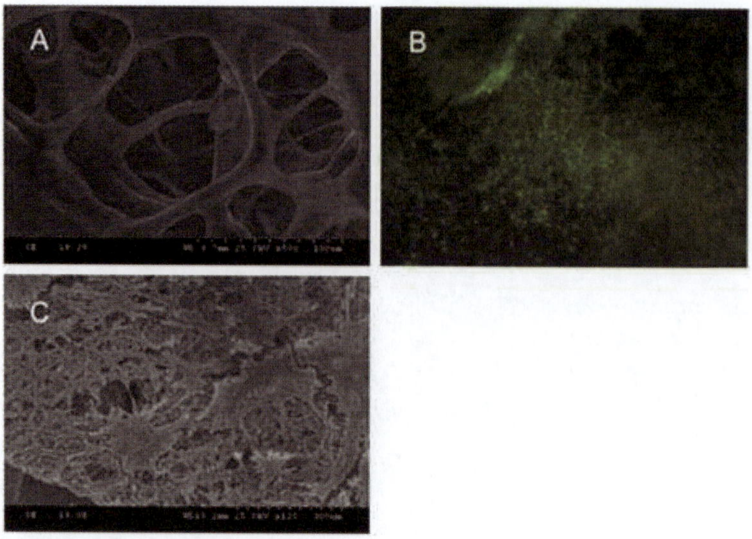

Figure 3.6 SEM images of scaffold surface indicate the presence of well defined pores (A) and even cell distribution of cells on the scaffold (B and C).

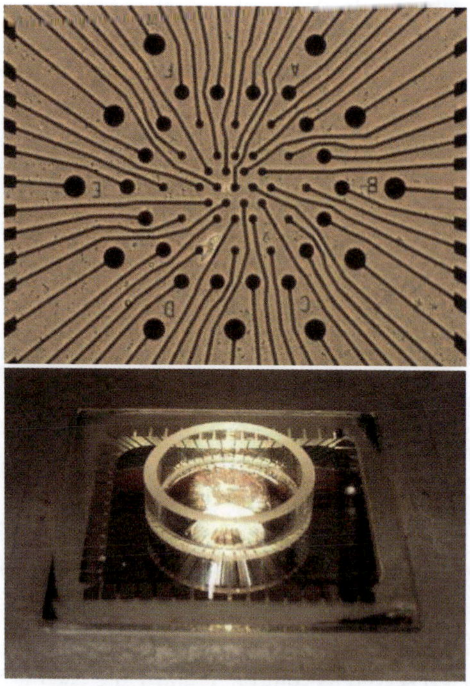

Figure 4.1 One example of a commercially available microelectrode array (Multichannel Systems Hexagonal Array). Neurons or neural tissues are plated on the array surface, top. The circular electrodes can be stimulated and recorded from, but the electrical traces are insulated from neurons plated above. The electrodes are found in the center of the well, which contains culture media to provide cells with nutrients, bottom.

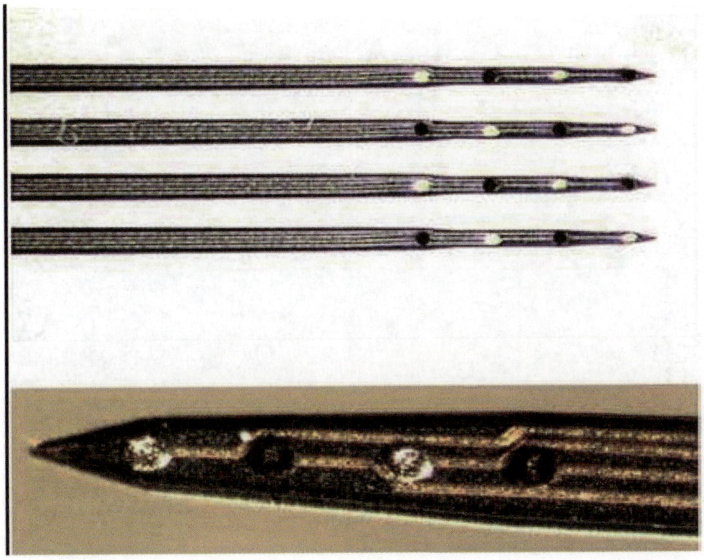

Figure 4.4 The four shanks of a Michigan probe, top; closer image of the four electrodes per shank, bottom. The darker electrode sites in the lower picture have been treated with the conductive polymer poly(3,4-ethylenedioxythiophene) (PEDOT).[32] Reprinted with permission from the *Journal of Neural Engineering*, Ludwig *et al.*, 2006, Vol 3, Issue 1, pp. 59–70.

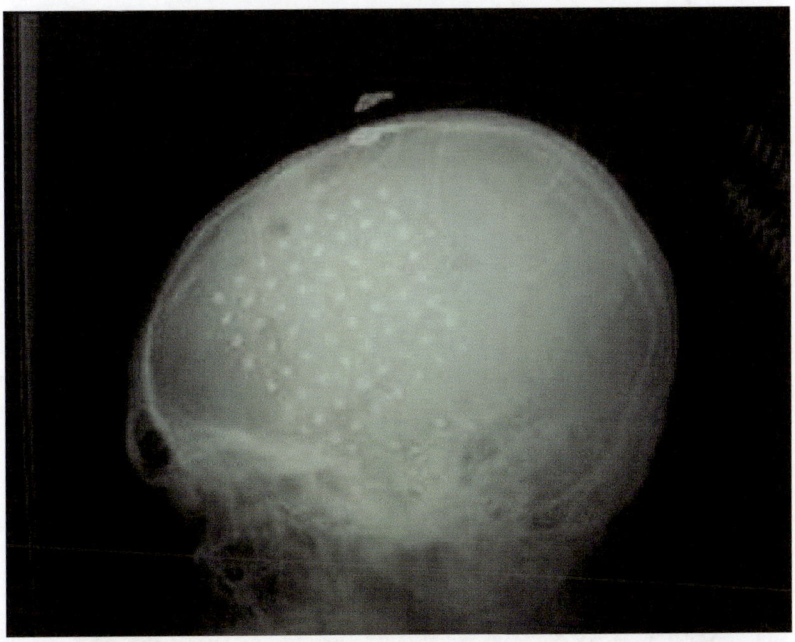

Figure 4.7 An ECoG electrode array located on top of the cortex but inside the skull.[91] Reprinted with permission from the *Journal of Neural Engineering*, Leuthanrdt *et al.*, 2004, Vol 1, Issue 2, pp 63–71.

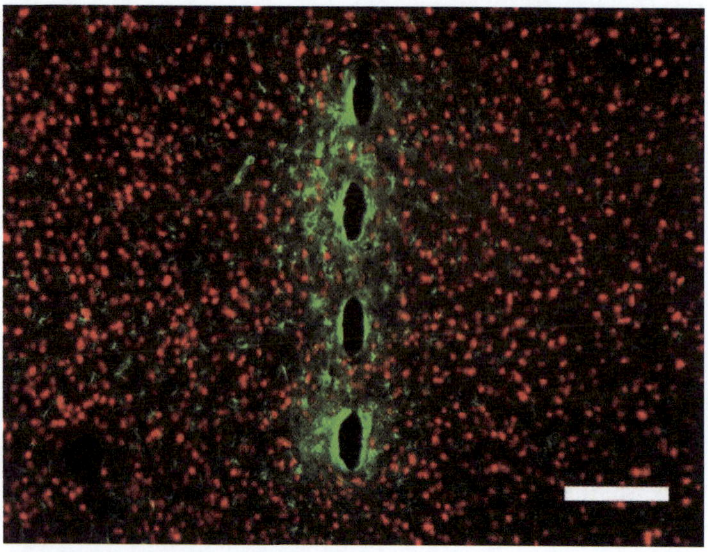

Figure 4.8 The response of rat cortical tissue 8 weeks after insertion. The green stain marks glial fibrillary acidic protein (GFAP), a marker for astrocytes, and the red NeuN stain marks neuron nuclei. The scale bar is 100 micrometers. Notice the astrocytes apparent at the electrode interface, and that the neurons appear further away. Image courtesy of George McConnell, Bellamkonda Lab, Georgia Institute of Technology.

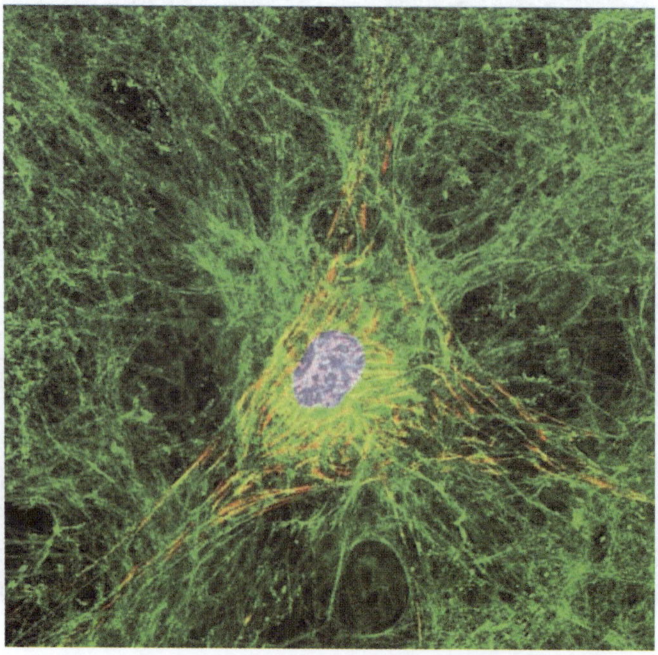

Figure 5.1 A cell in a three-dimensional culture forming links with the scaffold.[2]

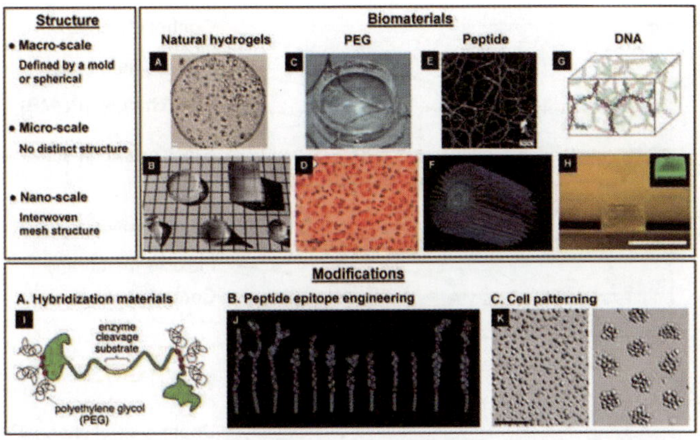

Figure 5.3 Overview of hydrogels used for 3-D cell entrapment.[14]

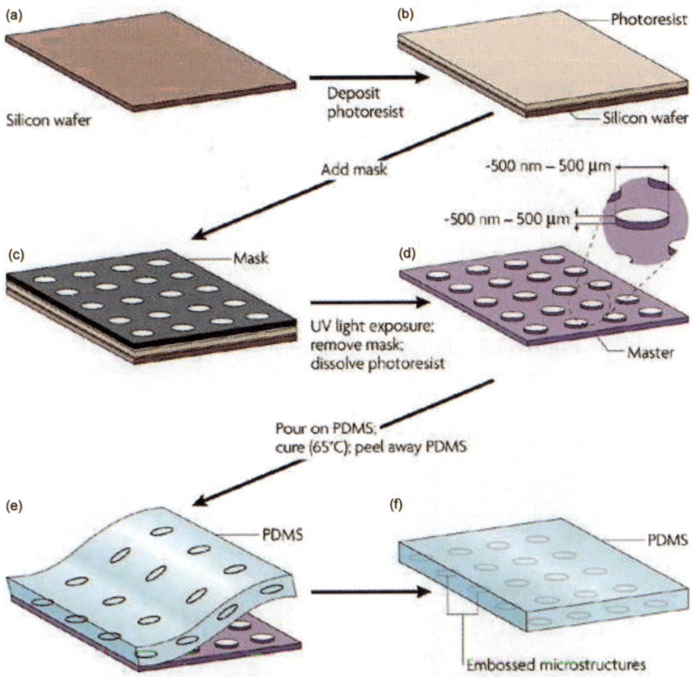

Figure 5.7 Microfabrication of PDMS structures. (a–b) Photoresist is spin-coated on a silicon wafer. (c) A mask is placed in contact with the layer of photoresist. (d) The photoresist is illuminated with ultraviolet (UV) light through the mask. An organic solvent dissolves and removes photoresist that is not crosslinked. The master consists of a silicon wafer with features of photoresist in bas-relief. An expanded view of one of the microfabricated structures with its characteristic critical dimensions is shown. (e) PDMS is poured on the master, cured thermally and peeled away. (f) The resulting layer of PDMS has microstructures embossed in its surface.[47]

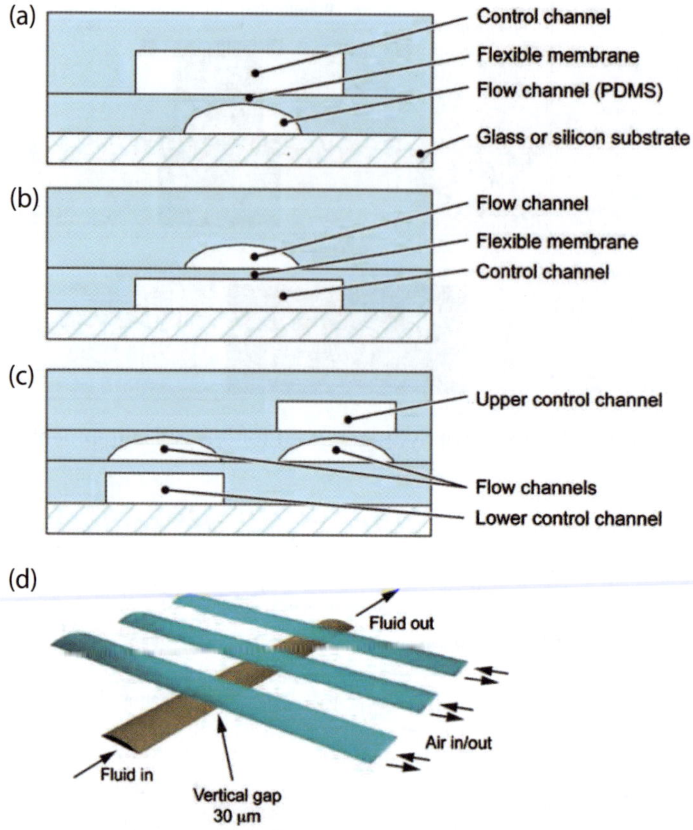

Figure 5.8 For color reference, see page 100.

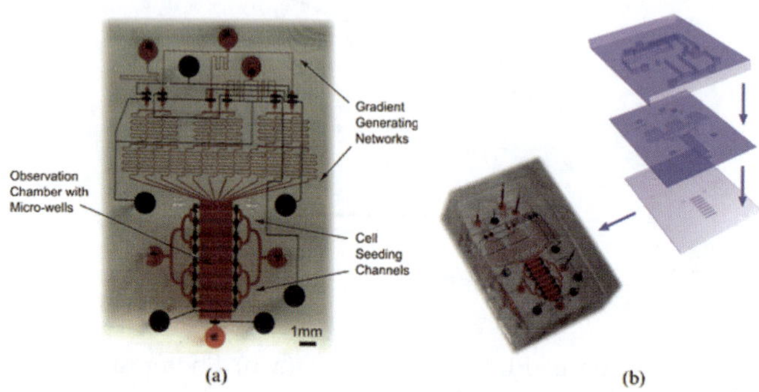

Figure 5.9 (a) A two-layer polydimethylsiloxane (PDMS) push-down microfluidic valve. An elastomeric membrane is formed where the flow channel is positioned orthogonal to the control channel directly above. Fluid flow is out of the page. (b) A two-layer PDMS push-up microfluidic valve where a control channel lies orthogonal to and below the flow channel. (c) A three-layer device with both push-up and push-down valves. (d) Schematic of a linear peristaltic pump using three membrane valves in a series.[53]

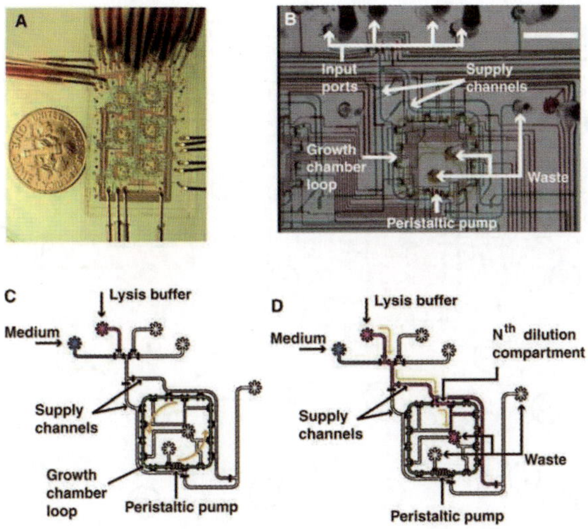

Figure 5.10 (A) Optical micrograph showing six microchemostats that operate in parallel on a single chip. Various inputs have been loaded with food dyes to visualize channels and sub-elements of the microchemostats. The coin is 18 mm in diameter. (B) Optical micrograph showing a single microchemostat and its main components. Scale bar, 2 mm. (C) Schematic diagram of a microchemostat in continuous circulation mode. Elements such as the growth loop with individually addressable connected segments, the peristaltic pump, supply channels, and input/output ports are labeled. (D) Isolation of a segment from the rest of the growth chamber during cleaning and dilution mode. A lysis buffer (indicated in red) is introduced into the chip through the lysis buffer port. Integrated microvalves direct the buffer through the segment, flushing out cells, including those adhering to chamber walls. The segment is then rinsed with fresh sterile medium and reunited with the rest of the growth chamber.[46]

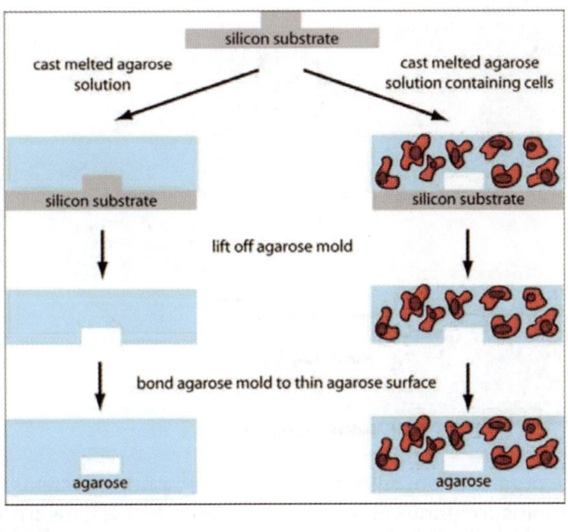

Figure 5.12 Schematic of the fabrication of agarose microfluidic devices with (right) and without (left) embedded cells.[68]

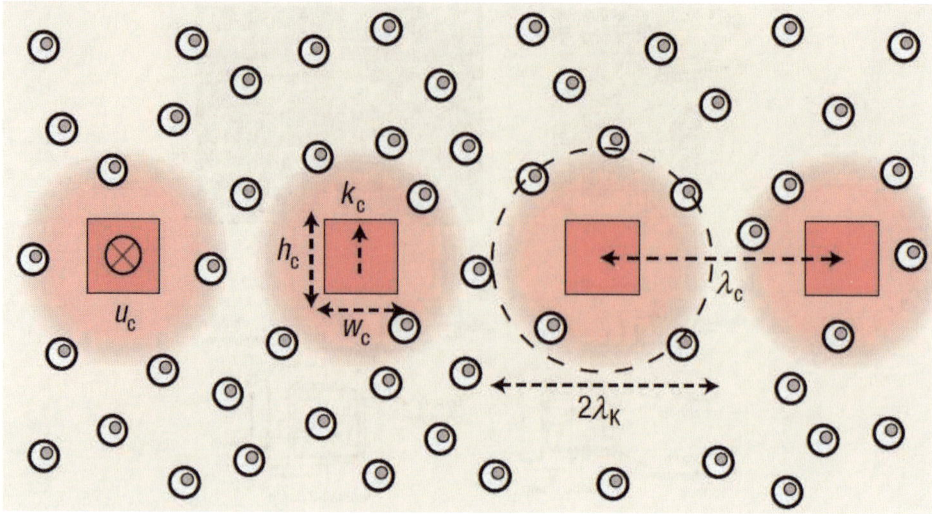

Figure 5.15 Schematic cross-sectional view of a cell-seeded microfluidic scaffold. The dispersed cells (circles) surround the microchannels (squares). The pink shading represents steady-state distributions of solutes. Here, a reactive solute is delivered via the channels and is consumed by the cells as it diffuses into the matrix. λ_K is the Krogh length, λ_c is the interchannel distance, w_c and h_c are the microchannel width and height, k_c is the mass transfer coefficient of the flow in the microchannels, and u_c is the speed of the flow in the microchannels.[94]

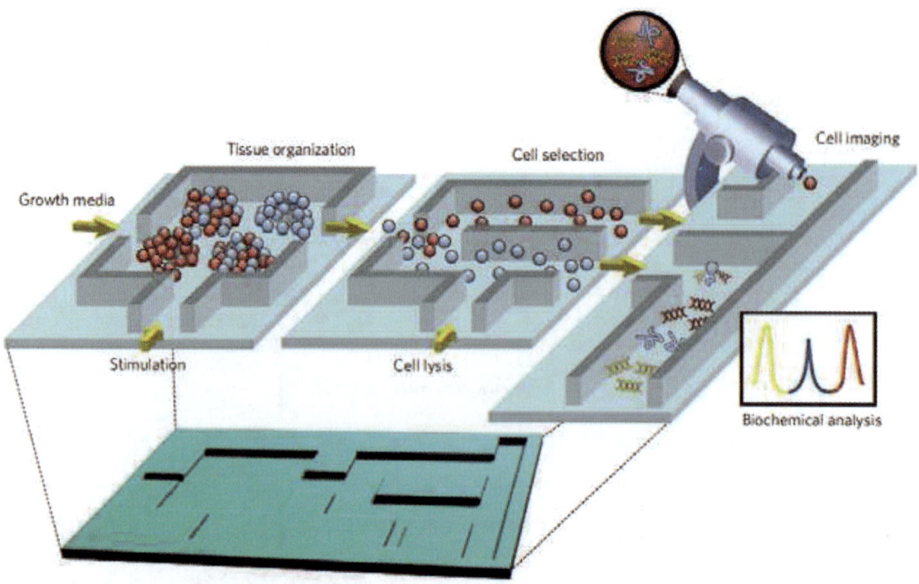

Figure 5.16 Tissue organization, culture and analysis in microsystems. Microsystems can incorporate 3D scaffolds to guide cell growth, microfluidic systems for nutrient transport, different techniques for biochemical analysis (such as image-based analysis), to give multiple functionalities on a single chip.[96]

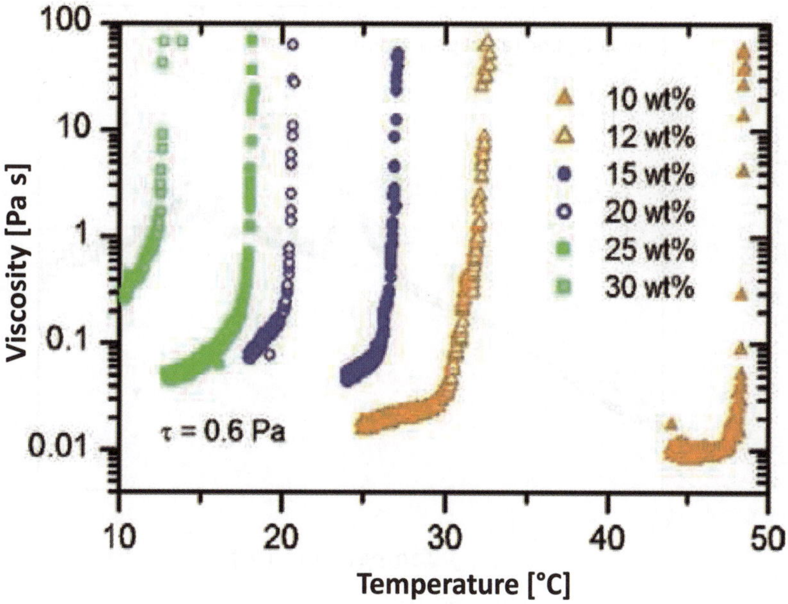

Figure 6.1 Viscosity as a function of temperature for different concentrations of Pluronic F127 in water from cone and plate viscometry at controlled shear stress (0.6 Pa s).[26]

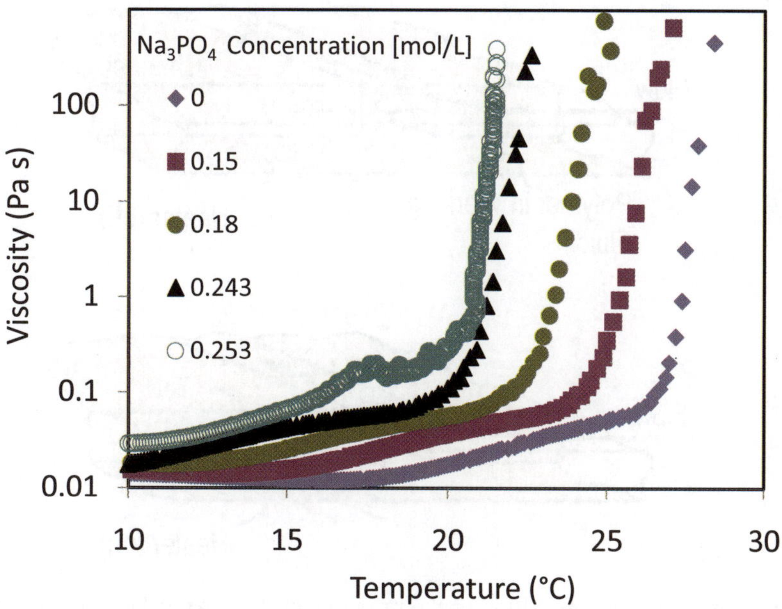

Figure 6.2 Viscosity as a function of temperature for different concentrations of sodium phosphate in a 15 wt% Pluronic F127 solution in water from cone and plate viscometry at controlled shear stress (0.6 Pa s).[110]

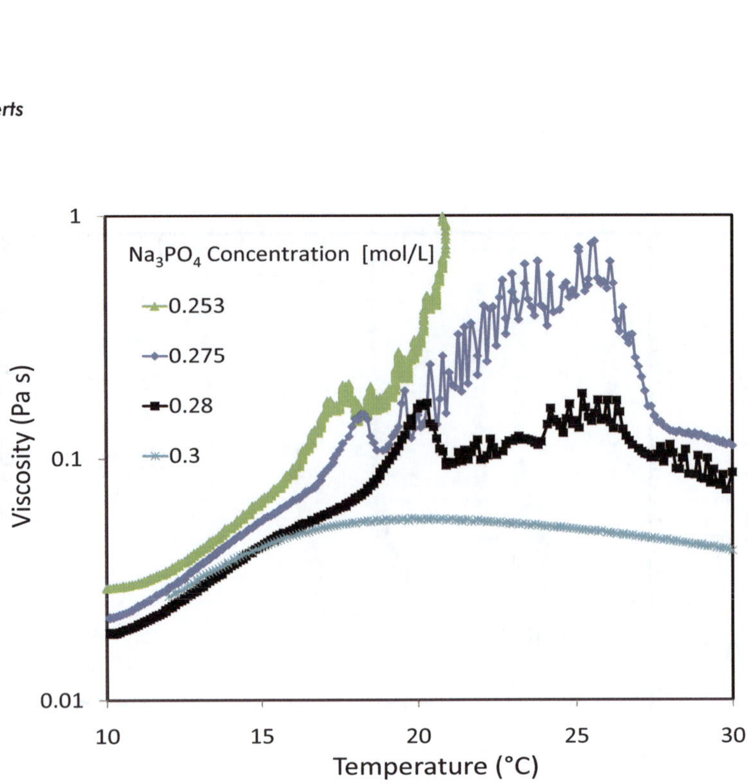

Figure 6.3 Viscosity as a function of temperature for high concentrations of sodium phosphate in a 15 wt% Pluronic F127 solution in water from cone and plate viscometry at controlled shear stress (0.6 Pa s).[110]

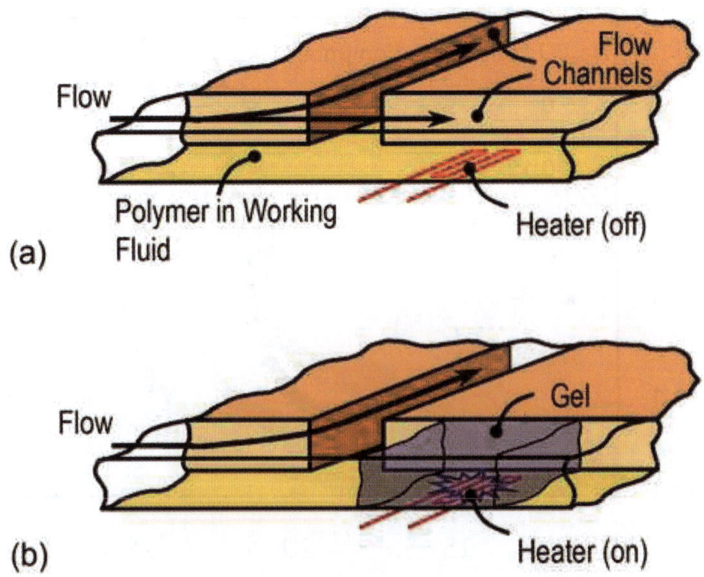

Figure 6.5 A schematic illustration of the principle of an active valve using a thermally responsive fluid in a microfluidic network. (a) Fluid from one channel is diverted into two channels at a channel bifurcation; (b) activating an integrated heater leads to localized gel formation in the corresponding microchannel, which subsequently blocks this channel to flow.

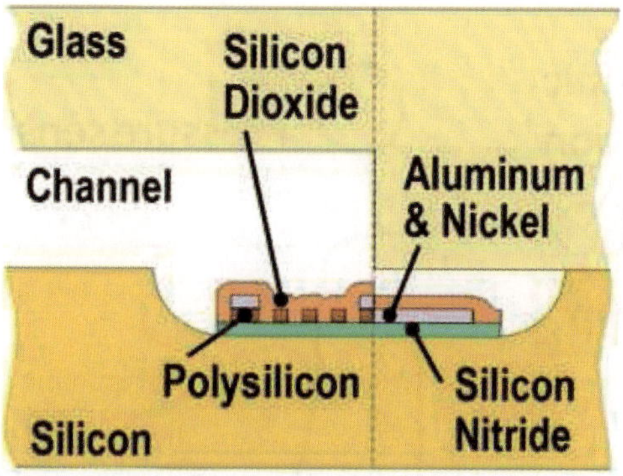

Figure 6.7 A schematic cross-section of the microfluidic device shown in Fig. 6.6. The left-hand side shows the cross-section across a heater in the channel, while the right-hand side shows a cross-section further away from the channel, through an aluminum lead.

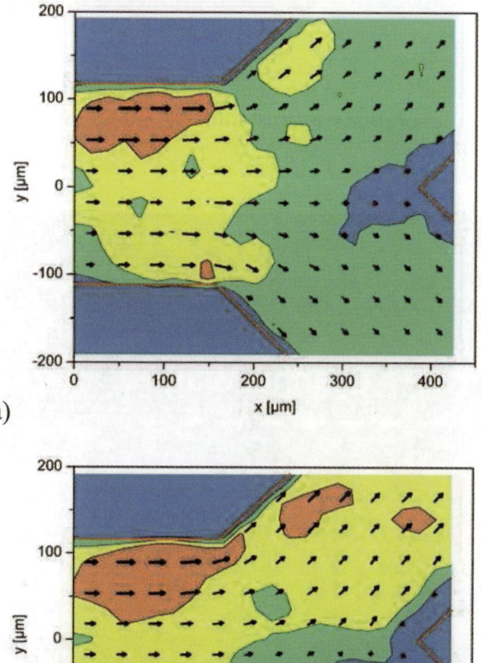

(a)

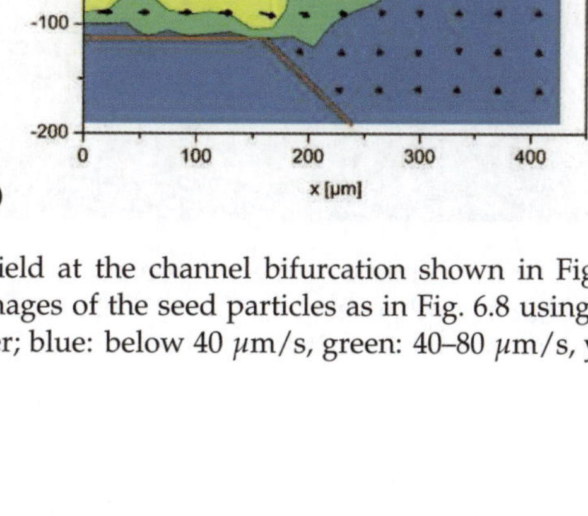

(b)

Figure 6.9 The flow field at the channel bifurcation shown in Fig. 6.6. The velocities were evaluated from images of the seed particles as in Fig. 6.8 using PIV; (a) before valve actuation, (b) 33 ms later; blue: below 40 μm/s, green: 40–80 μm/s, yellow: 80–120 μm/s, orange: 120–160 μm/s.

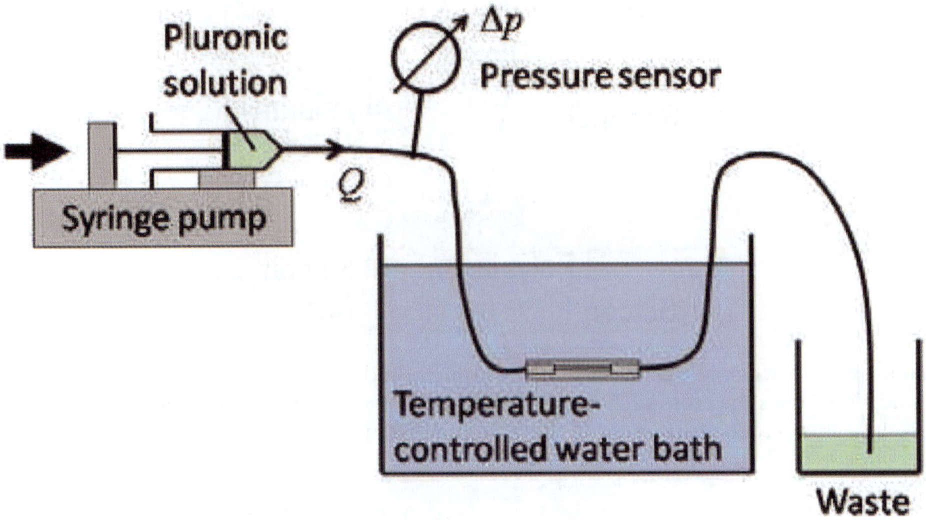

Figure 6.10 Experimental setup for the demonstration of passive flow control using thermally responsive Pluronic solutions at constant flow rate in microchannels.

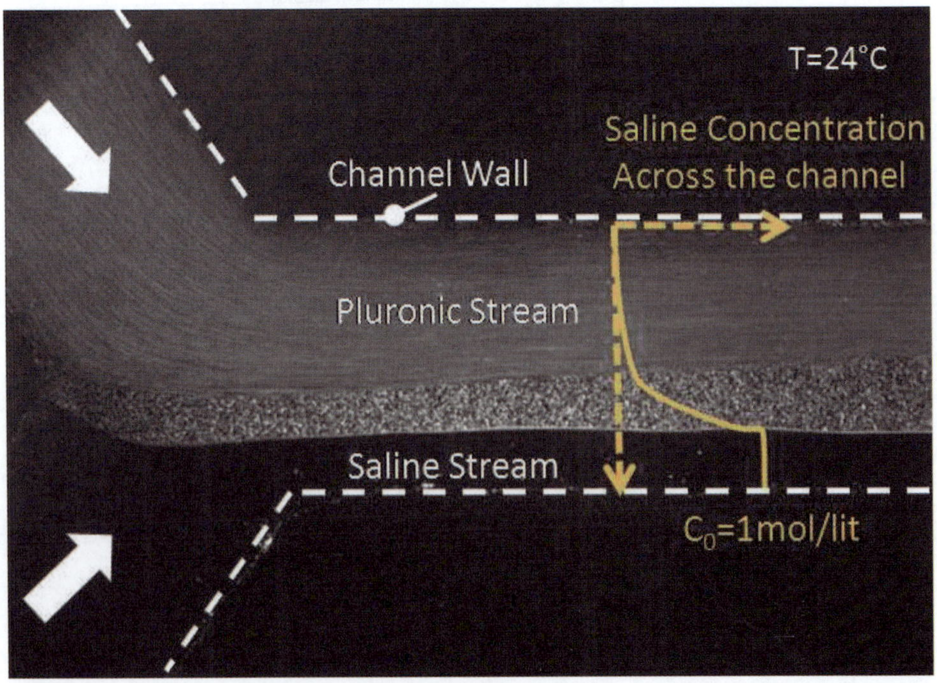

Figure 6.14 Top view of a microfluidic device. Pluronic solution and saline solution are introduced in a 10 μm high microchannel at T_a = 24°C. Gel formation occurs in the center of the channel. The Pluronic stream is seeded with fluorescent particles.

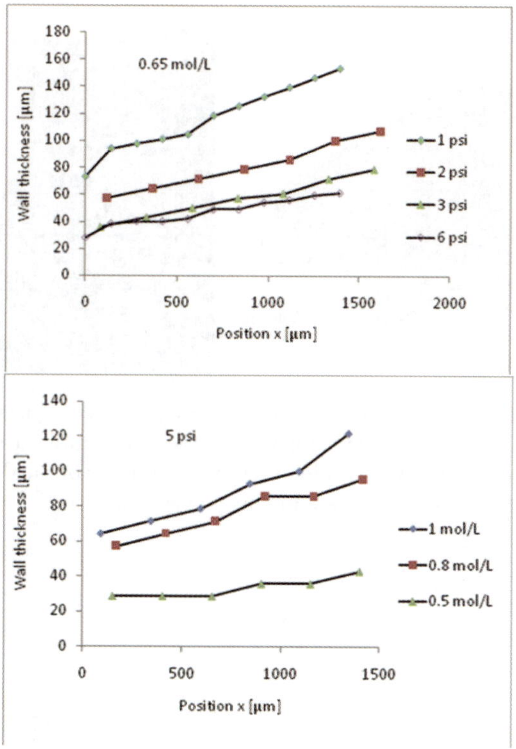

Figure 6.16 Thickness of the gel wall along the channel for different experimental conditions.

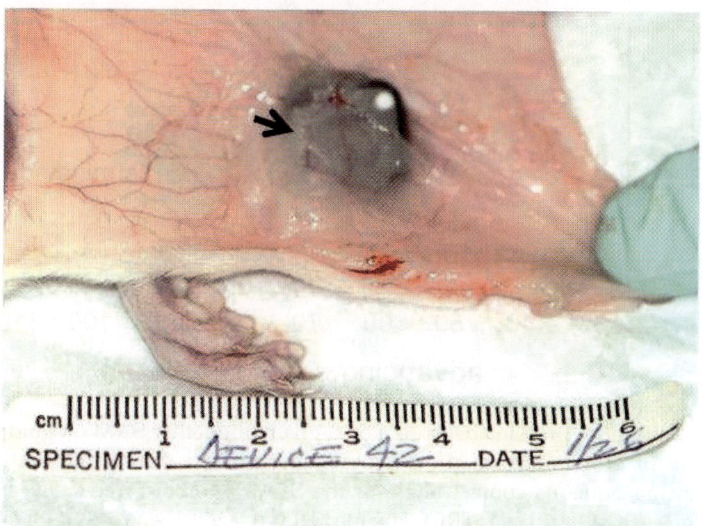

Figure 7.5 BioMEMS multi-array prototype implanted subcutaneously in a rodent model. The local environment had encapsulated the device in an attempt to isolate it from the body (28 days post-implantation).

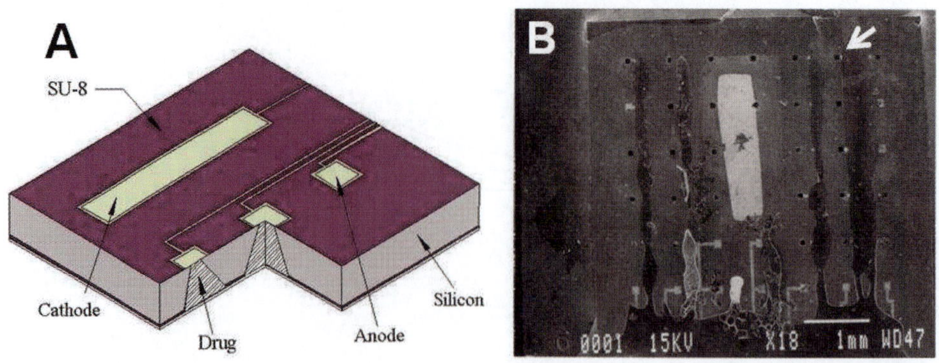

Figure 7.8 BioMEMS micro-array prototype. The inner-reservoir surface area is larger than it appears from a top view (A). The activation of the entire multi-array reservoir system leads to dissolution of the reservoir sealing membrane, release of the drug, and the onset of a "new" inflammatory response initiated by the exposed surface of the inner reservoir (B).

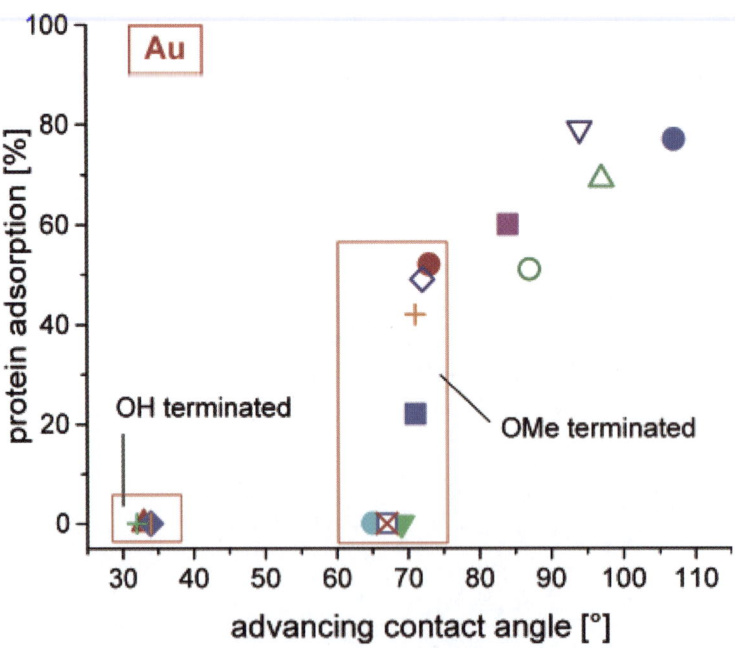

Figure 8.1 Amount of protein adsorption on a given oligoether SAM on gold normalized to the amount of protein adsorbed on a monolayer of hexa-decanethiol on gold (100%) versus advancing aqueous contact angle of the SAM. Symbols: red▲, EG_2OH; orange, EG_3OH; green+, EG_6OH; blue♦, TRI_3OH; blue ■, EG_1OMe; green▼, EG_2OMe; light blue •, EG_3OMe; red ×, EG_6OMe; blue□ , TRI_3OMe; red ×, PRO_2OMe; blue◊, PRO_3OMe; orange +, PRO_4OMe; purpl■, EG_3OEt; green○, EG_6OEt; blue▽, EG_3OPr; green△, EG_6OPr; blue •, EG_3OBu. Images adapted from Ref. 54.

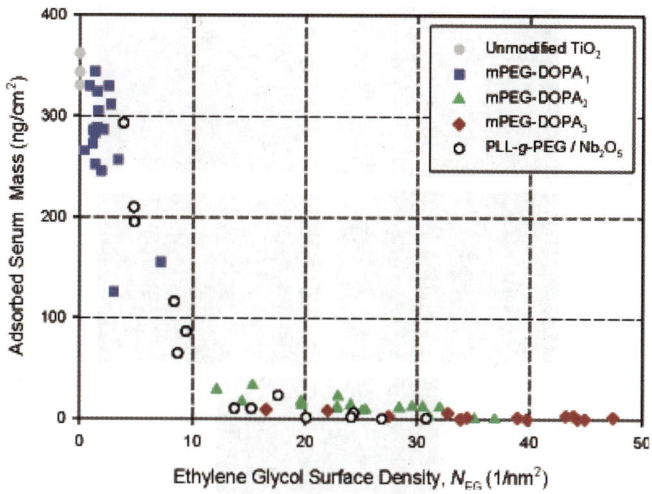

Figure 8.8 Ethylene glycol surface density versus adsorbed serum thickness. Images adapted from Ref. 78.

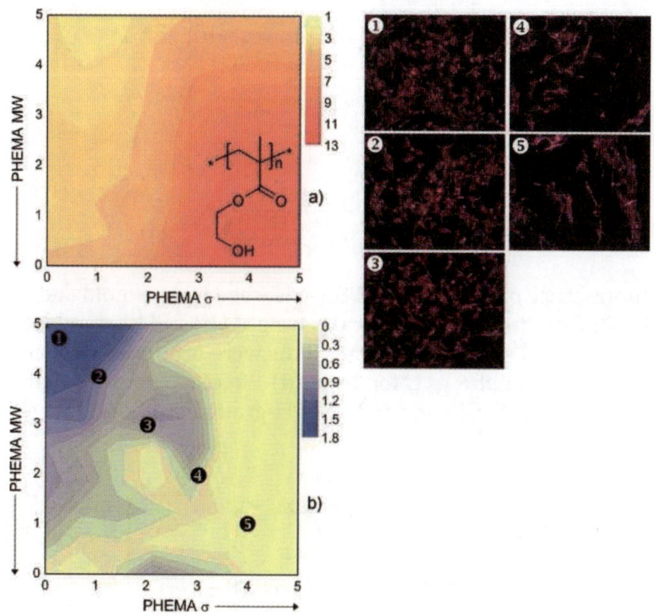

Figure 8.20 Left: contour plots of a) dry thickness of PHEMA in a MW-σ orthogonal PHEMA gradient (scale in nm); b) dry FN thickness in a MW-σ orthogonal PHEMA/FN gradient (scale in nm), the scales depicting the position on the substrate in parts (a) and (b) are in cm. Right: fluorescence microscopy (10x) images of fluorescently labeled MC3T3-E1 cells (nucleus: DAPI blue, cytoskeleton/actin: phalloidin red) cultured on PHEMA/FN gradient substrates. Images were taken at positions on the PHEMA/FN gradient sample marked with the numbers in part (b). The error bars associated with the thickness measurements ($\approx$ 0.1 nm) are smaller than the thickness increments in the contour plots. Images adapted from Ref. 109.

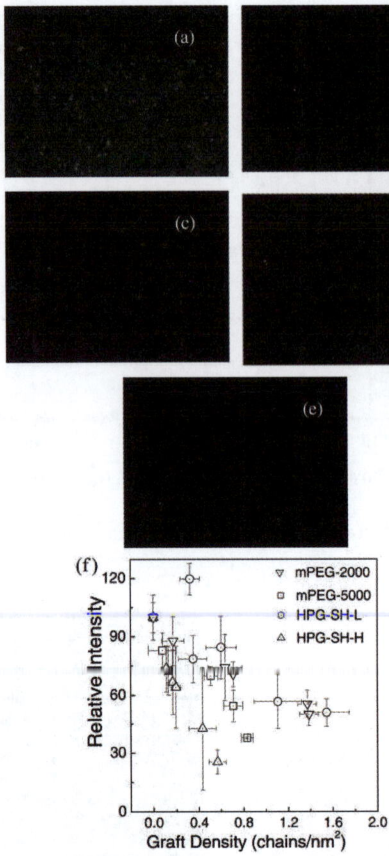

Figure 8.22 Fluorescence photographs of BSA-adsorbed (a) bare gold and (b) linear mPEG-2000-, (c) linear mPEG-5000-, (d) low molecular weight HPG-SH-L-, and (e) high molecular weight HPG-SH-H grafted surfaces. Polymer films were produced by incubating the gold surface in polymer solution at 6 g/L for 16 h. (f) Effect of the graft density on the BSA adsorption of mPEG-, HPG-SH-L-,and HPG-SH-H-grafted surfaces. Images adapted from Ref. 120.

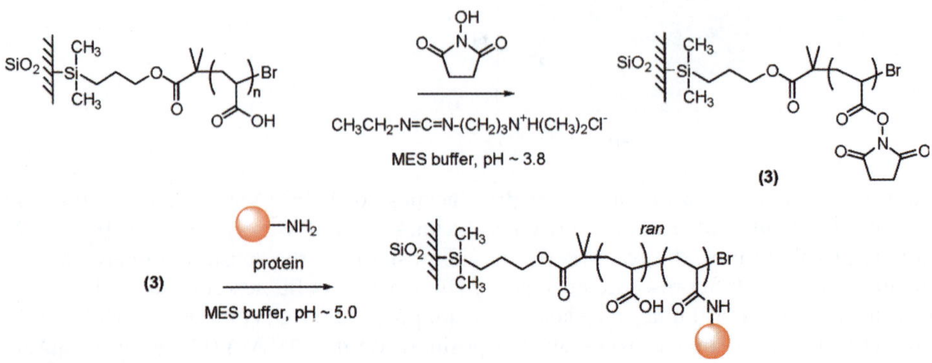

Figure 8.24 Formation of a peptide-modified poly(OEGMA) brush on titanium. Images adapted from Ref. 126.

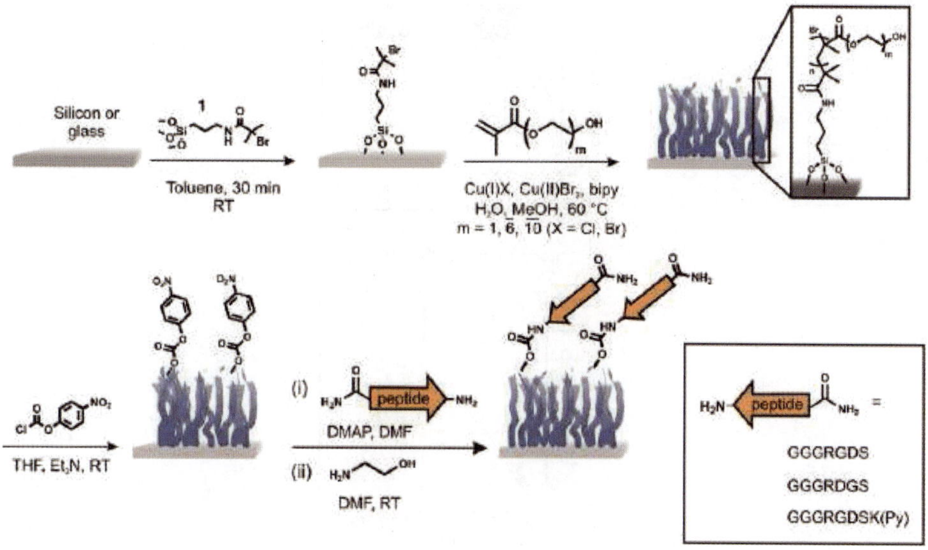

Figure 8.25 RGD functionalized PHEMA and PPEGMA brushes. Images adapted from Ref. 128.

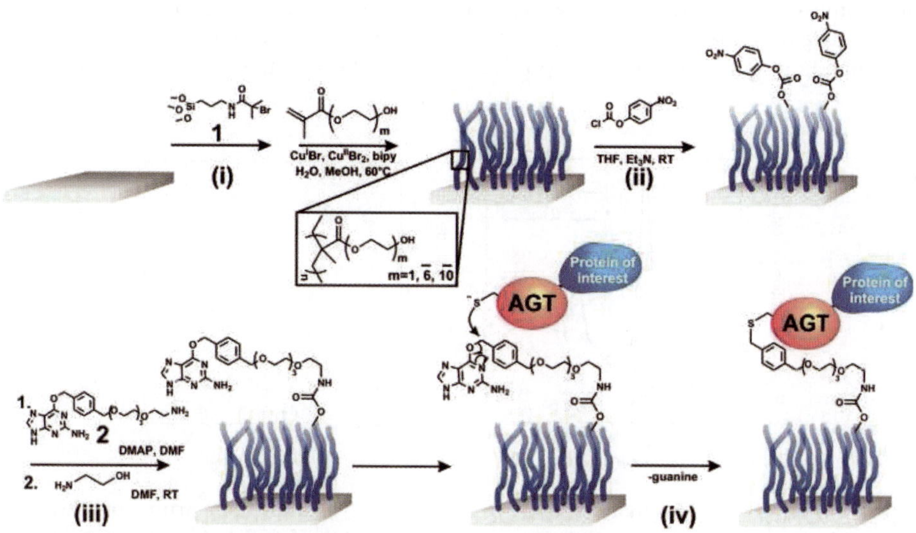

Figure 8.26 Fabrication of protein-functionalized polymer brushes: (i) grafting of ATRP initiator **1** and surface-initiated ATRP; (ii) activation of hydroxyl groups with p-nitrophenyl chloroformate (NPC); (iii) functionalization with benzylguanine derivative **2**and quenching of residual NPC groups; (iv) immobilization of AGT fusion proteins on benzylguanine-displaying surfaces. Images adapted from ref.[129]

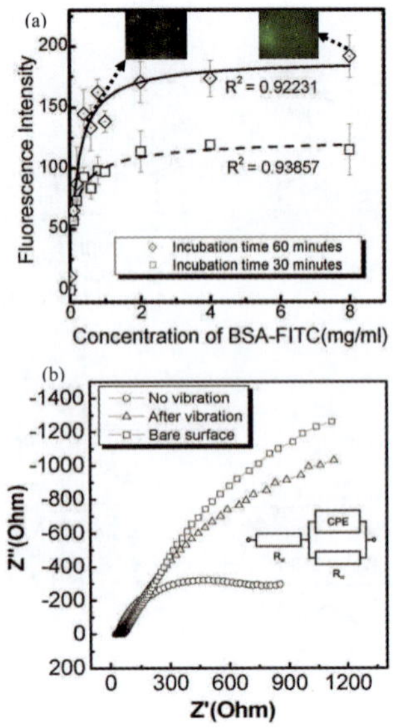

Figure 9.2 (a) BSA adsorption isotherms on a piezoelectric (PZT) surface incubated at 37°C for 1 hour and 30 minutes. The fluorescence intensity was transfer from tri-color to 256 gray scale by Adobe Photoshop 6.0. The intensity is linearly to the quantity of adsorbed BSA, and is a function, which follows the Langmuir isotherm, of BSA concentration. Two inserted photos are fluorescence photos in pointed experimental condition. (b) Nyquist plot of impedance between working electrodes (PZT) and platinum counter electrode. The voltage data (in Z measurement) is corresponding to the reference voltage. The inset is the corresponding equivalent circuit. Reproduced with the permission from Yeh *et al.* (2007).

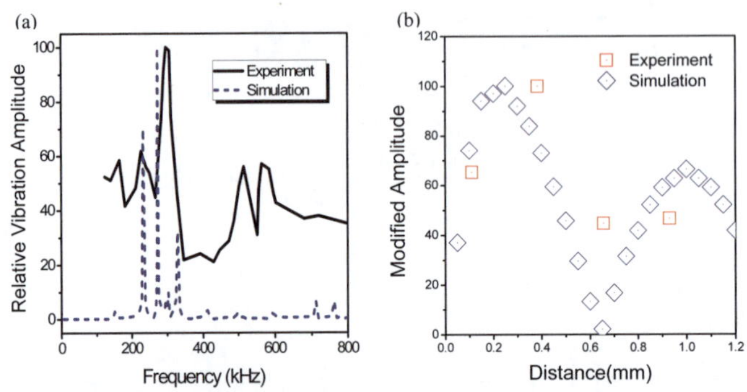

Figure 9.7 (a) The simulation and experimental vibration spectrum of the PZT plate/siliconmembrane. The vibration at 308, 320, 500, and 575 kHz correspond to 1.5, 2.5, 3.5 wavelength bending mode, and longitudinal mode, respectively. (b) The simulation and experimental vibration amplitude distribution across the Si/SiO_2 membrane along the membrane center line. The x axis is given as the distance from the center of the membrane to the left periphery. The half length of the membrane is 1000 μm. Reproduced with the permission from Yeh *et al.* (2008).

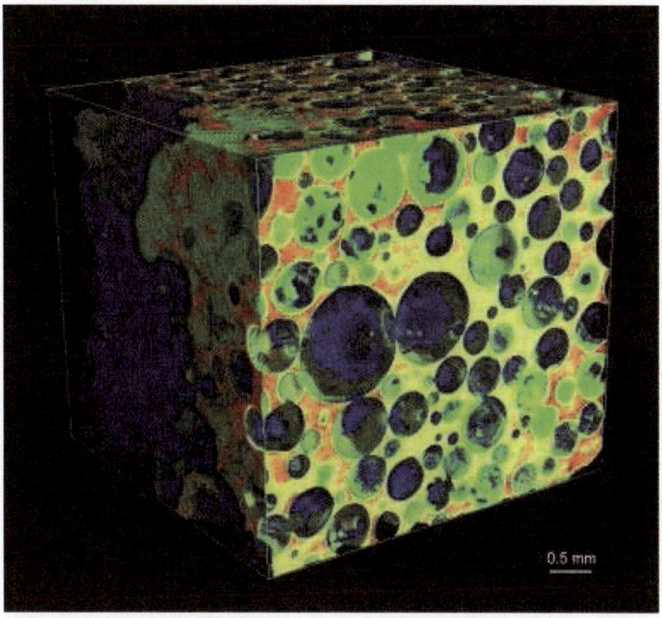

Figure 10.1 3D microCT image of hydroxyapatite scaffold (Engipore) loaded with sheep BMSC and implanted in an immunodeficient mouse, showing newly formed bone (green) onto the inner surface of the scaffold (yellow and red); the organic phase is blue. With permission from.[4]

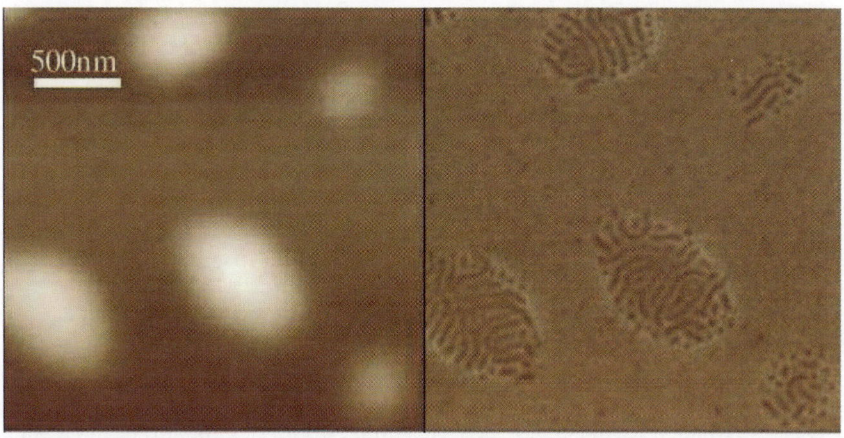

Figure 10.3 Topography and phase image of a poly (styrene-ethylene/butylene-styrene) (SEBS) copolymer treated by solvent annealing. (a) topography image, (b) phase image. The phase image in (b) shows unambiguous resolution of the two different phases of the copolymer. The phase angle contrast of two different copolymers results from the different probe–sample interaction forces. With permission from.[24]

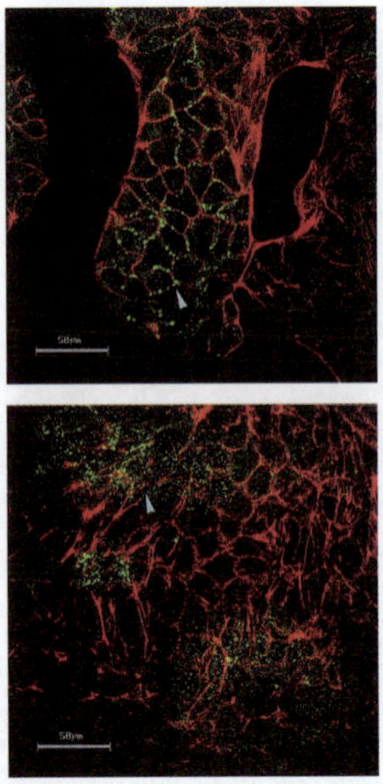

Figure 10.5 Fibronectin formation (green) and actin filament formation (red) in epithelial mHepR1 cells cultured on the biomaterial PHB (below) compared to control cultures (above). Note that in control cultures the actin filaments are structured in fibrils showing colocalization with fibronectin (arrow head) in contrast to PHB cultured cells, where the organization of actin filaments is more irregular and the colocalization of fibronectin with actin is lost (arrow head) (CLSM images taken with LSM 410, Carl Zeiss Jena). With permission from.[33]

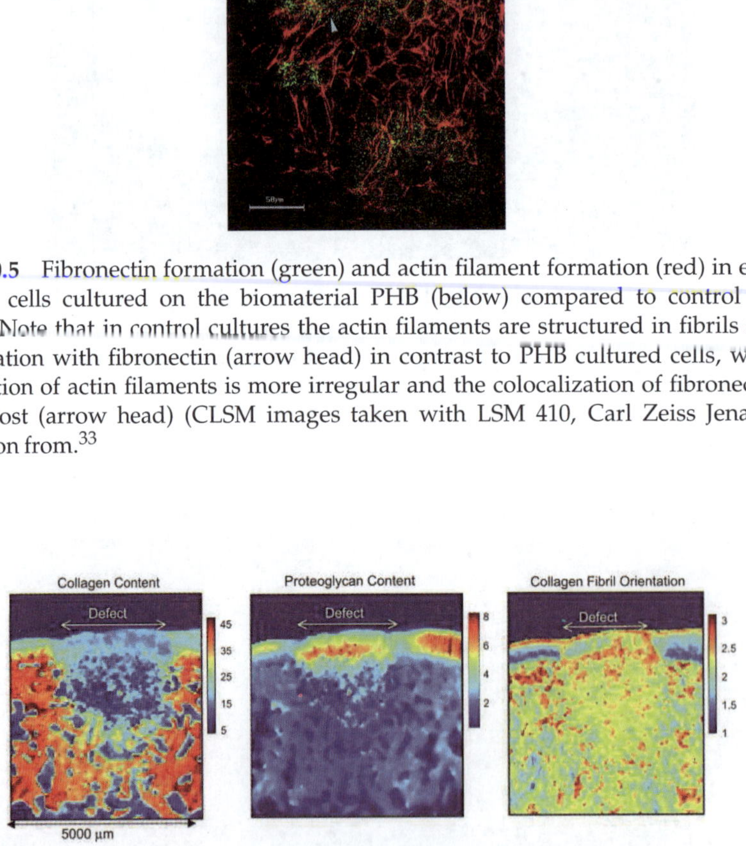

Figure 10.9 FT-IR images of rabbit osteochondral defect at 6 weeks post-repair with TP 508 protein. The defect repair cartilage has lower collagen content (amide I area) compared to the adjacent articular cartilage, but also has regions with PG content (PG sugar ring C-O absorbance (985–1140 cm^{-1}) of a similar magnitude to that in the native cartilage. There is some orientation present in the repair cartilage, as evidenced by the layer of fibrils parallel to the surface. For the orientation data, the boundaries on the color bar indicate the three collagen fibril orientation categories with respect to the articular surface, corresponding to amide I/amide II peak area ratios$>$2.7 (parallel fibrils), between 2.7 and 1.7 (random fibrils), and $<$1.7 (perpendicular fibrils), respectively. With permission from.[53]

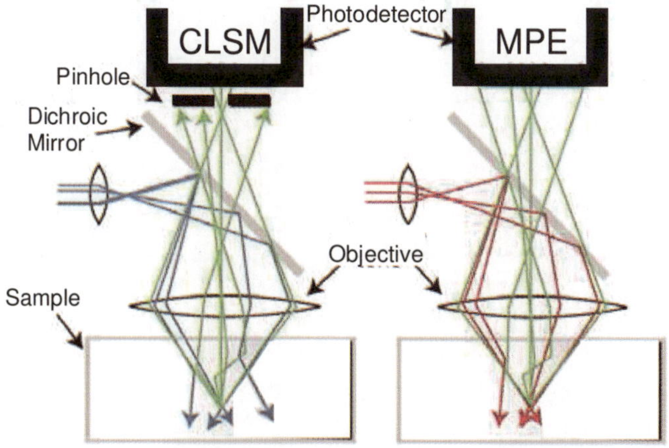

Figure 10.10 Comparison of a conventional confocal microscope (CLSM) with a multi-photon excitation (MPE) microscope.

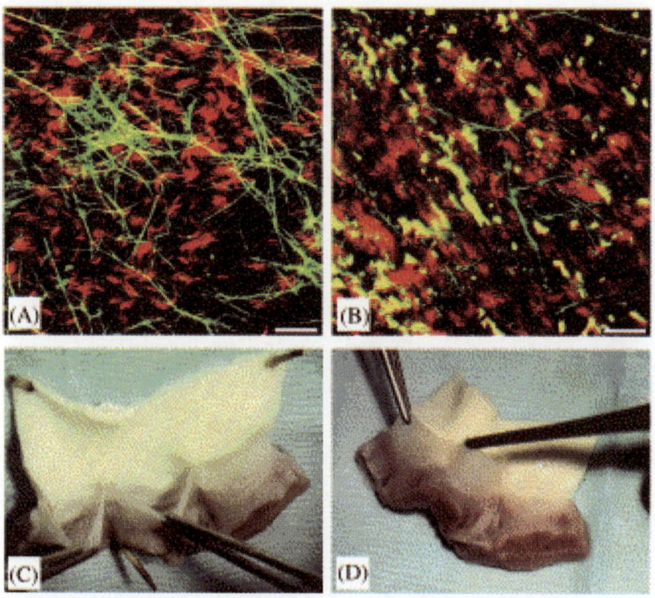

Figure 10.11 Overlay of optical sections showing elastic fibers (green) and collagen (red) of two TE valve leaflets, cultured under different conditions within the bioreactor system. The differences between the tissue states were only explicitly verifiable by the use of the MPE images. (A) TE heart valve, exposed to defined physical signals with a fixed frequency (1 Hz) and pressure conditions (3 l/min and 60/40 mm Hg), show normal extracellular matrix structures (bar: 20 μm). (C) Corresponding macroscopic appearance of the TE valves. (B) TE leaflet, exposed to supra-physiologic (>200 mm Hg) pressures, revealing an impaired matrix formation and a high amount of cell detritus (yellow). (D) Corresponding macroscopic pictures. With permission from.[72]

Index